Lecture Notes in Computer Scier

T0238516

Commenced Publication in 1973
Founding and Former Series Editors:
Gerhard Goos, Juris Hartmanis, and Jan van Leeuwen

Editorial Board

Kazuo Sakiyama   Masayuki Terada (Eds.)

Advances in Information and Computer Security

8th International Workshop on Security, IWSEC 2013
Okinawa, Japan, November 18-20, 2013
Proceedings

 Springer

Volume Editors

Kazuo Sakiyama
The University of Electro-Communications
Department of Informatics
1-5-1 Chofugaoka, Chofu
Tokyo 182-8585, Japan
E-mail: sakiyama@uec.ac.jp

Masayuki Terada
NTT DOCOMO, Inc.
Research Laboratories
3-6 Hikari-no-oka, Yokosuka
Kanagawa 239-8536, Japan
E-mail: teradam@nttdocomo.com

ISSN 0302-9743   e-ISSN 1611-3349
ISBN 978-3-642-41382-7   e-ISBN 978-3-642-41383-4
DOI 10.1007/978-3-642-41383-4
Springer Heidelberg New York Dordrecht London

Library of Congress Control Number: 2013949481

CR Subject Classification (1998): E.3, G.2, D.4.6, F.2, C.2, K.6.5, K.4.4

LNCS Sublibrary: SL 4 – Security and Cryptology

Typesetting: Camera-ready by author, data conversion by Scientific Publishing Services, Chennai, India

Printed on acid-free paper

Springer is part of Springer Science+Business Media (www.springer.com)

Kazuo Sakiyama Masayuki Terada (Eds.)

Advances in Information and Computer Security

8th International Workshop on Security, IWSEC 2013
Okinawa, Japan, November 18-20, 2013
Proceedings

 Springer

Volume Editors

Kazuo Sakiyama
The University of Electro-Communications
Department of Informatics
1-5-1 Chofugaoka, Chofu
Tokyo 182-8585, Japan
E-mail: sakiyama@uec.ac.jp

Masayuki Terada
NTT DOCOMO, Inc.
Research Laboratories
3-6 Hikari-no-oka, Yokosuka
Kanagawa 239-8536, Japan
E-mail: teradam@nttdocomo.com

ISSN 0302-9743   e-ISSN 1611-3349
ISBN 978-3-642-41382-7   e-ISBN 978-3-642-41383-4
DOI 10.1007/978-3-642-41383-4
Springer Heidelberg New York Dordrecht London

Library of Congress Control Number: 2013949481

CR Subject Classification (1998): E.3, G.2, D.4.6, F.2, C.2, K.6.5, K.4.4

LNCS Sublibrary: SL 4 – Security and Cryptology

Typesetting: Camera-ready by author, data conversion by Scientific Publishing Services, Chennai, India

Printed on acid-free paper

Springer is part of Springer Science+Business Media (www.springer.com)

Preface

IWSEC 2013, the 8th International Workshop on Security, was held at Okinawaken Shichouson Jichikaikan in Okinawa, Japan, during November 18–20, 2013. The workshop was co-organized by ISEC in ESS of IEICE (Technical Committee on Information Security in Engineering Sciences Society of the Institute of Electronics, Information and Communication Engineers) and CSEC of IPSJ (Special interest group on Computer Security of the Information Processing Society of Japan).

We received 63 submissions, of which 20 were accepted for publication. Each submission was anonymously reviewed by at least three reviewers, and these proceedings contain the revised versions of the accepted papers. There were also two keynote talks that were selected at the discretion of the general co-chairs and program co-chairs. The talks were given by Sebastian Faust and Nobuaki Hoshino. In addition to the presentations of the papers and the keynote talks, the workshop also featured a poster session.

The Best Paper Award was given to "Solving Google's Continuous Audio CAPTCHA with HMM-Based Automatic Speech Recognition," by Shotaro Sano, Takuma Otsuka, and Hiroshi G. Okuno, and the Best Student Paper Award was given to "Improvement of Faugère *et al.*'s Method to Solve ECDLP," by Huang Yun-Ju, Christophe Petit, Naoyuki Shinohara, and Tsuyoshi Takagi.

Our deepest appreciation goes to the Program Committee. The selection of the papers was a challenging and delicate task, and we are deeply grateful to the members of the Program Committee and the external reviewers for their in-depth reviews and detailed discussions.

A number of people contributed to the success of IWSEC 2013. We would like to thank all of the authors of submissions. Their great work made IWSEC 2013 a worthwhile conference. We are also grateful to Andrei Voronkov for developing EasyChair, which was used for the paper submission, reviews, discussions, and preparation of these proceedings.

Last but not least, we would like to thank the general co-chairs, Toshiaki Tanaka and Masakatsu Nishigaki, for leading the local Organizing Committee, and we also would like to thank the members of the local Organizing Committee for their dedicated efforts to ensure the smooth running of the workshop.

August 2013

Kazuo Sakiyama
Masayuki Terada

IWSEC 2013
8th International Workshop on Security

Okinawa, Japan, November 18–20, 2013

Co-organized by

ISEC in ESS of IEICE

(Technical Committee on Information Security in Engineering Sciences Society of the Institute of Electronics, Information and Communication Engineers)

and

CSEC of IPSJ

(Special interest group on Computer Security of the Information Processing Society of Japan)

General Co-chairs

Toshiaki Tanaka — KDDI R&D Laboratories Inc., Japan
Masakatsu Nishigaki — Shizuoka University, Japan

Advisory Committee

Hideki Imai — Chuo University, Japan
Kwangjo Kim — Korea Advanced Institute of Science and Technology, Korea
Günter Müller — University of Freiburg, Germany
Yuko Murayama — Iwate Prefectural University, Japan
Koji Nakao — National Institute of Information and Communications Technology, Japan
Eiji Okamoto — University of Tsukuba, Japan
C. Pandu Rangan — Indian Institute of Technology, Madras, India
Ryoichi Sasaki — Tokyo Denki University, Japan

Program Co-chairs

Kazuo Sakiyama — University of Electro-Communications, Japan
Masayuki Terada — NTT DOCOMO, Inc., Japan

Local Organizing Committee

Yuki Ashino	NEC, Japan
Takuro Hosoi	The University of Tokyo, Japan
Takehisa Kato	IPA, Japan
Akinori Kawachi	Tokyo Institute of Technology, Japan
Yuichi Komano	Toshiba, Japan
Koji Nuida	AIST, Japan
Anand Prasad	NEC, Japan
Kouichi Sakurai	Kyushu University, Japan
Yuji Suga	Internet Initiative Japan Inc., Japan
Mio Suzuki	National Institute of Information and Communications Technology, Japan
Alf Zugenmaier	Munich Universities of Applied Sciences, Germany

Program Committee

Rafael Accorsi	University of Freiburg, Germany
Toru Akishita	The University of Tokyo, Japan
Claudio Ardagna	Università degli Studi di Milano, Italy
Nuttapong Attrapadung	AIST, Japan
Andrey Bogdanov	Technical University of Denmark, Denmark
Sanjit Chatterjee	Indian Institute of Science, India
Koji Chida	NTT, Japan
Sabrina De Capitani di Vimercati	Università degli Studi di Milano, Italy
Bart De Decker	Katholieke Universiteit Leuven, Belgium
Isao Echizen	National Institute of Informatics, Japan
Sebastian Faust	EPFL, Switzerland
Dario Fiore	Max Planck Institute for Software Systems, Germany
Eiichiro Fujisaki	NTT, Japan
David Galindo	CNRS/LORIA, France
Dieter Gollmann	Hamburg University of Technology, Germany
Goichiro Hanaoka	AIST, Japan
Swee-Huay Heng	Multimedia University, Malaysia
Naofumi Homma	Tohoku University, Japan
Mitsugu Iwamoto	University of Electro-Communications, Japan
Tetsu Iwata	Nagoya University, Japan
Angelos Keromytis	Columbia University, USA
Hiroaki Kikuchi	Meiji University, Japan
Hyung Chan Kim	ETRI, Korea
Takeshi Koshiba	Saitama University, Japan
Noboru Kunihiro	The University of Tokyo, Japan
Kwok-Yan Lam	National University of Singapore, Singapore

Jian Guo

Koki Hamada

Yoshikazu Hanatani

Ryotaro Hayashi

Matthias Hiller

Takato Hirano

Masatsugu Ichino

Dai Ikarashi

Motohiko Isaka

Kenta Ishii

Takanori Isobe

Tadahiko Ito

Kangkook Jee

Mahavir Jhanwar

Christian Kahl

Satoshi Kai

Akira Kanaoka

Akinori Kawachi

Yutaka Kawai

Vasileios P. Kemerlis

Ryo Kikuchi

Minkyu Kim

Naoto Kiribuchi

Nobuaki Kitajima

Hiroki Koga

Masanobu Koike

Georgios Kontaxis

Sascha Koschinat

Sebastian Kutzner

Martin M. Lauridsen

Hyung Tae Lee

Zhenhua Liu

Atul Luykx

Changshe Ma

Takahiro Matsuda

Shin'ichiro Matsuo

Qixiang Mei

Kunihiko Miyazaki

Kirill Morozov

Pratyay Mukherjee

Sayantan Mukherjee

Debdeep Mukhopadhyay

Ivica Nikolic

Ryo Nishimaki

Ryo Nojima

Toshihiro Ohigashi

Akira Otsuka

Nguyen Phuong Ha

Michalis Polychronakis

Ahmad Sabouri

Minoru Saeki

Yusuke Sakai

Koichi Sakumoto

Masahito Shiba

Kyoji Shibutani

Koichi Shimizu

Seonghan Shin

Koutarou Suzuki

Mostafa Taha

Syh-Yuan Tan

Mehdi Tibouchi

Markus Tschersich

Shigenori Uchiyama

Berkant Ustaoglu

Daniele Venturi

Srinivas Vivek

Dai Watanabe

Lars Wolos

Jing Xu

Jun Yajima

Shota Yamada

Takashi Yamakawa

Naoto Yanai

Masaya Yasuda

Kenji Yasunaga

Wei-Chuen Yau

Shun'ichi Yokoyama

Hui Zhang

Zongyang Zhang

Table of Contents

Public Key Cryptosystems

Security Protocols

Secure Log Transfer by Replacing a Library in a Virtual Machine

Masaya Sato and Toshihiro Yamauchi

Graduate School of Natural Science and Technology, Okayama University,
3-1-1 Tsushima-naka, Kita-ku, Okayama, 700-8530 Japan
m-sato@swlab.cs.okayama-u.ac.jp, yamauchi@cs.okayama-u.ac.jp

Abstract. Ensuring the integrity of logs is essential to reliably detect
and counteract attacks, because adversaries tamper with logs to hide
their activities on a computer. Even though some research studies proposed different ways to protect log files, adversaries can tamper with
logs in kernel space with kernel-level malicious software (malware). In
an environment where Virtual Machines (VM) are utilized, VM Introspection (VMI) is capable of collecting logs from VMs. However, VMI is
not optimized for log protection and unnecessary overhead is incurred,
because VMI does not specialize in log collection. To transfer logs out
of a VM securely, we propose a secure log transfer method of replacing
a library. In our proposed method, a process on a VM requests a log
transfer by using the modified library, which contains a trigger for a log
transfer. When a VM Monitor (VMM) detects the trigger, it collects logs
from the VM and sends them to another VM. The proposed method provides VM-level log isolation and security for the mechanism itself. This
paper describes design, implementation, and evaluation of the proposed
method.

Keywords: Log transfer, log protection, virtual machine, digital
forensics.

1 Introduction

Logging information about activities and events in a computer is essential for
troubleshooting and for computer security. Logs are important not only for detecting attacks, but also for understanding the state of the computer when it was
attacked. The importance of logs for computer security is described in Special
Publication [1]. Adversaries tamper with logs to hide their malicious activities
and the installation of malwares on the target computer [2–4]. If logs related
to those activities are tampered with, detection of problems might be delayed,
and the delay could cause further damage to services. In addition, log tampering impedes the detection, prevention, and avoidance of attacks. With the
growth of cloud computing in recent years, security in VMs has become more
important [5, 6]. Especially, log forensics in cloud application has great importance [7]. However, existing logging methods are not designed for VMs or cloud
applications.

K. Sakiyama and M. Terada (Eds.): IWSEC 2013, LNCS 8231, pp. 1–18, 2013.

As described in a paper [8], secure logging using VMs provides integrity and completeness for logging. Boeck et al. proposed a method to securely transfer logs utilizing a trusted boot and a late launch [9]. While this method can prevent attacks to logging daemons, adversaries can still tamper with logs in kernel space. Logs must go through an Operating System (OS) kernel when transferred out of the computer. If malware is installed on, logs could be tampered with in kernel space. SecVisor [10] is a method that prevents the execution of illegal codes in kernel space. However, these methods depend on the structure of the OS kernel, making it difficult to adapt to various OSes. In a situation where a single machine provides many VMs, different OSes could be running on each VM. VMI [11] can be considered as a logging method for VM. However, VMI has problems including performance degradation and granularity of information.

These researches are considered as a method of log protection. However, even though the importance of logging for cloud application is increased [7], there is no method specialized for logging in VM environment. VM is commonly used for providing cloud computing environment. Providing services like logging hurts performance of APs on VMs [8]. Thus, reducing performance overhead incurred by additional services is an important challenge.

To collect logs from outside the VM securely, we propose a secure log transfer method using library replacement. To trigger a log transfer to a VMM, we embed an instruction in a library function to cause a VM exit. On Linux and FreeBSD, we modified the standard C library, libc, which contains standard logging function. When the VMM detects a VM exit, the VMM collects the logs generated by APs in the source VM and transfers them to the logging VM, which stores the logs to a file. We assumed that the modified library is secured in the memory by the method [12] that protects a specific memory area from being modified by kernel-level malware.

With the proposed method, adversaries cannot tamper with logs in kernel space because the VMM collects logs before they reach in kernel space. Because the modification to a library is kept minimal, adapting different OSes requires less effort. Performance degradation is minimal because the overhead incurs only when an AP calls a logging function. The proposed method replaces only a library, which includes a function to send logs to a syslog daemon. Therefore, we can make the possibility of bug inclusion low. Additionally, bugs in a library give less effect than that in a kernel.

This paper also describes evaluations of the proposed system. We evaluate the system with the standpoint of security of logs, adaptability to various OSes, and performance overhead. To evaluate the system with the standpoint of security of logs, we analyze the security of a logging path. Experiments to tamper with logs in the logging path are also described. Adaptability of the proposed system is provided with case studies to adapt to various OSes. Performance evaluations with APs commonly used in servers are described. As described in a paper [13], VMI causes large overhead. For practical use, performance degradation should be kept as small. With these evaluations, this paper presents how the proposed system is practical for generally-used APs and multi-VM environment.

The contributions made in this paper are as follows:

- We propose a secure log transfer method by replacing a library in a VM. With the proposed system, a kernel-level malware cannot delete or tamper with logs. Moreover, by comparing collected logs and tampered logs, we can identify the area that is tampered with.
- We design a tamper-resistant system using VMM. We implemented all of our system inside the VMM because of its attack-resistance.
- The proposed system is implemented with minimal modification to libc. Although no modification is preferable, modifying the library gives two advantages: slight overhead and ease of adaptation to varied OSes. This also reduces the possibility of bug inclusion, and makes the system more secure.

2 Method of Log Transfer

2.1 Existing Log Transfer Methods

In Linux and FreeBSD, syslog is a protocol for system management and security monitoring. Syslog consists of a syslog library and a syslog daemon. New syslog daemons and protocols [14–17] have been developed to achieve greater security. New syslog daemons can transfer logs to out of a computer and can encrypt syslog traffic using transport layer security (TLS). However, during log transfer, adversaries can delete or tamper with the log with a kernel-level attack [3]. Other methods using inter-process communications can be attacked in the same manner. Other malware tamper with logs by replacing syslog daemons [2].

VMI [11] inspects VMs by retrieving hardware information about the target VM and constructing a semantic view from outside the VM. ReVirt [18] collects instructions-level information for VM logging and replay. CloudSec [19] performs a fine-grained inspection of the physical memory used by VMs and detects attacks that modify kernel-level objects. While these methods enable us to collect information inside VMs, they increase complexity of semantic view reconstruction and performance overhead. In addition, the reconstruction of a semantic view strongly depends on the structure of the OS.

To overcome this problem, in-VM monitoring method [13], which inserts an agent into a VM, is proposed. It protects the agent from attacks from inside the VM. Inserting an agent is a practical and efficient way to collect information, however, it is difficult to adapt to various OSes because the implementation of an agent depends on the structure of the OS. The VMM-based scheme [20] can collect logs inside VMs without modifying a kernel or inserting agents. However, it has a large overhead and strong dependency to architecture of OS.

2.2 Problems of Existing Methods

Existing methods have the following four problems:

(1) Transferring log via inter-process communications can be preempted by kernel-level attacks.

(2) Collecting logging information inside a VM by monitoring the behavior of APs or OSes cause unnecessary performance overhead.

(3) Collecting logging information from various OSes requires efforts to adapt the method to a variety of OSes.

(4) Additional code increases the likelihood of bugs in the system.

No suitable method is currently available to transfer logs out of the VM. For security management, a secure logging method is required. Monitoring from outside the VM is a new approach, because the monitor itself is secured by VM-level separation. On the other hand, the information obtained by the method is difficult to translate into a semantic view or is too fine-grained. While VMI and other introspection methods securely collect information inside a VM, constructing the semantic view of the VM is strongly depends on structure of the target OSes. Adapting those methods to various OSes is nontrivial work. Inserting an agent into a VM can cause undesirable effects and make the VM unstable.

3 Secure Log Transfer by Replacing a Library in a VM

3.1 Scope and Assumptions

This paper covers the prevention of log tampering via attacks to the kernel, to the logging daemon, and to files that contain logs. Attacking specific APs requires nontrivial work and it cannot tamper with logs completely; therefore, adversaries attack the point where all logs go through. If we focus our attention on attacks to APs, preventing log tampering in kernel space and in a logging daemon is a reasonable challenge.

We assume attacks for a VMM is difficult because the conditions that allow attacks are limited. Therefore, we assume that a VMM can prevent those attacks.

3.2 Objectives and Requirements

The objectives of this paper are as follows:

Objective 1. To propose a fast and tamper-resistant log transfer method.
Objective 2. To propose a log transfer method that is easy to adapt to various OSes.

The objective of our research is to address problems detailed in Section 2.2. To address those problems, providing a tamper-resistant log transfer method is necessary. Specifically, we aim to prevent log tampering from kernel-level malware like adore-ng [3]. Moreover, low overhead is desired to implement the method to APs in the real world. Further, an OS-independent method is preferable, because it is assumed that various OSes are running on each VM.

To achieve the objectives, the followings are required.

Requirement 1. Transfer logs as soon as possible.
Requirement 2. Isolate logs from a VM.

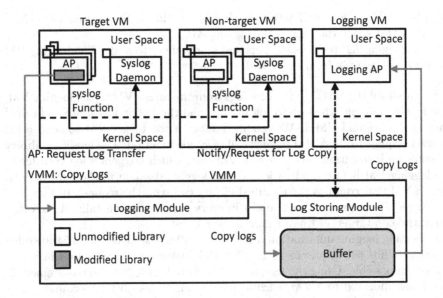

Fig. 1. Overview of the proposed system

Requirement 3. Secure the log transfer mechanism itself.

Requirement 4. Make the log transfer method OS-independent and small.

Requirement 5. Reduce unnecessary overhead related to log transfer.

In a logging path, logs generated by a process are passed to a kernel because the kernel provides the ability to send messages to other processes. Therefore, to prevent log tampering in kernel space, it is necessary to collect logs from outside the VM before the logs reach kernel space. To prevent tampering of log files, they must be isolated from the VM. To ensure the security of the log transfer method itself, install the method outside the VM. With low dependency on the OS, migration to other OSes becomes easy. Moreover, a smaller program size helps to reduce the possibility of bugs. A VM exit, which is a CPU-mode transfer between a VM and a VMM, can cause additional overhead. To adapt the method to APs in the real world, unnecessary VM exits must be removed.

3.3 Overview of the Proposed Method

The overall design of the proposed system is shown in Figure 1. In the proposed system, the target VM works on a VMM and the VMM collects logs from the VM. We assume that all of the VMs fully virtualized by Intel VT-x. An AP on the target VM can transfer logs with the proposed system as follows:

(1) An AP requests a log transfer to a VMM.
(2) The logging module inside the VMM receives the request and copies logs from the AP to the buffer inside the VMM.

(3) The VMM sends a notification to a logging AP inside the logging VM. Then, the VMM sends the logs to the logging AP.

(4) The logging AP receives the logs and stores them to a file. The logging VM accepts logs only from the VMM.

We modified the VMM to transfer logs from the target VM to the logging VM. The logging module, the log storing module and the buffer VMM are additional part to the original VMM. We modified `libc` in the target VM to send a log transfer request to the VMM in each call of syslog function. The modified library executes an instruction that causes a VM exit, which triggers a log transfer to the logging VM before sending logs to the logging daemon in the current VM. Only a VM that contains the modified library can send the request. In Figure 1, the target VM requests a log transfer in every syslog function call; on the other hand, the non-target VM never makes the request.

Collecting logging information immediately after the invocation of the syslog function fulfills the requirement 1. With this feature, tampering logs in kernel space is impossible. Using the logging VM to store logs fulfills the requirement 2. Resources allocated to a VM, such as memory, network, disk space, and others are separated from resources allocated to another VM, therefore, it is difficult to tamper with logs outside the VM being attacked. It is also difficult to attack a VMM from inside a VM; therefore, using a VMM and modifying a library fulfill the requirement 3. Library modification also makes OS-adaptation easier and fulfills the requirement 4. Finally, VM exits occur only when a syslog library function is called; therefore, the requirement 5 is fulfilled.

3.4   Comparison between the Proposed Method and VMI

The proposed method and VMI are similar from the standpoint of collecting information inside VM. However, there are following differences between them:

- Security of logs.
- Dependency to a data structure in a VM.
- Overhead.

The proposed system can achieve greater security of logs than VMI. VMI collects information of VMs by monitoring hardware states and some events. However, it is difficult to detect log generation by monitoring hardware states or events. Even if VMI can detect log generation, when VMI detects it after a mode transition to kernel space, logs are tampered by kernel-level malware. By contrast, kernel-level malware cannot tamper with logs because the trigger of log transfer is given by a library in the user space of each VM.

To inspect a state of a VM, VMI collects some information strongly related to a data structure in a VM. Thus, VMI must have enough knowledge of layout of data structure in the VM. Additionally, to inspect a state of a VM, VMI must collect a lot of information (e.g. process list, process descriptor). This creates strong dependency to version of OSes in VMs.

As just described above, VMI can inspect a state of a VM with fine-grained information; however, it creates strong dependency of data structure in a VM and some overheads. On the other hand, however the proposed system cannot collect much information of a VM; it achieves weak dependency of data structure in a VM and low overheads. VMI has large overhead because it monitors a state of a VM with various and fine-grained information. A research [13] shows that VMI causes 690% overhead in monitoring of process creation. On the other hand, in-VM monitoring causes only 13.7% overhead in that monitoring. Thus, the approach of the proposed system is efficient because the system can be considered as one of an in-VM monitoring. Additionally, our proposed system only monitors invocation of syslog function. Therefore, overhead related to the proposed system arises only when an AP invokes syslog function.

4   Implementation

4.1   Flow of Log Transfer

Transferring logs from a VM to a VMM takes place in two phases: requesting the log transfer and copying the log. This section describes the implementation of each phase in Section 4.2 and Section 4.3. The modified code to libc library is shown as Figure 4 and explained at Section 5.5. The overall flow is as follows:

(1) An AP in the target VM requests a log transfer.
(2) A VM exit occurs and the VMM receives the request.
(3) The VMM copies logs from the AP to the VMM buffer.
(4) The VMM sends a log storing request to the logging VM.
(5) The logging VM receives the request and notifies the VMM that it is ready to receive the logs.
(6) The VMM copies the logs to the logging VM.
(7) The logging VM stores the logs to a file.

4.2   Request of Log Transfer

We embed a cpuid instruction in a library to request a log transfer to the VMM from an AP. The instruction does not affect the CPU state; however, if executed in a virtualized environment, the instruction causes a VM exit. Therefore, we embedded the instruction into a library to request the copying of logs to the external VMM before sending the logs to a logging daemon. The interface of log transfer request is shown in Table 1. The embedded codes set the appropriate values to the registers and execute the cpuid instruction. Additional codes are shown in Section 5.5. We utilize cpuid instruction to counteract detection of our approach that scans memory or a library file. One of a typical instruction to call a VMM is vmcall. If we use vmcall instruction as a trigger of log transfer, adversaries easily detect our approach by scanning a memory because the instruction is not used in regular APs. To make detection of our approach harder, we utilize cpuid instruction.

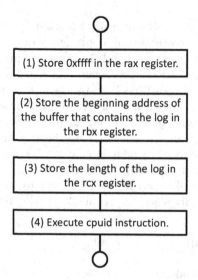

Fig. 2. Flow of a log transfer request

Table 1. Interface of log transfer

Register	Explanations
rax	0xffff: the value represents a log transfer request.
rbx	Address of the buffer that contains logs to transfer.
rcx	Length of logs to transfer.

Figure 2 depicts the flow of a log transfer request. At first, the AP on the target VM stores 0xffff in the rax register, the beginning address of the buffer in the rbx register, and the length of the buffer in the rcx register. Then, the AP executes `cpuid` instruction to request a log transfer.

4.3   Log Copying from a VM to a VMM

Figure 3 depicts the flow of log copying by a VMM. A `cpuid` instruction is a trigger for log transfer. After detected the instruction, the VMM copies logs from the AP and notifies to the logging VM if the value contained in the guest's rax register is 0xffff. If not, the VMM do not copy logs and only emulates the instruction. The buffer inside the VMM is implemented as a ring buffer to reduce the loss of logs in a high-load situation. Step (4) only sends notification. Log copying to the logging VM is made asynchronously. Thus, the time of log copying is kept as short as possible.

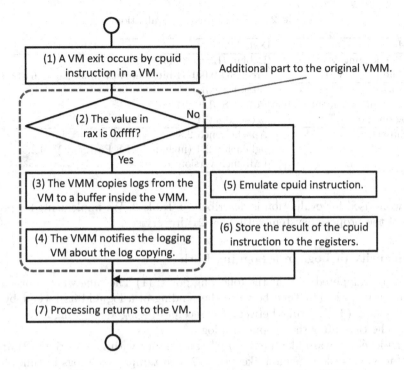

Fig. 3. Flow of log copying from the AP to the VMM

5 Evaluation

5.1 Purpose and Environment

We evaluated the proposed system at following standpoints:

- Security of logs in a logging path
  To evaluate the ability of prevention of log tampering, we inserted a malware into the kernel running on a VM.
- Prevention of log tampering and loss
  To check whether the proposed method can prevent log tampering and loss, we tried to prohibit log storing procedure with malware and some attacks.
- Completeness of log collection
  By sending a massive number of log transfer requests from an AP in the target VM, we tested the system in a high-load environment.
- Efforts for adapting various OSes
  Ease of adaptation to various OSes was also evaluated.
- Performance evaluation
  Performance overhead in Database Management System (DBMS) is also evaluated.
- Performance in multi-VM environment
  We measured performance of a web server with many VMs to clarify performance overhead incurred by the proposed system in multi-VM environment.

Table 2. Software used for evaluation

VMM	Xen 4.2.0
OS (The logging VM)	Debian (Linux 3.5.0 64-bit)
OS (The target VM)	FreeBSD 9.0.0 64-bit, Debian (Linux 2.6.32 64-bit)
Web server	thttpd 2.25b
Database management system	PostgreSQL 9.2.4
Syslog daemon	rsyslogd 4.6.4
Benchmark	ApacheBench 2.3
	pgbench 9.2.4 (included with PostgreSQL 9.2.4)
	LMbench version 3

Software used for evaluation is described in Table 2. We implemented a prototype of the proposed system with Xen [21] hypervisor.

5.2 Security of Logs in a Logging Path

Logs can be tampered with at the following point: (1) The time when a process generates a log, (2) The time between the sending of a log and its receipt by a syslog daemon, (3) The time between the receipt of a log and storing it to a file, and (4) The time after the output of a log.

Kernel-level malware like *adore-ng* [3] can tamper with logs in time (2) and (3). Attacks for syslog daemon like *tuxkit* [2] can tamper with logs in time (3). Adversaries who have privileges to write to the log file can tamper with logs in time (4). Our proposed method can prevent attacks in time (2), (3), and (4) because logs are transferred to outside of the VM before it reaches in a kernel.

Without hypervisor-based software runtime memory protection mechanism [12], we cannot prevent log tampering in time (1). A kernel-level malware can manipulate memory of user processes; therefore, it tampers with logs before they are transferred to out of a VM. With the memory protection mechanism [12], a kernel-level malware cannot tamper with logs of user processes. Thus, to prevent log tampering in time (1), the memory protection mechanism [12] is necessary.

5.3 Prevention of Log Tampering and Loss

To check whether the proposed method can prevent log tampering or not, we tried to tamper with logs. First, we used *adore-ng* [3], which is a kernel-level malware that tamper with logs sent to the syslog daemon, to check if the proposed system can prevent log tampering in kernel space. The *adore-ng* patches runtime memory of kernel code to tamper with logs. The *adore-ng* monitors inter-process communication using socket function and deletes a message if it contains disadvantageous words for the adversary. This experiment proves that the proposed method can prevent log tampering by the kernel-level malware. Logs sent to the VMM with the proposed method were not tampered with while logs stored in the target VM is tampered with. Moreover, we can find log tampering by comparing logs between the target VM and the logging VM. With this comparison, we can estimate a purpose of the adversary.

Third, we tampered with a policy file of syslog daemon as no logs are written to files. The policy file is loaded by the syslog daemon at a start-up. By this attack, no logs are written even if a syslog daemon is running. In this situation, we confirmed that the proposed system collects logs with no modification or loss. This result shows that the proposed system is resistant to attack for policy file of syslog daemon. The result also shows that log tampering by replacing a syslog daemon has no effect in a log collected by the proposed system. Thus, the proposed system is resistant to attacks like *tuxkit* [2].

Fourth, we stopped a syslog daemon on a target VM to prevent logging. Obviously, no logs are transferred to the syslog daemon. We also confirmed that the proposed system can collect logs completely. However, its completeness depends on a flow of the log transfer. In GNU libc, a syslog function aborts log transfer when the establishment of a connection is failed. Our prototype used for evaluation requests log transfer before establishing a connection to the syslog daemon; therefore we can collect logs completely. This implies that logs might be lost if the library requests log transfer after establishing connection.

Finally, we tampered with a log file. This type of attack is used in *LastDoor* backdoor [4]. It wipes specific entries in log files. Because the logs written to the file are already transferred to the logging VM, while logs in the target VM are tampered with, there is no effect to the log file in the logging VM.

These results show that the proposed system can collect almost all logs and collected logs are not affected by attacks on the target VM. Additionally, adversaries tend to install log tampering malware to a place where all logs go through. For example, *adore-ng* [3] is installed to a kernel function and *tuxkit* [2] is installed to a syslog daemon. All logs sent by syslog library function go through that kernel function and syslog daemon. From the reason, we can estimate that log tampering attacks to an AP, which is a source of logs, is rare.

5.4 Completeness of Log Collection

To ensure that the proposed system can collect all logs in the target VM with no loss, we tested the proposed system in a high-load environment. In an experiment, we sent a log transfer request 10,000 times within approximately 0.26 seconds. The length of the log in each request was approximately 30 bytes. All logs were successfully transferred to the logging VM. No logs were incomplete or lost. This result shows that our proposal is sufficient in terms of completeness of log collection in a high-load environment.

5.5 Efforts for Adapting Various OSes

In the prototype, we implemented the proposed method with FreeBSD and Linux as a target VM and Xen as a VMM. To adapt to various OSes, modification to the target VM must be minimal. We added 20 additional lines of codes to libc on FreeBSD and Linux. Figure 4 shows the result of diff command. As shown is Figure 4, we can adapt the proposed system to the libc library by inserting cpuid_logxfer() function before invocation of a send system call. The rest of

```
void
  __vsyslog_chk(int pri, int flag, const char *fmt, va_list ap)
  {
          int saved_errno = errno;
          char failbuf[3 * sizeof (pid_t) + sizeof "out of memory []"];

+         reg_t regs;
+         regs.rax = 0xffff;
+
  #define  INTERNALLOG LOG_ERR|LOG_CONS|LOG_PERROR|LOG_PID
          /* Check for invalid bits. */
          if (pri & ~(LOG_PRIMASK|LOG_FACMASK)) {
***************
*** 278,283 ****
--- 297,308 ----
          if (LogType == SOCK_STREAM)
            ++bufsize;

+         regs.rbx = (unsigned long)buf;
+         regs.rcx = bufsize;
+
+         cpuid_logxfer(regs.rax, &regs);
+         regs.rax = regs.rbx = regs.rcx = 0;
+
          if (!connected || __send(LogFile, buf, bufsize, send_flags) < 0)
            {
            if (connected)
```

Fig. 4. The result of diff command between source codes of the unmodified library and the modified library

the additional codes are definition of the **regs** structure and the **cpuid_logxfer** function. These additional lines consist of (1) setting the registers with the appropriate values and (2) executing the **cpuid** instruction. Based on the size of the additional code, adapting the proposed system to various OSes would be a small effort.

5.6   Performance Evaluation

Measured Items and Environment. We measured the performance of the syslog function, some system calls, and an AP. We also measured performance overhead in multi-VM environment. The performance measurements of both the syslog function and an AP show the additional overhead incurred by the proposed system. On the other hand, the performance measurement of some system calls shows that the proposed system causes additional overhead only when the syslog function is called.

We measured the performance with a computer, which has Core i7-2600 (3.40 GHz, 4-cores) and 16 GB memory. In each measurement, one virtual CPU (VCPU) is provided and 1 GB memory is allocated to each VM. Hyper-threading is disabled. Each VCPU is pinned to physical CPU core to avoid the instability of measurement. If many VMs work on one physical CPU, performance of APs on those VMs would be instable. Each VM has one VCPU and 1 GB memory.

Table 3. Performance comparison of the syslog function

	Time (μs)	Overhead (μs (%))
Xen	31.47	–
Proposed system	33.38	1.91 (6.08%)

Table 4. Frequency of library function calls when providing a web page with thttpd web server

Function name	Count	Rate (%)	Function name	Count	Rate (%)
strncasecmp	1600	17.77	strftime	200	2.22
strlen	1400	15.55	accept	200	2.22
strcpy	800	8.89	gmtime	200	2.22
vsnprintf	600	6.67	__errno_location	200	2.22
memmove	400	4.44	time	100	1.11
strchr	400	4.44	close	100	1.11
select	301	3.34	read	100	1.11
gettimeofday	301	3.34	getnameinfo	100	1.11
strstr	300	3.33	strcat	100	1.11
fcntl	300	3.33	readlink	100	1.11
strpbrk	300	3.33	strrchr	100	1.11
strcasecmp	200	2.22	syslog	100	1.11
__xstat	200	2.22	writev	100	1.11
strspn	200	2.22			

Syslog Function and System Call. In the proposed system, the modified library requests log transfer when an AP called syslog function. To clarify the overhead incurred by the proposed system, we measured and compared the performance of syslog function with unmodified Xen and the proposed system. Table 3 compares the performance of the syslog function between Xen and the proposed system. In the proposed system, the additional overhead of the syslog function is 1.91 μs (6.08%), which is small enough, because the function is not called frequently.

Table 4 shows counts of function call in thttpd accessed by ApacheBench for 100 times. We measured the number of counts of library function call by ltrace. Table 4 shows the ratio of syslog function call in thttpd is about 1%. Additionally, we measured a performance impact of library functions in thttpd with the same workload. Table 5 shows the result of measurement. These results are measured in Ubuntu 13.04. The function named __syslog_chk is same as syslog. As shown in Table 5, performance impact of syslog function is only 0.18%, thereby it can be considered as 6.08% of overhead in syslog function has limited impact of the performance of APs.

Additionally, we measured the performance of some system calls by LMbench, which measures the performance of file creation and deletion, process creation, system call overhead, and other processes. In this measurement, the additional overhead is not significant.

Table 5. Performance impact of library functions in thttpd

Function name	Rate (%)	Function name	Rate (%)
writev	76.90	memmove	0.13
poll	17.71	gmtime	0.10
strncasecmp	0.82	strcasecmp	0.10
strlen	0.81	strftime	0.10
strcpy	0.51	strspn	0.09
close	0.30	strcat	0.09
__vsnprintf_chk	0.30	read	0.08
__xstat	0.23	getnameinfo	0.05
strchr	0.22	memcpy	0.05
fcntl	0.22	time	0.05
__syslog_chk	0.18	strrchr	0.05
accept	0.17	__strcpy_chk	0.04
readlink	0.15	malloc	0.02
strpbrk	0.14	mmap	0.00
strstr	0.14	open	0.00
__errno_location	0.14	realloc	0.00
gettimeofday	0.14		

Table 6. Performance comparison of a PostgreSQL

tmpfs	VMM	TPS	Relative performance
disabled	Xen	400.37	–
	Proposed system	395.76	0.99
enabled	Xen	1,448.80	–
	Proposed system	1,372.60	0.95

Performance of AP. We measured performance overhead by the proposed system on a DBMS. To measure the performance overhead caused by the proposed system in DBMS, we used PostgreSQL as a DBMS. We configured PostgreSQL to call syslog function in each transaction. We used pgbench to measure performance of PostgreSQL. The workload with pgbench includes five commands per transaction. The benchmark measures transactions per second (TPS) of a DBMS. The concurrency of transactions is set to one.

Table 6 shows the comparison of a performance of the PostgreSQL DBMS. Higher TPS is better. Performance degradation with the proposed method is less than 1%. The proposed method degrades performance of a CPU intensive process. Because PostgreSQL accesses to disk heavily, the overhead incurred with the proposed method becomes small. To clarify that the proposed system is CPU intensive, we measure the performance with tmpfs, which provides a memory file system. Transactions do not require access to disk; therefore, performance overhead with the proposed method would be higher. Table 6 shows that the relative performance to unmodified Xen with tmpfs is about 5%. The performance degradation is higher than that in the case without tmpfs. If a processing is I/O intensive, performance degradation with the proposed method becomes

Table 7. Throughputs of a web server (request/s) in multi-VM environment

File size	VMM	Number of VM						
		0	2	4	6	8	10	12
	Xen	1396.9	1329.27	1295.61	1225.22	1171.51	1231.72	1172.15
1 KB	Proposed system	1231.06	1150.54	1057.95	1017.53	987.24	1015.69	946.61
	Relative performance	0.88	0.87	0.81	0.83	0.84	0.82	0.8
	Xen	680.61	658.15	639.76	627.9	628.56	609.45	615.64
10 KB	Proposed system	664.48	626.12	612.93	559.02	582.24	578.89	589.58
	Relative performance	0.98	0.95	0.89	0.92	0.93	1.00	0.96
	Xen	11.41	11.41	11.4	11.39	11.38	11.39	11.39
1,000 KB	Proposed system	11.41	11.41	11.4	11.39	11.37	11.39	11.06
	Relative performance	1.00	1.00	1.00	1.00	1.00	1.00	0.98

less. Thus, the proposed method is suitable for I/O intensive APs. In this measurement, we configured PostgreSQL to call syslog in each transaction, however, logging frequency in general use of DBMS becomes less. Thus, the performance degradation can be assumed as almost negligible in normal use.

Performance in Multi-VM Environment. To examine the ability of our proposal to scale to its target of many domains, we measured a performance of a web server in a VM with many other VMs. These VMs have a process that sends logs using syslog function every second. This evaluation is experimented with the machine that has four CPU cores; the logging VM is placed on the core 0, a VM that has a web server is placed on core 1, and other VMs are placed on core 2 and core 3 to measure the pure performance changes of the web server. We placed 2, 4, 6, 8, 10 and 12 VMs on core 2 and core 3. The number of VMs on core 2 and core 3 is same. Scheduling priority of each VM is configured as same. The performance is measured by ApacheBench on a remote machine with 1 Gbps network.

Table 7 shows performance in each environment. Figure 5 shows changes of performance in each environment. If the number of VM increases, the performance of the web server degrades. Performance degradation with the proposed system is less than about 10% when the file size is larger than 10 KB. Especially, when the file size is 1,000 KB, performance degradation is nearly 0. From the result, we can estimate that change of relative performance related to the number of VM is small enough. Despite the number of VM changes, change of relative performance is approximately same. For this reason, the proposed system is efficient in multi-VM environment.

6   Related Works

6.1   Secure Logging

Accorsi classified and analyzed secure logging protocols [22]. In that paper, extensions of syslog, including syslog-ng [15], syslog-sign [16], and reliable syslog

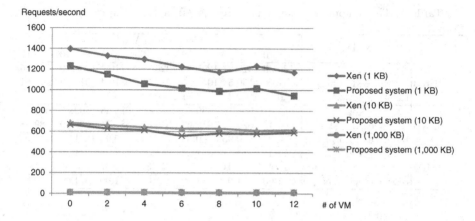

Fig. 5. Performance comparison in multi-VM environment. Horizontal axis shows the number of other VM. Vertical axis shows throughput of a web server in request/s; higher measurements are better.

[17] are distinguished as a protocol that provides security in transmission of log messages, not for storage phase. We focus on transmission phase because our proposal is highly related to that phase. Accorsi described that only reliable syslog fulfills security requirements that guarantee the authenticity of audit trails. Even if those protocols can detect and verify log message as not tampered, they cannot prevent deletion or tampering of logs. At this point, those protocols are different from our proposal. Therefore, this paper proposes a protection of log messages from a viewpoint of system security. By combining our proposal and existing secure logging protocol, we can increase security of logged data.

6.2   Logging with Virtual Machine

ReVirt [18] logs non-deterministic events on a VM for replay. Because it logs events for analysis of attacks, types of data are different from our proposal. While ReVirt logs instruction-level information, our proposal collects log messages for syslog. With our proposal, we can easily monitor the target VM without deep analysis of logged information because those logs are already formatted.

Virtual machine is also used to separate logged information [23]. While reference [23] separates information about file system logs, our proposal separates logs for syslog. They utilized split device driver model of Xen and it is provided for para-virtualization, thus, their proposal can be applied only for para-virtualized environment. Our prototype is implemented with fully virtualized environment; however, implementing in para-virtualized environment is easy.

VMI [11] and other introspection method [13] can be considered as a logging method with a VM. In that regard, these methods are similar to our proposal. However, information gathered by those methods are not formatted like syslog, therefore, to analyze these data, existing tools are unavailable. In contrast, with our proposal, existing tools work well without modification because the format of information gathered by our proposal is same as messages produced by syslog. Our previous work [20] can gather information from a VM without modification to a library in that VM. However, to adapt to various OSes, it requires modification to a VMM. Modification to a VMM requires restart of all VMs on that VMM. Besides, it causes measurable overheads. By contrast, although modification to a library on a VM is required, our proposal in this paper requires no modification to a VMM to adapt to various OSes and has less overhead.

7 Conclusions

The secure log transfer method by replacing a library in a VM provides processes on a VM with an ability to transfer logs without involving the VM kernel. Thus, even though kernel-level malware tamper with logs on that VM, logs gathered by our proposal have no effect. In addition, we implemented the proposed system with VMM, therefore, attacking the proposed system from a target VM is difficult enough because of the property of a VMM. Further, adapting the method to various OSes is easy because of its implementation with library modifications. Evaluation of resistance for log tampering shows that tampering of logs from the target VM is difficult enough. From the experiment of adapting different OSes showed that an effort of adaptation is only 20 lines of additional code to libc library. Performance evaluation shows that performance degradation of syslog function is only about 6%. Performance degradation is negligible if a processing of an AP is I/O intensive. Performance evaluation in multi-VM environment shows that the proposed system has enough performance with many VMs.

References

1. Kent, K., Souppaya, M.: Guide to computer security log management, special publication 800-92 (September 2006)
2. spoonfork: Analysis of a rootkit: Tuxkit,
   http://www.ossec.net/doc/rootcheck/analysis-tuxkit.html
3. stealth: Announcing full functional adore-ng rootkit for 2.6 kernel,
   http://lwn.net/Articles/75991/
4. Symantec: Backdoor.lastdoor, http://www.symantec.com/security_response/
   writeup.jsp?docid=2002-090517-3251-99
5. Subashini, S., Kavitha, V.: A survey on security issues in service delivery models of cloud computing. Journal of Network and Computer Applications 34(1), 1–11 (2011)
6. Grobauer, B., Walloschek, T., Stocker, E.: Understanding cloud computing vulnerabilities. IEEE Security & Privacy 9(2), 50–57 (2011)
7. Marty, R.: Cloud application logging for forensics. In: Proceedings of the 2011 ACM Symposium on Applied Computing, SAC 2011, pp. 178–184 (2011)
8. Chen, P.M., Noble, B.D.: When virtual is better than real. In: Proceedings of the Eighth Workshop on Hot Topics in Operating Systems, HOTOS 2001, pp. 133–138. IEEE Computer Society (2001)

9. Boeck, B., Huemer, D., Tjoa, A.M.: Towards more trustable log files for digital forensics by means of "trusted computing". In: International Conference on Advanced Information Networking and Applications, pp. 1020–1027 (2010)
10. Seshadri, A., Luk, M., Qu, N., Perrig, A.: Secvisor: a tiny hypervisor to provide lifetime kernel code integrity for commodity oses. SIGOPS Oper. Syst. Rev. 41(6), 335–350 (2007)
11. Garfinkel, T., Rosenblum, M.: A virtual machine introspection based architecture for intrusion detection. In: Proceedings of the Network and Distributed Systems Security Symposium, pp. 191–206 (2003)
12. Dewan, P., Durham, D., Khosravi, H., Long, M., Nagabhushan, G.: A hypervisor-based system for protecting software runtime memory and persistent storage. In: Proceedings of the 2008 Spring Simulation Multiconference, SpringSim 2008, pp. 828–835 (2008)
13. Sharif, M.I., Lee, W., Cui, W., Lanzi, A.: Secure in-vm monitoring using hardware virtualization. In: Proceedings of the 16th ACM Conference on Computer and Communications Security, CCS 2009, pp. 477–487 (2009)
14. Adiscon: rsyslogm, http://www.rsyslog.com/
15. Security, B.I.: Syslog server | syslog-ng logging system, http://www.balabit.com/network-security/syslog-ng
16. Kelsey, J., Callas, J., Clemm, A.: Signed syslog messages (May 2010), http://tools.ietf.org/html/rfc5848
17. New, D., Rose, M.: Reliable delivery for syslog (November 2001), http://www.ietf.org/rfc/rfc3195.txt
18. Dunlap, G.W., King, S.T., Cinar, S., Basrai, M.A., Chen, P.M.: Revirt: enabling intrusion analysis through virtual-machine logging and replay. SIGOPS Oper. Syst. Rev. 36(SI), 211–224 (2002)
19. Ibrahim, A., Hamlyn-Harris, J., Grundy, J., Almorsy, M.: Cloudsec: A security monitoring appliance for virtual machines in the iaas cloud model. In: 5th International Conference on Network and System Security, pp. 113–120 (September 2011)
20. Sato, M., Yamauchi, T.: Vmm-based log-tampering and loss detection scheme. Journal of Internet Technology 13(4), 655–666 (2012)
21. Barham, P., Dragovic, B., Fraser, K., Hand, S., Harris, T., Ho, A., Neugebauer, R., Pratt, I., Warfield, A.: Xen and the art of virtualization. SIGOPS Oper. Syst. Rev. 37(5), 164–177 (2003)
22. Accorsi, R.: Log data as digital evidence: What secure logging protocols have to offer? In: Proceedings of the 33rd Annual IEEE International Computer Software and Applications Conference, vol. 02, pp. 398–403 (2009)
23. Zhao, S., Chen, K., Zheng, W.: Secure logging for auditable file system using separate virtual machines. In: 2009 IEEE International Symposium on Parallel and Distributed Processing with Applications, pp. 153–160 (2009)

Static Integer Overflow Vulnerability Detection in Windows Binary

Yi Deng, Yang Zhang, Liang Cheng, and Xiaoshan Sun

Institute of Software, Chinese Academy of Sciences, Beijing, China
{dengyi,zhangyang,chengliang,sunxs}@tca.iscas.ac.cn

Abstract. In this paper, we present a static binary analysis based approach to detect integer overflow vulnerabilities in windows binary. We first translate the binary to our intermediate representation and perform Sign type analysis to reconstruct sufficient type information, and then use dataflow analysis to collect suspicious integer overflow vulnerabilities. To alleviate the problem that static vulnerability detection has high false positive rate, we use the information how variables which may be affected by integer overflow are used in security sensitive operations to compute priority and rank the suspicious integer overflow vulnerabilities. Finally the weakest preconditions technique is used to validate the suspicious integer overflow vulnerabilities. Our approach is static so that it does not run the software directly in real environment. We implement a prototype called EIOD and use it to analyze real-world windows binaries. Experiments show that EIOD can effectively and efficiently detect integer overflow vulnerabilities.

Keywords: Binary analysis, Integer overflow, Priority ranking, Weakest Precondition.

1 Introduction

Integer overflows are dangerous: while the integer overflow itself is usually not exploitable, it may trigger other classes of vulnerabilities, including stack overflows and heap overflows. The number of integer overflow vulnerabilities has been increasing rapidly in recent years. Common Vulnerability and Exploit (CVE) shows that more and more integer overflows have been recorded[1].

In the past few years, some tools have been presented to detect or prevent integer overflows in source code, such as CCured[2], Cyclone[3], BLIP[4], RICH[5], LCLint[6], IntPatch[7], they either translates the program into type safe language, or checks the code for certain operations during compiling. However, as for many programs like common off-the-shelf(COTS) programs, source code is not available to users, the state-of-the-art techniques have to detect integer overflow vulnerabilities at binary level, several approaches are proposed. IntScope[8] leverages symbolic execution and taint analysis to identify the vulnerable points of integer overflow, and reports suspicious integer overflow vulnerabilities. IntFinder[10] decompiles x86 binary code and use type analysis and

K. Sakiyama and M. Terada (Eds.): IWSEC 2013, LNCS 8231, pp. 19–35, 2013.

taint analysis to create the suspect instruction set, then dynamically inspects the instructions in the suspect set. They both use static analysis to find suspicious integer overflow vulnerabilities, then dynamically check each suspicious vulnerability. This mechanism has low efficiency because of high false positive rate of static analysis, detectors have to spend a large amount of time on checking each suspicious vulnerability to find real vulnerabilities.

Integer overflow is caused by arithmetic instruction, but not all arithmetic instructions could cause integer overflow vulnerabilities. By observing most known integer overflow vulnerabilities, we find some characteristics of integer overflow vulnerabilities, incomplete or improper input validation, overflow values using in security sensitive operations, no integer overflow checking, etc. Only the arithmetic instruction with these characteristics may cause integer overflow vulnerability, and the possibility is related to these characteristics. Based on the observation, we present EIOD, a static binary analysis based approach for detecting integer overflow vulnerabilities with a suspicious vulnerabilities ranking algorithm. In EIOD, we first find all the arithmetic instructions in the program, in order to check whether each arithmetic instruction could overflow, we collect the information related to these arithmetic instructions, such as how the overflowed value are used and whether there is existing checking for inputs and arithmetic results. If the values came from integer overflow points are used in security sensitive operations, we treated them as suspicious vulnerabilities, and use the information collected in first step to calculate the vulnerability priority of each suspicious integer overflow vulnerability which represented their vulnerability possibility and rank the suspicious integer overflow vulnerability by their priorities. This ranking strategy can help users to check those potential vulnerabilities that are most likely to be real vulnerabilities at first. Moreover, users can ignore suspicious vulnerabilities with low vulnerability priorities. Finally, unlike existing tools, we use weakest precondition(WP) computation to check suspicious vulnerabilities. The advantage of WP is that it is static so that we do not have to run the binary in real environment.

We have implemented our system EIOD on Microsoft Windows platform. EIOD first use IDA Pro[11] to disassemble binary executables, then translate to intermediate code, our analysis and detection are perform on the intermediate code. We use it to detect integer overflow vulnerabilities in some real world programs from Microsoft platform, and we have got encouraging experimental results: EIOD can effectively and efficiently detect integer overflow vulnerabilities.

Our paper makes three major contributions:

1. We propose a static systematic approach based on dataflow analysis and weakest precondition to detect integer overflow vulnerabilities in windows executables.
2. An effective suspicious vulnerability ranking scheme based on vulnerability likelihood has been used to alleviate the problem of high false positive rate of static analysis.

3. We implement a prototype EIOD and use it to analyze real world binaries. Experiments show that our approach can effectively and efficiently detect integer overflow vulnerabilities.

The rest of this paper is organized as follows: The background of integer overflow detection is described in section 2. In section 3, we present the overview of our approach. The design details are described in section 4. Section 5 gives the evaluation of our tool. Section 6 discusses related work on detecting integer overflow vulnerabilities. Section 7 concludes this paper.

2   Background

In this section, we will describe the characteristics of integer overflow vulnerabilities, and then discuss the challenges of integer overflow detection at binary level.

2.1   Characteristics of Integer Overflow

By studying more than 200 integer overflow case on CVE, we concluded the following characteristics of integer overflow vulnerabilities:

1. Incomplete or improper input validation. Almost all the integer overflow vulnerabilities are related to incomplete or improper input validation, because if input values have been completely checked, they could be safely used in program and will not cause integer overflow.
2. Integer overflowed values are used in sinks. Not all the integer overflow is harmful, but depends on where and how the program uses the overflowed value. If an overflowed value is used in some security sensitive points, it will be dangerous, because it may lead to other vulnerabilities, such as buffer overflow vulnerability. We call these security sensitive points as sinks. From our case studies, we summarize the sinks as following:
   - Memory allocation function: Memory allocation functions directly manipulate the memory space, if overflowed value is used as the size argument, the allocated memory will be insufficient, which may be used by attackers to make buffer overflow attacking.
   - Memory copy function: Memory copy functions copy memory from source to destination, if overflowed value is used as the size argument, it will lead to buffer overflow.
   - Memory offset: Memory offset is often used with base address to access the memory. if overflowed value is used as memory offset, such as array index, it may cause arbitrary bytes memory read or overwritten.
   - Branch statement: if the overflowed value is used in a branch statement, and the branch statement is not designed to catch the integer overflow, it could lead to bypass security checks or result in an undesirable execution.
3. Absence of integer overflows checking. Integer overflow checking is usually used after integer overflow points to detect integer overflows and prevent dangerous operation on overflow values. Almost all the integer overflow vulnerabilities have no integer overflow checking.

2.2   Challenges

As we use static approach to detect integer overflow vulnerabilities in binaries, it may encounter some challenges:

1. Lack of high level semantics. Binary is different from source code, there is no high level semantics in binary, such as function information, type of variables which is important to integer overflow detection.
2. High false positive rate of static analysis. It is not simple to rule out false positive report, especially in binary code.
3. Distinguish harmless integer overflow. There are some benign integer over-flow existing in programs[8], for example, GCC compiler use integer overflow to reduce a comparison instruction, we cannot treat them as vulnerabilities. Another harmless integer overflow with integer overflow checking, such as:

```
add eax, ebx
cmp eax, ebx
ja  target
```

In this case, programmer has checked whether the sum is overflow, so we should treat it as harmless integer overflow.

3   System Overview

In this section, we will describe the architecture and working process of EIOD.

The main disadvantage of static analysis based vulnerability detection ap-proaches is the high false positive rate, too many false alarm reported by static vulnerability detection restrict static vulnerability detection to be used in large program, users have to spend much time on identifying real integer overflow vulnerabilities. We found not every suspicious vulnerabilities reported by static vulnerability detection have same possibility to be real vulnerability, but existing methods treat them as same. To solve this problem, we presented a suspicious in-teger overflow vulnerabilities ranking algorithm which use the information about how the suspicious integer overflow vulnerabilities are used in program to evalu-ate their possibilities, then rank the suspicious vulnerabilities by their possibili-ties. This algorithm can help identifying real vulnerabilities from false alarm.

EIOD consists of four components: (1) Binary lifting component. (2) Integer overflow vulnerability finder. (3)Integer overflow vulnerability validation compo-nent. The workflow of EIOD is as following:

First, to obtain high level semantics in binary, the Binary lifting component decompiles binary, and translates it into our designed intermediate representa-tion which includes enough information to detect integer overflow vulnerabilities mean while remove information which no need for integer overflow detection.

Second, Integer overflow vulnerability detection component finds all the arith-metic and treat them as potential integer overflow points. and collects informa-tion related to these points, including variables which store the arithmetic result and variables which store arithmetic operands, security sensitive instructions where the variables are used, the potential integer overflow points which value used in security sensitive instruction are recognized as suspicious integer overflow vulnerabilities, then use our priority algorithm to calculate the priority of each integer overflow point, the integer overflow points with high priority are considered be more possible to be real vulnerability than others.

Finally, we validate the potential integer overflow vulnerabilities by weakest precondition computation in the order of their priorities.

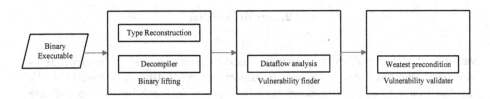

Fig. 1. Architecture of EIOD

4   Design Details

We first present the design of our intermediate representation in Section 4.1 and describe type analysis in Section 4.2. We discuss how to find potential integer overflow points and collect relating information in Section 4.3 and how to get suspicious integer overflow vulnerabilities in Section 4.4. In Section 4.5, we present integer overflow vulnerabilities ranking algorithm. Finally in Section 4.6, we give the validation method of suspicious vulnerabilities.

4.1   Intermediate Representation

It is complicated to directly analyze x86 binary instructions as the x86 instruction set is very complex. It has hundreds of different instructions and each instruction can have complex semantics. Moreover, instructions lack higher-level semantics, such as functions' information and variables' type. So it is necessary to design an intermediate language to simplify the representation of x86 instructions. By studying existing intermediate language such as Vine[13] IL, BIL[14], we found they not only have some grammar which is useless for detection of integer overflow vulnerability, and also lack necessary sign type information. Based on Vine IL, we defined our IR in SSA-like form, whose grammar is shown in Figure 2.

There are 6 different kinds of statement in our IR: (1) `Assign(`*var*`, `*exp*`)` assigns variable with expression *exp*. (2) `Jmp(`*label*`)` jumps to the statement *label*. (3) `Cjmp(`*exp*`, `*label1*`, `*label2*`)` is conditional jump, when *exp* is true, it jumps to *label1*, else it jumps to *label2*. (4) `Call(`*exp*`)` call function at address *exp*. (5) `Label(`*label*`)`. The variables in our IR are divided into two classes: memory and register, and memory variables include stack variables and heap

```
statement ::= Assign(var, exp) | Jmp(label) | Cjmp(exp, label1, label2)
              | Label(label) | Call(exp)
label     ::= string
exp       ::= Binop(exp, exp) | Unop(exp) | var | integer
Binop     ::= Add | Sub | Mul | Div ...
Unop      ::= Minux
var       ::= Mem(string, id, exp, l, s) | Reg(string, id, l, s)
l         ::= 64| 32 |16 |8 |1
s         ::= Signed | Unsigned
```

Fig. 2. Our IR grammar

variables. Stack variables are frequently used in program to store local variables and pass function arguments, so it is important to identify stack variables. We use the function name with stack variables' offset on stack bottom to indicate them, the offset is computed by data flow analysis. Memory and register variables both have two types: length type and sign type, length types are 64, 32, 16, 8, 1, denote the variable's length, and sign types are signed and unsigned, denote variable's sign information. Vine IL has provided variable's length type, we need to perform extended type analysis to get variables' sign type.

4.2 Sign Type Analysis

Variables' type information, include length and sign is necessary for generation of integer overflow condition. Existing variable type analysis in Vine only provides variables' length information, so we extend the type analysis. We use both control flow analysis and data flow analysis to reconstruct sign type.

Our sign type system include:

- $\top$, which corresponds to a variable being "any" sign, and $\bot$, which corresponds to a variable being used in a sign-inconsistent manner.
- *Signed*, *Unsigned*, which corresponds to a variable being signed or unsigned.
- Intersection types ($S1 \cap S2$) and union types ($S1 \cup S2$).

sign information reconstruction consists of two phases: sign type initialization and sign type propagation.

Sign Type Initialization. At first, we initialize the sign of every variable with $\top$, then traverse program, assign variables with sign type based on how the variables are used, there are some hints to initialize variables' sign type:

- For most memory allocation functions, the argument used as size is unsigned.
- Array index should be unsigned type.
- X86 conditional jump instructions give some sign information, e.g., variables compared by JG, JNLE, JGE, JNL, JNGE, JLE, JNG, JE and JNE has signed type, and variables compared by JA, JNBE, JAE, JNB, JB, JNAE, JBE, JNA, JE and JNE has unsigned type.

This step is performed by traversing program, when variable used in above manner is found, it assigned to corresponding sign type.

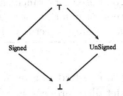

Fig. 3. Sign type lattice

Sign Information Propagation. We first propagate the sign information within a basic block, then traverse the control flow graph, and propagate the sign information to other basic blocks. Sign type propagation rule specify how sign information is propagate in IR. We use the following rule: sign type propagates by statement `Assign(var, exp)`, there are two situation:

- If *exp* is a arithmetic, i.e. `Binop(e1, e2)`, then the sign of *var* is $(S1 \cup S2)$.
- If *exp* is variable *var*, then the sign type of *var* is equal to *exp*.

The traversal is stopped when there is no new updated sign information.

4.3   Potential Integer Overflow Points

Definition. A vulnerability is a type of bug that can be used by attackers to alter the behaves of the software in a malicious way, such as overwrite critical data. Integer overflow is a category of vulnerability which caused by arithmetic's result being overflowed. Given P is a program, integer overflow point i is a program statement where integer overflow occurred, c is the condition which is necessary for integer overflow occurred at i, the vulnerability is defined as a tuple (P, i, c). Since we analyze at binary level, P is binary program and i is assembly instruction.

Potential Integer Overflow Points Searching. We consider arithmetics as potential integer overflow points. A potential integer overflow point p is denoted by a tuple (i, r, o), where i is program point which contain arithmetic, r is the arithmetic result which could overflow and o is the operand of the arithmetic. Potential integer overflow variable is the variable which could influenced by the arithmetic results, it is represented by a map from variable to the set of potential integer overflow points, denoted as $v \rightarrow (p_j, pc_j)$, $j = 1, 2, ...$, where pc is the potential integer overflow point p_j's propagation count before reaching the variable. We use function $F(v)$ represents v's potential integer overflow points.

Table 1. Type of sinks

type	description
Memcopy	used for memory copy
MemAlloc	used for memory alloc
BranchCond	used for branch select
MemAccess	used for memory access
OvCheck	integer overflow check

First, we traverse over program to find potential integer overflow points and store them in a set P. The detailed process is when encounter a statement `Assign(var, exp)` and exp is arithmetic, i.e. `Binop(e1, e2)`, where `Binop` is `Add`, `Sub` or `Mul`, we consider it as a potential integer overflow point, given the instruction address corresponding to this statement is a, the integer overflow point is denoted as $(a, var, \{e1, e2\}, 1)$.

Then forward data flow analysis is leveraged to collect potential integer overflow variables which are stored in a hash table H. The process detail is as following:

- For statement `Assign(var, exp)`, there are some situation:

  1. If exp is `Binop(v1, v2)`, we add $F(v1) \cup F(v2)$ with pc add 1 to var's potential integer overflow set.
  2. If exp is variable $rvar$, then add $F(rvar)$ to var's potential integer overflow set.

4.4 Suspicious Integer Overflow Vulnerability

Not every integer overflow points are dangerous, but only the ones which used in sinks could possibly cause vulnerability, we call them suspicious integer overflow vulnerability. Furthermore, the suspicious integer overflow vulnerabilities do not have the same possibility to be real vulnerability. Therefore, to estimate the likelihood that an integer overflow point be real vulnerability, we collect information that how the overflowed value at the integer overflow point is used, especially used in the sinks. We use pair $(s, type)$ to record a overflowed value produced at instruction i used in sink s, and $type$ represents the sink type. Table 2 lists types of sinks. Since overflowed value produced in a integer overflow point could used in many sinks, so we use a set of the pair $(s, type)$ to describe the information of how the integer overflow value is used, the set is called sinks set. For example, a overflowed value is used as a memory allocation function's size argument at program point $s1$, and also used in branch condition statement at program point $s2$, thus its sinks set is $((s1, \text{MemAlloc}), (s2, \text{BranchCond}))$.

We traverse over the program again to get the information about how the variables collected in above step are used. The detail is as following:

- For statement `Cjmp(var, label1, label2)`, we search var in hash table H, assume $F(var)$ is (p, pc), $weadd(s, \text{BranchCond})$ to p's sinks set, where s is address of the statement. In addition, we find the variables which the `Cjmp` related to, If they contain operand of an arithmetic, and the `Cjmp` statement is before the arithmetic in cfg, then we add $(s, \texttt{BranchCheck})$ to p's sinks set. If the variables are a operand of an arithmetic and the arithmetic's result, and the statement `Cjmp` is after the arithmetic in cfg, then we add $(s, \texttt{OvCheck})$ to p's sinks set.

- For statement `Call`(exp), if the function is memory allocation function or memory copy function, we find its size argument, denoted as sv, and search sv in hash table H, assume $F(sv)$ is (p, pc), then we add $(s, \texttt{MemAlloc})$ or $(s, \texttt{MemCopy})$ to p's sinks set, where s is address of the statement.
- For any statement using `Mem`$(\texttt{string}, id, var, l, s)$, we search var in the hash table H, assume $F(var)$ is (p, pc), then we add $(s, \texttt{MemAccess})$ to p's sinks set, where s is address of the statement.

After this step, we got the suspicious integer overflow points and their information about how they used in sinks.

4.5   Suspicious Integer Overflow Vulnerability Ranking

EIOD use the information collected from above step to estimate the suspicious integer overflow vulnerabilities' possibility being real vulnerability. We have obtained suspicious integer overflow vulnerabilities and their corresponding sinks set, they are used to compute priority of each integer overflow vulnerability which is representation of the vulnerable possibility of the integer overflow vulnerability. By studying existing integer overflow vulnerabilities, we proposed priority computing algorithm: each sink type is given a weight depending on how much the sink type using in existing integer overflow vulnerabilities. The specific rules are as follows:

- Memory alloc functions are main cause of integer overflow vulnerabilities, if memory alloc function use overflowed value as its size argument, it causes less memory to alloc than expected and leads to buffer overflow, i.e. integer overflow to buffer overflow vulnerability(IO2BO). According to NVD, more than a half of integer overflow vulnerabilities belong to this type. Therefore, we give the MemoryAlloc type high weight.
- Overflowed value using as memory index will cause arbitrary memory access, but it is few in existing integer overflow vulnerabilities, so it has low weight.
- When overflowed value used as condition statement, the security checking may be bypassed, there are some existing integer overflow vulnerabilities belongs to this type, so we give it middle weight.
- If programmers have checked whether arithmetic is overflowed, then the arithmetic is very likely not to cause integer overflow vulnerability, so it has a large negative weight.
- Checking for arithmetic operand reduces the possibility of integer overflow, but can't avoid the integer overflow, so we give it negative weight.

– Propagation count pc represent how many times a overflowed value has been transformed(like Add, Sub), each transform will reduce its threat, so for a sink, its weight equal to initial weight subtract $pc \times q$, where q is weaken factor.

In addition, we consider the situation when a overflowed value used in many sinks with same sink type, its vulnerable possibility will increase, but not in linear. Therefore, for a suspicious integer overflow vulnerability, assume its sinks set contains n_i of sinks with sink type s_i, $i = 1, 2, ..., n$, the priority is equal to:

$$p = \sum_{i=1}^{n}(1 + 1/2 + ... + 1/2^{n_i})(w_{s_i} - pc \times q) \tag{1}$$

The suspicious vulnerability ranking algorithm is configurable that users can customize the weight for each sink type for different application. We have tested our ranking algorithm in experiments to detect some windows executables and the results show the effectiveness of the ranking algorithm. It should be pointed out that users can also configure detection threshold for different scenarios, if they need high detection effectiveness and don't care some false negative rate, they can set a high threshold, otherwise, if they need balance on false positive rate and false negative rate, they can choose a middle threshold.

4.6 Suspicious Integer Overflow Vulnerability Validation

In this section, we describe our method for validating the suspicious integer overflow vulnerabilities.

In EIOD, we have got suspicious integer overflow vulnerability position, in this situation, the weakest precondition is a suitable technique to validate vulnerability which can compute weakest precondition from suspicious vulnerability position to entry along backward direction and thus avoid path explosion problem.

The validation takes suspicious integer overflow vulnerability as input, find the arithmetic cause this vulnerability and generates its vulnerable condition c according to table 2, then convert our IR to GCL to compute the weakest precondition $s = wp(G, c)$ over the GCL, where c is overflow condition and s is the weakest precondition at program beginning. Finally, STP[16] is used to solve s and get input stratifying s, if such input doesn't exist, we can treat the suspicious vulnerability as false alarm.

It should be noted that other technique such as fuzzing test can also be used to validate the suspicious integer overflow vulnerabilities.

5 Implementation

We have implemented our tool EIOD in Microsoft Windows, which includes the following components:

- Binary lifting component: We makes use of IDA Pro[11] as our decompiler. IDA Pro provides a lot of useful information: function boundaries and library functions' name. hence we can identify memory allocation functions and memory copy functions by their names. We translate assemble codes to our IR on top of Valgrind[12] and Vine, the Sign type analysis is implemented in 1k of OCaml codes.
- Integer overflow vulnerability detection component: we reuse some module of Vine, and add about 2.3k of OCaml codes to implement it.
- Vulnerability validation component: we implement our weakest precondition computation by using existing Vine's module with adding 1.2k OCaml codes. The condition checker is built on top of STP[16], a decision procedure for bit-vectors and arrays.

6   Evaluation

To verify the effectiveness and efficiency of EIOD, we have conducted a number of experiments. All the experiments are performed on an Intel i7 cpu (2.6 GHz) with 4GB memory running the Microsoft Windows XP. We first evaluate the effectiveness in Section 4.1, then measure the efficiency in Section 4.2.

6.1   Effectiveness

We use EIOD to detect comctl32.dll and gdi32.dll, they both exist known integer overflow vulnerability. EIOD successfully detected both known integer overflow vulnerabilities. We presented the result in the following.

DSA_SetItem Integer Overflow Vulnerability[15]. DSA_SetItem is a function in comctl32.dll used to set the contents of an item in a dynamic structure array. DSA_SetItem has three parameters: *hdsa* is a pointer to a dynamic structure array, *index* is an index for the item in *hdsa* to be set, and *pItem* is a pointer to a new item data which will replace the item specified by *index*. If *index* is greater than ($hdsa{\rightarrow}nMaxCount$), DSA_SetItem calls ReAlloc to allocate a new buffer. A large *index* can trigger a integer overflow in nNewItems $hdsa{\rightarrow}nItemSize$, where $hdsa{\rightarrow}nItemSize$ is the size of an item, resulting return a smaller size than expected.

Table 2 is the detection result of DSA_SetItem. There are 272 potential integer overflow points, Table 2 only list 19 points which has highest priority. The CVE-2007-3034 vulnerability is at integer overflow point 0x5d1aca26, from the table, we can see that it has priority 55, its ranking position is 2, higher than other 270 potential overflow points.

Figure 4 is the validation result of DSA_SetItem integer overflow vulnerability, when the condition in figure 4 is satisfied, integer overflow will be triggered, it means integer overflow vulnerability at 0x5d1aca26 is real vulnerability.

Table 2. Detection result of comctl32.dll

address of suspicious integer overflow point	sign	prority
0x5d190146	Signed	41
0x5d187fe0	Signed	41
0x5d174bef	Unsigned	42
0x5d19f1a0	Signed	42
0x5d1b08a9	Unsigned	43
0x5d180c80	Signed	43
0x5d1b088f	Unsigned	43
0x5d1acd54	Unsigned	44
0x5d180c82	Signed	45
0x5d19f1ca	Signed	53
0x5d1aca23	Signed	53
0x5d190d36	Signed	53
0x5d19f1cc	Signed	55
0x5d1aca26	Signed	55
0x5d1a32f4	Unsigned	65

```
ASSERT( mem_b_1065427[0hex00000000FFFFFFFD] = 0hex00 );
ASSERT( mem_b_1065427[0hex00000000FFFFFFFA] = 0hex00 );
ASSERT( mem_b_1065427[0hex00000000FFFFFFFE] = 0hex00 );
ASSERT( mem_b_1065427[0hex0000000000000002] = 0hex00 );
ASSERT( mem_b_1065427[0hex0000000000000006] = 0hex00 );
ASSERT( mem_b_1065427[0hex0000000000000005] = 0hex00 );
ASSERT( mem_l_1065428[0hex000000000000000C] = 0hexFFFFFFFF );
ASSERT( mem_b_1065427[0hex00000000FFFFFFF6] = 0hex00 );
ASSERT( mem_l_1065428[0hex0000000000000000] = 0hex00000000 );
ASSERT( mem_b_1065427[0hex00000000FFFFFFF7] = 0hex00 );
ASSERT( mem_b_1065427[0hex0000000000000000] = 0hex00 );
ASSERT( mem_b_1065427[0hex00000000FFFFFFF9] = 0hex00 );
ASSERT( mem_b_1065427[0hex00000000FFFFFFF5] = 0hex00 );
ASSERT( mem_l_1065428[0hex0000000000000008] = 0hex00000000 );
ASSERT( R_ESP_1 = 0hex00000004 );
ASSERT( R_EBX_6 = 0hex00000000 );
```

Fig. 4. Validation Result of DSA_SetItem Integer Overflow Vulnerability

```
ASSERT( mem_l_614033[0hex0000000000000008] = 0hex00000000 );
ASSERT( mem_b_614032[0hex00000000FFFFFFF1] = 0hex00 );
ASSERT( mem_b_614032[0hex00000000FFFFFFF2] = 0hex00 );
ASSERT( mem_l_614033[0hex000000000000000C] = 0hex00000000 );
ASSERT( mem_b_614032[0hex00000000FFFFFFF3] = 0hex00 );
ASSERT( mem_b_614032[0hex00000000FFFFFFED] = 0hex00 );
ASSERT( mem_b_614032[0hex00000000FFFFFFEE] = 0hex00 );
ASSERT( mem_b_614032[0hex00000000FFFFFFEF] = 0hex00 );
ASSERT( R_ESP_6 = 0hex00000000 );
ASSERT( mem_b_614032[0hex0000000000000011] = 0hex00 );
ASSERT( mem_b_614032[0hex0000000000000010] = 0hex00 );
ASSERT( mem_b_614032[0hex00000000FFFFFFF9] = 0hex00 );
ASSERT( mem_b_614032[0hex00000000FFFFFFFA] = 0hex00 );
ASSERT( mem_l_614033[0hex0000000000000004] = 0hex00000000 );
ASSERT( mem_b_614032[0hex00000000FFFFFFFB] = 0hex00 );
ASSERT( mem_b_614032[0hex0000000000000000] = 0hex00 );
ASSERT( mem_b_614032[0hex0000000000000001] = 0hex00 );
ASSERT( mem_b_614032[0hex0000000000000002] = 0hex00 );
ASSERT( mem_b_614032[0hex00000000FFFFFFF5] = 0hex00 );
ASSERT( mem_b_614032[0hex00000000FFFFFFF6] = 0hex00 );
ASSERT( mem_b_614032[0hex00000000FFFFFFF7] = 0hex00 );
ASSERT( R_EBP_0 = 0hex00000000 );
ASSERT( R_EAX_615075 = 0hexC0000000 );
ASSERT( R_ESI_2 = 0hex00000000 );
ASSERT( R_EDI_3 = 0hex00000000 );
ASSERT( R_ESI_615079 = 0hexC0000000 );
ASSERT( R_ESP_1 = 0hex00000000 );
```

Fig. 5. Validation Result of GDI AttemptWrite Integer Overflow Vulnerability

Table 3. Detection result of gdi32.dll

address of suspicious integer overflow point	sign	priority
0x77ef5b4a	Unsigned	30
0x77f0427d	Unsigned	30
0x77f04288	Unsigned	30
0x77f0c929	Unknown	31
0x77efdb18	Unknown	31
0x77ef5eee	Unknown	31
0x77ef6c19	Unknown	31
0x77f06c03	Unsigned	32
0x77f1d0ae	Unsigned	33
0x77ef6216	Unsigned	36
0x77f1ba0a	Unsigned	38
0x77f1d11b	Unsigned	46
0x77efd5c8	Uncertain	47
0x77f03220	Uncertain	51
0x77f0dacf	Unsigned	51
0x77f075da	Unsigned	53
0x77f03c34	Unsigned	53
0x77f0d7da	Unsigned	59
0x77f1b63c	Unsigned	61

GDI AttemptWrite Integer Overflow Vulnerability. AttemptWrite is a function in gdi32.dll used to copy some data to a buffer which named *Buffer*, whose capacity is *Buffer_Capacity*. AttemptWrite performs memory management in following way:

```
if(NumberOfBytesWritten + NumberOfBytesToWrite < Buffer_Capacity)
memcpy(Buffer, file_data, NumberOfBytesToWrite)
```

where *NumberOfBytesWritten* denotes the number of bytes that has been written to *Buffer*, and *NumberOfBytesToWrite* denotes the number of bytes still to be written to *Buffer*. To avoid copying too much data, AttemptWrite checks the bound of *NumberOfBytesToWrite*. But a large *NumberOfBytesToWrite* will cause an addition overflow and bypass the bounds check, resulting in a heap overflow in the subsequent call to memcpy.

Table 3 is the detection result of GDI AttemptWrite Integer Overflow Vulnerability. There are 341 potential integer overflow points, Table 3 only list 19 points which has highest priority. The CVE-2007-3034 vulnerability is at integer overflow point $0x77f0427d$, from the table, we can see that it has priority 30, its ranking position is 19, higher than other 322 potential overflow points. The result shows our priority ranking scheme can save much time on validate vulnerability.

Figure 5 is the validation result of DSA_SetItem integer overflow vulnerability, when the condition in figure 5 is satisfied, integer overflow will be triggered, it means integer overflow vulnerability at $0x77f0427d$ is real vulnerability.

6.2   Efficiency

In this section, we measure the performance of our system. Table 3 shows the result of efficiency evaluation. We measured the time that EIOD spent translating x86 assembly into our IR, the time EIOD spent detecting and ranking suspicious integer overflow vulnerabilities and the time EIOD spent validating the vulnerabilities. We can see that detecing and ranking potential overflow vulnerabilities and validating the vulnerabilities is time-consuming part.

Table 4. Result of efficiency evaluation

File	File Size	Translating time(sec)	Finding time(sec)	Validating time(sec)
Comctl32.dll	597KB	150	634	1322
Gdi32.dll	271KB	94	455	943

7   Related Work

Integer Vulnerability Prevention and Detection. To prevent integer vulnerabilities, many techniques have been proposed. Given program source code, there are three methods:

1. Language based method, this method either translates the C program into type safe language such as CCured[2], Cyclone[3] or uses safe class such as SafeInt[30], IntSafe[31].
2. compiler based method, this method inserts checking code for certain operations when compile the source code such as BLIP[4] RICH[5].
3. Static source code analysis, this method inspects the whole program to find the suspect instruction such as LCLint[6], integer bug detection algorithm proposed by Sarkar et al.

Given program at binary level, there are some tools to detect integer vulnerabilities. UQBTng[17] is a tool to automatically find integer overflow vulnerabilities in Windows binaries. UQBTng first translates binaries into C code by UQBT [18], then inserts assert statements before the calls to memory allocation functions, finally, UQBTng uses a Bounded Model Checker CBMC[19] to verify the program. UQBTng is limited by the binary translator, because the automatic decompilation of binary executables to C code is very challenging task.

IntScope[8] is proposed by Wang to detect integer overflow vulnerabilities in x86 binaries, it leverages symbolic execution and taint analysis to identify the vulnerable point of integer overflow, and reports suspicious integer overflow vulnerabilities. Finally, to confirm the suspicious vulnerability, they use dynamic vulnerability test case generation tool[9].

IntFinder[10] can automatically detect integer bugs in x86 binary programs. It first decompiles x86 binary code and creates the suspect instruction set, Second,

IntFinder dynamically inspects the instructions in the suspect set and confirms which instructions are actual Integer bugs with the error-prone input.

These tools all use static analysis to find suspicious integer overflow vulnerabilities, then dynamically check each suspicious vulnerability. However, static analysis face the problem of high false positive rate of suspicious vulnerability reports, researchers have to spend a large amount of time on carefully checking each suspicious vulnerability. Unlike them, Our approach focus on how to rank the suspicious vulnerabilities and try to present a priority ranking scheme.

Binary Analysis. Vine[13] is a static analysis component of the BitBlaze[20] project. Vine is divided into a platform-specific front-end and a platform-independent back-end. The front-end can accurately translate each x86 instruction into a RISC-like IR and the back-end supports a variety of core program analysis utilities. CodeSurfer/x86[21] is a static binary analysis platform which make use of IDA Pro and the CodeSurfer system. CodeSurfer/x86 uses the value-set analysis and aggregate-structures identification to recover binary information and translates x86 binary code into IR, then analyze on IR by CodeSurfer. Chevarista[23] is a tool to perform vulnerability analysis on binary program of SPARC. Chevarista translates binary code into SSA form IR and can detect buffer overflows or integer overflows vulnerability.

Error Ranking. Several research work focused on error ranking of static checkers for C and Java programs. Kremenek and Engler proposed z-ranking algorithm[24] to rank errors, based on the observation that true error reports tend to issue few failed checks while false positives always generate lots of failed checks. Shen et al presented EFindBugs[25] to employ an effective two-stage error ranking strategy that suppresses the false positives and ranks the true error reports on top.

8   Conclusion

In this paper, we proposed a static binary analysis based approach for detecting integer overflow vulnerabilities with an potential vulnerabilities ranking strategy. We have implemented our system EIOD on the Microsoft Windows platform and evaluated it with some real world programs from Microsoft platform, the experimental results shows EIOD can effectively and efficiently detect integer overflow vulnerabilities. The limitation of EIOD is that it can reduce the false positive rate of detection, but it also has some false positive reports. In the future, we plan to combine some dynamic analysis methods with EIOD to detect integer overflow vulnerabilities and remove false positive report at all.

Acknowledgement. We are grateful to the anonymous reviewers for their insightful comments and suggestions. This research was supported in part by National Natural Science Foundations of China (Grant No. 60970028 and 61100227), and the National High Technology Research and Development Program (863Program) of China under Grant No. 2011AA01A203.

References

1. Vulnerability type distributions in cev. CVE (2007),
   http://cve.mitre.org/docs/vuln-trends/vuln-trends.pdf
2. Necula, G.C., McPeak, S., Weimer, W.: Ccured: Type-safe retrofitting of legacy code. In: Proceedings of the Principles of Programming Languages, pp. 128–139 (2002)
3. Jim, T., Morrisett, G., Grossman, D., Hicks, M., Cheney, J., Wang, Y.: Cyclone: A safe dialect of c. In: Proceedings of the Annual Conference on USENIX Annual Technical Conference (2002)
4. Horovitz, O.: Big loop integer protection. Phrack Inc. (2002),
   http://www.phrack.org/issues.html?issue=60&id=9#article
5. Brumley, D., Chiueh, T., Johnson, R., Lin, H., Song, D.: Rich: Automatically protecting against integer-based vulnerabilities. In: Proceedings of the 14th Annual Network and Distributed System Security, NDSS (2007)
6. Evans, D., Guttag, J., Horning, J., Tan, Y.M.: Lclint:a tool for using specification to check code. In: Proceedings of the ACM SIGSOFT 1994 Symposium on the Foundations of Software Engineering, pp. 87–96 (1994)
7. Zhang, C., Wang, T., Wei, T., Chen, Y., Zou, W.: IntPatch: Automatically fix integer-overflow-to-buffer-overflow vulnerability at compile-time. In: Gritzalis, D., Preneel, B., Theoharidou, M. (eds.) ESORICS 2010. LNCS, vol. 6345, pp. 71–86. Springer, Heidelberg (2010)
8. Wang, T., Wei, T., Lin, Z., Zou, W.: Intscope: Automatically detecting integer overflow vulnerability in x86 binary using symbolic execution. In: Proceedings of the 16th Annual Network and Distributed System Security Symposium, NDSS 2009 (2009)
9. Lin, Z., Zhang, X., Xu, D.: Convicting exploitable software vulnerabilities: An efficient input provenance based approach. In: Proceedings of the 38th Annual IEEE/IFIP International Conference on Dependable Systems and Networks (DSN 2008), Anchorage, Alaska, USA (June 2008)
10. Chen, P., Han, H., Wang, Y., Shen, S., Yin, X., Mao, B., Xie, L.: INTFINDER: automatically detecting integer bugs in x86 binary program. In: Proceedings of the International Conference on Information and Communications Security, Beijing, China, pp. 336–345 (December 2009)
11. Ida pro, http://www.hex-rays.com/idapro/
12. Nethercote, N., Seward, J.: Valgrind: A Program Supervision Framework. In: Third Workshop on Runtime Verification, RV 2003 (2003)
13. Vine: BitBlaze Static Analysis Component,
    http://bitblaze.cs.berkeley.edu/vine.html
14. BAP: The Next-Generation Binary Analysis Platform, http://bap.ece.cmu.edu/
15. Brumley, D., Poosankam, P., Song, D., Zheng, J.: Automatic patch-based exploit generation is possible: Techniques and implications. In: Proceedings of the 2008 IEEE Symposium on Security and Privacy (May 2008)
16. Ganesh, V., Dill, D.L.: A decision procedure for bit-vectors and arrays. In: Damm, W., Hermanns, H. (eds.) CAV 2007. LNCS, vol. 4590, pp. 519–531. Springer, Heidelberg (2007)

17. Wojtczuk, R.: Uqbtng: a tool capable of automatically finding integer overflows in win32 binaries. In: 22nd Chaos Communication Congress (2005)
18. UQBT: A Resourceable and Retargetable Binary Translator, http://www.itee.uq.edu.au/cristina/uqbt.html
19. Clarke, E., Kroning, D., Lerda, F.: A tool for checking ANSI-C programs. In: Jensen, K., Podelski, A. (eds.) TACAS 2004. LNCS, vol. 2988, pp. 168–176. Springer, Heidelberg (2004)
20. BitBlaze: The BitBlaze Binary Analysis Platform Project, http://bitblaze.cs.berkeley.edu/index.html
21. Balakrishnan, G., Gruian, R., Reps, T., Teitelbaum, T.: CodeSurfer/x86—A platform for analyzing x86 executables. In: Bodik, R. (ed.) CC 2005. LNCS, vol. 3443, pp. 250–254. Springer, Heidelberg (2005)
22. Microsoft. Phoenix framework, http://research.microsoft.com/phoenix/
23. Automated vulnerability auditing in machine code, http://www.phrack.com/issues.html?issue=64id=8
24. Kremenek, T., Engler, D.R.: Z-ranking: Using statistical analysis to counter the impact of static analysis approximations. In: Cousot, R. (ed.) SAS 2003. LNCS, vol. 2694, pp. 295–315. Springer, Heidelberg (2003)
25. Zhang, C., Xu, H., Zhang, S., Zhao, J., Chen, Y.: Frequency Estimation of Virtual Call Targets for Object-Oriented Programs. In: Mezini, M. (ed.) ECOOP 2011. LNCS, vol. 6813, pp. 510–532. Springer, Heidelberg (2011)
26. Godefroid, P., Levin, M., Molnar, D.: Automated whitebox fuzz testing. In: Proceedings of the 15th Annual Network and Distributed System Security Symposium (NDSS 2008), San Diego, CA (February 2008)
27. Aho, A.V., Lam, M.S., Sethi, R., Ullman, J.D.: Compilers: Princiles, Techniques, and Tools, 2nd edn. Addison- Wesley (2006)
28. Balakrishnan, G., Reps, T.: Analyzing memory accesses in x86 executables. In: Duesterwald, E. (ed.) CC 2004. LNCS, vol. 2985, pp. 5–23. Springer, Heidelberg (2004)
29. Balakrishnan, G., Reps, T.: DIVINE: DIscovering Variables IN Executables. In: Cook, B., Podelski, A. (eds.) VMCAI 2007. LNCS, vol. 4349, pp. 1–28. Springer, Heidelberg (2007)
30. LeBlanc, D.: Integer handling with the c++ safeint class (2004), http://msdn.microsoft.com/library/default.asp?url=/library/en-us/dncode/html/secure01142004.asp
31. Howard, M.: Safe integer arithmetic in c (2006), http://blogs.msdn.com/michaelhoward/archive/2006/02/02/523392.aspx
32. Dipanwita, S., Muthu, J., Jay, T., Ramanathan, V.: Flow-insensitive static analysis for detecting integer anomalies in programs. In: Proc. SE, pp. 334–340. ACTA Press, Anaheim (2007)

Solving Google's Continuous Audio CAPTCHA with HMM-Based Automatic Speech Recognition

Shotaro Sano, Takuma Otsuka, and Hiroshi G. Okuno

Graduate School of Informatics, Kyoto University, Kyoto, Japan
{sano,ohtsuka,okuno}@kuis.kyoto-u.ac.jp

Abstract. CAPTCHAs play critical roles in maintaining the security of various Web services by distinguishing humans from automated programs and preventing Web services from being abused. CAPTCHAs are designed to block automated programs by presenting questions that are easy for humans but difficult for computers, e.g., recognition of visual digits or audio utterances. Recent audio CAPTCHAs, such as Google's audio reCAPTCHA, have presented overlapping and distorted target voices with stationary background noise. We investigate the security of overlapping audio CAPTCHAs by developing an audio reCAPTCHA solver. Our solver is constructed based on speech recognition techniques using hidden Markov models (HMMs). It is implemented by using an off-the-shelf library HMM Toolkit. Our experiments revealed vulnerabilities in the current version of audio reCAPTCHA with the solver cracking 52% of the questions. We further explain that background stationary noise did not contribute to enhance security against our solver.

Keywords: audio CAPTCHA, human interaction proof, reCAPTCHA, automatic speech recognition, hidden Marcov model.

1 Introduction

CAPTCHAs (Completely Automated Public Turing tests to tell Computers and Humans Apart) are programs that distinguish humans from automated programs by presenting task that humans can easily solve but computers cannot [1]. Many websites use CAPTCHAs to prevent their services from being abused such as when services are flooded with spam accounts. While they have been widely used in recent Web services, even well known commercial CAPTCHAs (such as Google's, Microsoft's, and Yahoo's) are sometimes easily compromised by simple machine learning algorithms [2] [3], which immediately adversely affect the quality of services by providing malicious programs with unauthorized access. Thus, there has been huge demand to assess vulnerabilities in the design of current CAPTCHAs.

Our research has especially focused on the security of audio CAPTCHAs. While most CAPTCHAs display images of characters and numerals and users have to input the same texts as those in images, some of them provide audio versions for accessibility reasons. Because a user that solves either a visual or audio question is authorized by the CAPTCHA, sometimes audio CAPTCHA systems provide another loophole for malicious programs.

K. Sakiyama and M. Terada (Eds.): IWSEC 2013, LNCS 8231, pp. 36–52, 2013.

Audio CAPTCHAs are mainly divided into two classes of non-continuous and continuous. There are several target voices (e.g., digits, letters of the alphabet, and words), mixed with irrelevant background noise in non-continuous audio CAPTCHAs, and the target voices do not overlap. Previous researchers on attacks [4] [5] have mainly aimed at solving non-continuous audio CAPTCHAs where a solver decodes CAPTCHAs through two stages of segmenting the target voices separately and labeling each of them with a certain method of supervised classification. These studies have revealed that the security of non-continuous audio CAPTCHAs depends on the difficulty of the segmentation stage, since given a perfect segmentation, a machine's accuracy of classification is often superior to that of a human's [6] [7].

On the other hand, continuous CAPTCHAs present overlapping target voices to make automated segmentation even more difficult, which has been demonstrated in Google's audio reCAPTCHA [8]. They cannot be solved with conventional methods of segmentation and classification, since these methods have been designed on the assumption that the target voices do not overlap. Due to their assumed effectiveness in security, continuous audio CAPTCHAs have been used even though no formal security assessments of these types of audio CAPTCHAs have been undertaken.

Here, we discuss a system that automatically solves continuous audio CAPTCHAs and assesses their vulnerability. The solver was able to crack the current version of audio reCAPTCHA (as of April 2013) with 52% accuracy, which means this type of audio CAPTCHA is no longer safe. Since the labeling process of our solver is formulated with a well-known method of automatic speech recognition (ASR) that is based on the hidden Markov model (HMM) [9], the solver can easily be implemented with an off-the-shelf library called the HMM Toolkit (HTK) [10]; therefore, our method may further threaten the security of CAPTCHAs.

Section 2 presents the current version of the audio reCAPTCHA scheme. Section 3 outlines how an HMM-based method of ASR works. Section 4 describes the implementation of our solver system. Section 5 discusses several experiments we conducted to evaluate our solver's accuracy and presents our assessment of how efficient reCAPTCHA's security techniques are. Section 6 discusses better audio CAPTCHAs and stronger solvers based on the experimental results, and finally we conclude the paper in Section 7.

2 Audio reCAPTCHA

Figure 1 shows the waveform of an audio clip, which we refer to as a *challenge*, from the current version of audio reCAPTCHA. A challenge consists of three clusters with distinct intervals, and a *cluster* contains three or four *digit* utterances spoken in English. Digit utterances in a cluster overlap at random intervals. When all digits in a challenge are correctly estimated, the CAPTCHA is solved (or equivalently broken using a certain algorithm), i.e., the audio reCAPTCHA confirms that the listener is a human.

Audio reCAPTCHA also prevents automated programs with two types of distortions: *challenge-distortion* and *digit-distortion*. Challenge-distortion is additive stationary noise that covers the entire audio signal of the challenge, which prevents both clusters from being detected and recognized. On the other hand, digit-distortion

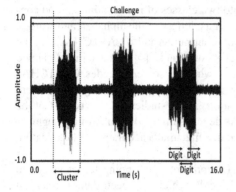
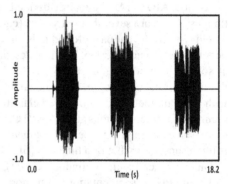

Fig. 1. Waveform of audio reCAPTCHA challenge. Challenge consists of three clusters, and each cluster contains three or four overlapping digits. If all digits in challenge are correctly identified, audio reCAPTCHA is solved.

Fig. 2. Waveform of former version of audio reCAPTCHA challenge. Intervals between clusters are completely silent.

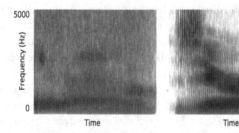

Fig. 3. Comparison of spectrograms for distorted digital voice of reCAPTCHA (left) and clear digital voice (right). Both of them are pronounced "zero."

is convolutive non-additive noise and applied for each digit. Figure 3 shows an example spectrogram of a distorted digit voice of a reCAPTCHA that is pronounced "zero" (left), comparing it to that of a clear digit voice (right). Some of the distorted digit's features collapse especially in its high frequency range. Although digit-distortion effectively seems to prevent clusters or digits from being recognized, distortion degrades usability since it is often too strong even for humans to hear, in the author's opinion.

An audio reCAPTCHA regards a response as correct even when one of the digits in a challenge is deleted or replaced to increase usability. For example, a challenge whose correct answer is "012 345 6789" may be labeled as "012 345 678" or labeled as "112 345 6789." Current audio reCAPTCHA allows deletion or substitution errors but not insertion errors, and the challenge should not be mislabeled as "0012 345 6789."

In summary, audio reCAPTCHA adopt four defensive techniques. These are:

- Overlap of target voices,
- Random number of target voices in a cluster,
- A stationary noise signal that entirely covers the challenge, and
- Filtering that collapses high frequency features of digits.

They also adopt an additional idea to ensure usability by:

- Allowing off-by-one error to label a challenge.

Former Version of Audio reCAPTCHA

Figure 2 shows an audio clip from the former version of reCAPTCHA, which had been used until February 2013. As this version did not adopt challenge-distortion, the intervals between clusters were completely silent. In Section 5, we evaluate the solver's accuracy both for the former and current versions to assess the efficiency of challenge-distortion.

3   Preliminaries

Our reCAPTCHA solver decodes a challenge in three steps by segmenting it into clusters, extracting feature vectors from each cluster, and labeling each cluster. The labeling stage is carried out with a method of ASR based on HMM. This section overviews how an ASR system recognizes an audio signal, after it describes an effective acoustic feature called Mel-frequency cepstral coefficient (MFCC) [11] and the mechanism for HMM. The methods described in this section have been black-boxed and are easily available in the HTK or related documents.

3.1   MFCC

Before the source audio signal is recognized by ASR, it is transformed into a sequence of feature vectors. MFCC is one of the best transformation techniques successfully used in recent ASR systems that is based on the mechanism for human auditory perception.

An MFCC vector is extracted in the four steps for each short time window of the source audio signal. These are:

1. Calculate the fast Fourier transform of the short time signal.
2. Reduce the dimensionality of the power spectrum obtained in Step 1 using a Mel-scale filter bank [12].
3. Map the Mel powers obtained in Step 2 onto the logarithmic scale.
4. Calculate the discrete cosine transform of Mel log powers obtained in Step 3.

See [11] for details.

The first and second derivative of MFCC are called a delta MFCC and a delta-delta MFCC, both of which are also effective temporal representations [13]. Thus, a feature vector is composed of a combination of MFCC, Delta MFCC, and Delta-Delta MFCC in most ASR methods.

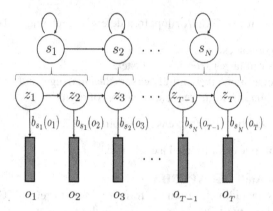

Fig. 4. N-state left-to-right HMM. Observed sequence O is generated by sequence of hidden states Z. Z has form of Markov chain in states S.

3.2　HMM

HMM is a probabilistic model for a sequential observation. Given an observed sequence, $O = o_1, \ldots, o_T$, an HMM, $\lambda = \{\pi, A, B\}$ calculates the likelihood, $P(O|\lambda)$. O is assumed to be generated by a sequence of hidden states $Z = z_1, \ldots, z_T$, which has the form of a Markov chain in states $S = \{s_1, \ldots, s_N\}$, i.e., $z_t \in S$. An observation value, o_t, is generated by a state, s_n, with probability $b_{s_n}(o_t)$. Thus, an HMM λ is defined with three parameters:

- An initial probability vector of hidden states: $\pi = [\pi_n | 1 \leq n \leq N]$.
- A transition matrix of hidden states: $A = \{a_{i,j} | 1 \leq i, j \leq N\}$ where each element $a_{i,j}$ corresponds to $P(s_j|s_i)$, which means the transition probability from state s_i to s_j.
- Observation likelihood functions: $B = \{b_s(o) | s \in S\}$ where o may be a continuous value by defining $b_s(o)$ as a continuous density function.

When the elements of transition matrix A satisfy the conditions in Equation (1), HMM is called left-to-right HMM (strict left-to-right HMM) where a hidden state, s_n, is supposed to transit to s_n itself or the next state s_{n+1}.

$$a_{ij} = 0 \text{ if } i \neq j \text{ and } i+1 \neq j \tag{1}$$

The probability that hidden states Z generates observation O can be calculated with the parameter of HMM λ:

$$P(O, Z|\lambda) = P(O|Z, \lambda)P(Z|\lambda) \tag{2}$$

$$= \{\prod_{t=1}^{T} b_{z_t}(o_t)\}\{\pi_{z_1} \prod_{t=1}^{T-1} a_{z_t, z_{t+1}}\} \tag{3}$$

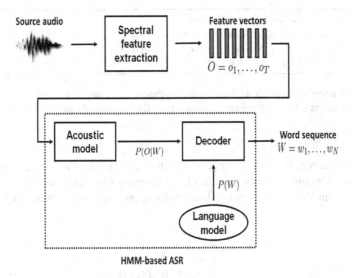

Fig. 5. Overview of ASR system based on HMM. Given sequence of sentence's feature vectors O, ASR outputs sentence's word sequence W.

Thus, $P(O|\lambda)$ is obtained as:

$$P(O|\lambda) = \sum_Z P(O,Z|\lambda) \tag{4}$$

$$= \sum_Z \{\prod_{t=1}^{T} b_{z_t}(o_t)\}\{\pi_{z_1} \prod_{t=1}^{T-1} a_{z_t,z_{t+1}}\} \tag{5}$$

We can efficiently calculate the summation over Z in Equation (5) by using the forward algorithm [9].

The parameters of an HMM are usually trained through an unsupervised training called the Baum-Welch algorithm [14] where we provide only a set of observed sequences. While this algorithm is effective for the training of a single HMM, multiple HMMs should be trained in the ASR method. This is because the ASR method integrates multiple HMMs into the network of words and phones. In order to train multiple HMMs simultaneously, we use a semi-supervised training called concatenated training [15]. In this setup, the training data is provided as pairs of not only a sequence of audio features, but also the corresponding text transcription. Note that the pairwise data do not necessarily require a strict alignment. Indeed, the training of our model is carried out with a sequence of audio features of a cluster and annotated three or four digits in the cluster.

3.3 HMM-Based ASR

As shown in Figure 5, the input of an ASR is a sequence of feature vectors $O = o_1,\ldots,o_T$ that is obtained by pre-processing the source audio signal. ASR finds the

> $<cluster>::=<cluster_3> \mid <cluster_4>;$
> $<cluster_3>::=<digit><digit><digit>;$
> $<cluster_4>::=<digit><digit><digit><digit>;$
> $<digit>::= \text{`0`}|\text{`1`}|\text{`2`}|\text{`3`}|\text{`4`}|\text{`5`}|\text{`6`}|\text{`7`}|\text{`8`}|\text{`9`};$

Fig. 6. BNF to generate language L. Start symbol is cluster and terminal symbols are digit labels. This grammar meets schema of cluster that consists of three or four digits.

most likely sentence, $\hat{W} = w_1, \ldots, w_N$, for O out of all sentences in a language, a set of possible word sequences, denoted by L. This problem is formulated as Equations (6) to (8), which means the problem can be broken down into the computations of $P(W)$ and $P(O|W)$:

$$\hat{W} = \arg\max_{W \in L} P(W|O) \tag{6}$$

$$= \arg\max_{W \in L} \frac{P(W)P(O|W)}{P(O)} \tag{7}$$

$$= \arg\max_{W \in L} P(W)P(O|W), \tag{8}$$

where Equation (7) is obtained with Bayes' rule. The denominator, $P(O)$, may be left out because it is always the same for given feature vectors O. Thus, we can obtain Equation (8). The computation of $P(W)$ and $P(O|W)$ is called language model and acoustic model respectively.

Language L describes the sequences that can be recognized. Language L in most cases is determined in either of two ways: (1) defining it with a specific grammar or (2) constructing a statistical language model from a corpus. The former grammatical model suits tasks that involve structured sentences whereas the latter statistical model suits tasks of handling arbitrary utterances of various topics.

We can assume that L is represented by a Backus-Naur form (BNF) in our problem that is listed in Figure 6, and $P(W)$ is determined as:

$$P(W) = \begin{cases} \frac{1}{2 \cdot |cluster_3|} & \text{if } d = 3, \\ \frac{1}{2 \cdot |cluster_4|} & \text{if } d = 4, \\ 0 & \text{otherwise,} \end{cases} \tag{9}$$

where d is the number of digits in a cluster, and $|symbol|$ is the number of sentence patterns generated from non-terminal symbol $<symbol>$ in Figure 6's BNF.

The computation of $P(O|W)$ is enabled by a phone-based HMM network [15] where each phone is represented by an HMM; each word is represented by several consecutive phones; and the phone HMMs are integrated into a network of HMMs to recognize the sentences. A certain sentence W corresponds to a path in the HMM network. Equation (5) is evaluated for each HMM along the path so as to determine the best path that maximizes the likelihood in Equation (8).

4 reCAPTCHA Solver

Our reCAPTCHA solver is depicted in Figure 7. The input to our solver is a challenge's audio signal of reCAPTCHA, and the solver outputs the challenge's answer. The system solves a challenge in three steps:

1. The input challenge is segmented into three clusters with a voice activity detection algorithm (cluster segmentation).
2. Each cluster's audio signal is converted to feature vectors (spectral feature extraction).
3. The feature vectors of each cluster are labeled with the HMM-based ASR (cluster labeling).

The spectral feature extraction component and cluster labeling component are implemented with HTK.

The cluster labeling component is trained with the actual audio signals of audio reCAPTCHA. As outlined in Figure 8, the training set of challenges is downloaded and stored into the challenge database (DB). The data in the challenge DB is manually labeled for each cluster by the solver's user.

4.1 Cluster Segmentation

This component segments a challenge audio signal into three clusters. Clusters are extracted with a volume-based algorithm for voice activity detection. The challenge audio signal, $s_1, \ldots, s_N$, is split into segments of length l and is subsampled as $Volume(t)$:

$$Volume(t) = \frac{1}{l} \sum_{n=t}^{t+l-1} \{\bar{s} - s_n\}^2, \tag{10}$$

where $\bar{s}$ is the mean of $s_t, \ldots, s_{t+l-1}$.

Figure 9 plots the volume analysis of a challenge. There are three cluster segments between four noise segments in which every volume value is less than a threshold, θ. First, this component removes the four longest segments in which every window has a lower volume than the threshold, θ, and it then returns the remaining three segments as clusters. We set $l = 512$ and $\theta = 0.01$ where the sampling rate of the challenge audio signal is 16 kHz and the amplitude of the input waveform is normalized to 1.0.

4.2 Spectral Feature Extraction

A feature vector consists of a 13-dimensional MFCC, a 13-dimensional Delta MFCC, and a 13-dimensional Delta-Delta MFCC, and is in total a 39-dimensional vector. It is extracted from each short-time window. We set the window size to 25 ms and the frame shift to 10 ms for short-time Fourier transform.

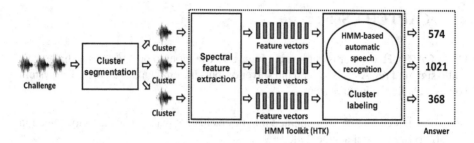

Fig. 7. Audio reCAPTCHA solver. Input challenges are decoded through three stages of (1) segmenting challenge into three clusters, (2) extracting feature vectors from clusters, and (3) labeling each sequence of feature vectors with HMM-based ASR.

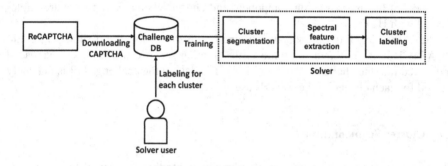

Fig. 8. Training process for audio reCAPTCHA solver

4.3 Cluster Labeling

The input of the cluster labeling component is a spectral feature sequence of a cluster, and this component outputs the cluster's label. This component labels a cluster with the HMM-based ASR method that is described in Section 3.3. A sentence for the ASR corresponds to a cluster in our problem, and a word corresponds to a digit.

The grammar in Figure 6 is applied to the language model. This grammar satisfies a cluster's schema for audio reCAPTCHA: a cluster consists of three or four digits.

We assume that each digit consists of seven consecutive phones. A phone HMM is a left-to-right HMM and has three states whose observation likelihood function is a 39-dimensional Gaussian distribution. We adopted the triphone model [16]; thus, one HMM denotes each triplex pattern of phones.

All HMMs are simultaneously trained with concatenated training where the training data are given as pairs of a cluster's feature vectors and the cluster's label. As outlined in Figure 8, the HMMs are trained in four steps using data from the challenge DB as follows:

1. Segment data from the challenge DB into clusters with the cluster segmentation component.

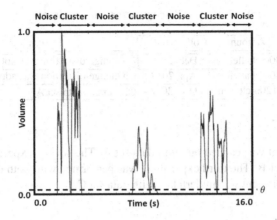

Fig. 9. Volume analysis of challenge. Volume does not reach threshold θ in non-utterance sections.

2. Input each cluster obtained in Step 1 into the spectral feature extraction component.
3. Label each feature sequence obtained in Step 2 as the source cluster's label.
4. With the training set obtained in Step 3, obtain the HMMs' parameters by using concatenated training.

5 Experiments

We carried out four experiments to evaluate our solver's performance as well as assess the security of audio reCAPTCHA. We evaluated the performance of our solver in the former version of audio reCAPTCHA in the first experiment, and that of the current version in the second experiment. In the third experiment, we investigated how uncertainty in the number of digits in each cluster contributed to the security of reCAPTCHA where we evaluated accuracy by giving the ground truth number of digits in a cluster. In the fourth experiment, we examined the robustness of the ASR method in terms of challenge-distortion. There, we evaluated the performance of the cluster labeling component with several simulation CAPTCHAs that were generated by adding several types of noise to the former version of audio reCAPTCHA and varying the strength of noise.

5.1 Data

As listed in Table 1, the experiments were performed with three data sets. Data set A was a set of challenges downloaded from the former version of audio reCAPTCHA, and data set B was that from the current version. We downloaded 400 challenges both from the former and the current versions. (As described in Section 2, the main difference between the former and current versions of reCAPTCHA was that the current version adopted stationary noise that entirely covered a challenge, which we refer to as challenge-distortion.) Data set C was obtained by segmenting data set A into clusters.

Table 1. Data set description for our experiments

Data set	Version	Amount	Collected date	Description
A	Former	400 challenges	Dec. 2012	Challenge-distortion was not adopted.
B	Current	400 challenges	Apr. 2013	Challenge-distortion was adopted.
C	Former	1200 clusters	Dec. 2012	Clusters of data set A.

The first experiment was carried out with data set A. The second experiment was conducted with data set B. The third experiment was performed with both data sets A and B. The fourth experiment was undertaken with data set C.

5.2   Metrics

Accuracies.   We used five metrics to evaluate our system:

- *Off-by-one accuracy* evaluates the actual vulnerability of the reCAPTCHA. It is defined as "the number of challenges correctly answered" divided by "the number of challenges" where an output of the system is regarded as correct when the Levenshtein distance between the output and the correct answer is less than two excluding the case of inserted error.
- *Strict accuracy* is the ratio of strictly correct challenges. It is defined as "the number of challenges correctly answered" divided by "the number of challenges" where an output of the system is regarded as correct only when it is exactly the same as the correct answer.
- *Per-cluster accuracy* is defined as "the number of clusters correctly answered" divided by "the number of clusters".
- *Per-digit accuracy* is defined as "the number of digits correctly answered" divided by "the number of digits". If the number of digits in the cluster is misestimated, all the digits in the cluster are regarded as incorrect.
- *N-segment accuracy* evaluates the success rate of segmentation. It is defined as "the number of clusters estimated with the correct number of digits" divided by "the number of clusters".

For example, if there are two CAPTCHA challenges whose correct answers are "000 000 0000" and "111 111 1111" while the solver outputs "000 000 0001" and "111 112 111", the solver's off-by-one accuracy is $\frac{1}{2} = 0.50$, strict accuracy is $\frac{0}{2} = 0.00$, per-cluster accuracy is $\frac{3}{6} = 0.50$, per-digit accuracy is $\frac{14}{20} = 0.70$, and N-segment accuracy is $\frac{5}{6} = 0.83$.

Note that the actual vulnerability of audio reCAPTCHA is represented by off-by-one accuracy, because audio reCAPTCHA regards an output of the system as correct even when it has off-by-one error in terms of the Levenshtein distance.

Closed Test and Open Test.   The solver is trained with four-fifth of the entire data set. *Open test* uses the rest one-fifth data for the evaluation. This test measures the practical

Table 2. Results from performance evaluations of former and current versions of reCAPTCHA

	Former reCAPTCHA		Current reCAPTCHA	
	Closed test	Open test	Closed test	Open test
Off-by-one	0.54	0.51	0.54	0.52
Strict	0.21	0.19	0.20	0.17
Per-cluster	0.59	0.57	0.61	0.59
Per-digit	0.75	0.74	0.77	0.76
N-segment	0.84	0.84	0.85	0.85

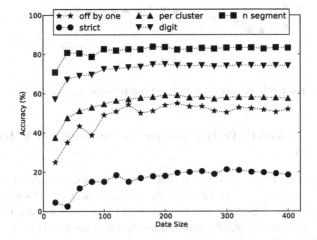

Fig. 10. Relationship between accuracy and data size (former version of reCAPTCHA)

performance of the system because it uses unknown data excluded from the training. *Closed test* uses the same data as used in the training phase. We perform this five times to obtain the average performance, which is called *five-fold cross validation*.

5.3   Experiment 1: Solver's Performance on Former version of Audio reCAPTCHA

We evaluated our solver's performance with 400 challenges downloaded from the former version of audio reCAPTCHA (data set A in Table 1). Five times, we trained the cluster labeling component with four-fifths of data set A, and evaluated Section 5.2's metrics with the rest of the data. We also evaluated the transition in performance by changing the numbers of training data in increments of 20.

The left side of Table 2 lists the results when all of data set A was used. Our system solved audio reCAPTCHA with 51% accuracy for the former version of audio reCAPTCHA. Figure 10 plots the transition in performance. We can see the solver's performance saturates when the data size reaches around 200.

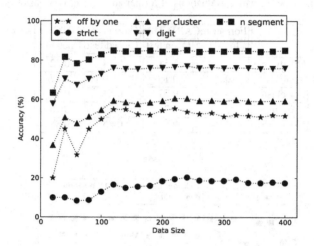

Fig. 11. Relationship between accuracy and data size (current version of reCAPTCHA)

5.4   Experiment 2: Solver's Performance for Current version of Audio reCAPTCHA

We evaluated our solver's performance for the current version of audio reCAPTCHA with 400 challenges (data set B in Table 1). We evaluated Section 5.2's metrics as we did in Experiment 1 by five-fold cross validation and also evaluated the transition in performance by changing the numbers of training data in increments of 20.

The right side of Table 2 lists the results when all of data set B was used. Our solver attained 52% accuracy for the current version of audio reCAPTCHA. Its performance was almost the same as that in Experiment 1, which means challenge-distortion was hardly effective against our solver. Figure 11 plots the transition in performance. Performance saturates when the data size reaches around 200 similarly to that in Experiment 1.

5.5   Experiment 3: Giving Number of Target Voices

This experiment assessed the efficiency of security when there was uncertainty in the number of target voices. We evaluated our solver's performance, assuming that the number of target voices in each cluster was known, and compared its performance with what we obtained in Experiments 1 and 2. We could conduct this experiment by properly changing Figure 6's BNF according to the number of digits in the cluster because we annotated the correct answer for each cluster. We evaluated Section 5.2's metrics both for data sets A and B by five-fold cross validation.

Table 3 lists the results for performance in this experiment. Compared to the results for Experiments 1 and 2, these results suggest that uncertainty in the number of digits greatly decreases the solver's performance both for the former and current versions of audio reCAPTCHA; hence, it contributes to the security of reCAPTCHA.

Table 3. Results for performance with ground truth number of digits in cluster

	Former reCAPTCHA		Current reCAPTCHA	
	Closed test	Open test	Closed test	Open test
Off-by-one	0.62	0.60	0.60	0.58
Strict	0.27	0.26	0.30	0.27
Per-cluster	0.66	0.64	0.68	0.67
Per-digit	0.87	0.87	0.87	0.86
N-segment	-	-	-	-

5.6 Experiment 4: Robustness of Recognition for Various Additive Noise

The results obtained from Experiments 1 and 2 proved that our solver's recognition did not suffer from stationary noise adopted in the current version of audio reCAPTCHA. We evaluated the robustness of the ASR method against various kinds of other additive noise in Experiment 4 where the cluster labeling component decoded several simulated CAPTCHA clusters.

Simulation clusters were generated by adding one of the noise signals listed in Table 4 to each cluster signal of data set C in Table 1. We tested five kinds of noise that could be divided into two classes: stationary noise, such as white noise, and semantic noise. Semantic noise has more similar characteristics to CAPTCHA's target voices like those in spoken audio. This is known to be an effective technique that defends against the classification of non-continuous audio CAPTCHAs [5]. We tested white noise, brown noise, and pink noise [17] as stationary noise, and spoken sentences and music as semantic noise.

We also examined how the solver's performance was affected with various noise levels. The noise level was controlled by changing signal-to-noise ratio (SNR) calculated as:

$$SNR = 10\log_{10} \sqrt{\frac{\sum_{t=1}^{T} s_t}{\sum_{t=1}^{T} n_t}}, \tag{11}$$

where $s_1, \ldots, s_T$ is the source audio signal and $n_1, \ldots, n_T$ is the noise signal. Note that the lower SNR is, the stronger noise becomes. The SNR ranged from -25 to 25 at 5 interval.

We conducted five-fold cross validation for each type of noise and for each value of SNR where the cluster labeling component was trained with four-fifths of data set C and evaluated per-cluster accuracy with the rest of the data.

Figure 12 plots the relationship between SNR and the performance of the cluster labeling component for each noise. Semantic noise results in lower accuracy for each value of SNR, which means semantic noise enables more secure distortion without increasing the strength of noise.

6   Discussion

Our solver cracks the current version of audio reCAPTCHA with 52% accuracy. Chellapilla et al. stated that "depending on the cost of the attack and value of the service,

Table 4. Description of noise signals used in Experiment 4. The noise can be classified into two classes: stationary and semantic.

Class	Name	Description
Stationary	White	White noise.
	Brown	Brown noise.
	Pink	Pink Noise.
Semantic	Speech	Spoken audio signal. Noise audio is selected for each cluster from corpus of spontaneous Japanese [18].
	Music	Music audio signal. Noise audio is randomly clipped from "I Saw Her Standing There" by the Beatles for each cluster.

automatic scripts should not be more successful than 1 in 10,000" [19]; therefore, we conclude on this basis that our solver discloses the vulnerability of the current audio reCAPTCHA system. The security of reCAPTCHA is also threatened by the fact that the solver can easily be implemented by using an off-the-shelf library, HTK, and its performance saturates only with 200 annotated challenges for training.

6.1   Toward Better CAPTCHAs

Our experimental results suggest that the CAPTCHA's security may be improved in the following ways:

Forcing strict evaluation. We can conclude that applying strict evaluations to user responses drastically enhances the security of CAPTCHAs by comparing off-by-one accuracy and strict accuracy in the experimental results in Tables 2 and 3. However, this somewhat decreases usability because the strict evaluation is also difficult for human users. The trade-off should be carefully considered.

Increasing uncertainty in number of digits. The results from Experiment 3 demonstrates that uncertainty about the number of digits in a cluster efficiently prevents our solver from solving CAPTCHAs. We speculate that the more this uncertainty increases, the more secure CAPTCHAs becomes. We also speculate that randomly providing the number of clusters also enhances the security of CAPTCHAs.

Adopting semantic noise. The experimental results in Figure 12 indicate that semantic noise is better at decreasing the performance of ASR than stationary noise. In addition, humans can easily handle such noise even at low SNRs when there are semantic differences between the target audio and noise, which is known as the cocktail party effect [20]. Thus, we can expect that adopting proper semantic noise will enhance the security of CAPTCHAs as well as retain excellent usability.

6.2   Toward Stronger Solver

This paper did not refer to the workaround for digit-distortion (described in Section 2), since the specific filtering process of digit-distortion is unknown other than that it

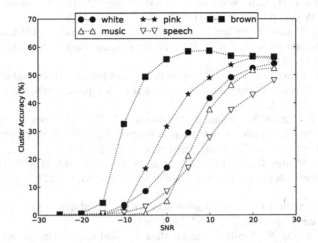

Fig. 12. Relationship between SNR and performance of cluster labeling component for several kinds of additive noise

collapses the high frequency range of the original signal's power spectrum. Identifying the filtering process and designing more appropriate features for distortion are problems that remain to be solved.

7 Conclusion

We developed and evaluated an audio CAPTCHA solver that could solve one of the most modern defensive techniques — overlapping target voices. Our solver demonstrated 52% accuracy for the current version of audio reCAPTCHA, which threatens the security of the CAPTCHA system. We also assessed several defensive techniques used in audio reCAPTCHA and demonstrated that increasing uncertainty in the number of target voices and adopting proper semantic noise can enhance the security of CAPTCHAs.

Future work will involve analysis of digit-distortion for a stronger solver as well as the design of more secure and user friendly algorithms to generate CAPTCHAs.

References

1. Von Ahn, L., Blum, M., Langford, J.: Telling humans and computers apart automatically. Communications of the ACM 47(2), 56–60 (2004)
2. Mori, G., Malik, J.: Recognizing objects in adversarial clutter: Breaking a visual CAPTCHA. In: Proceedings of IEEE Computer Society Conference on Computer Vision and Pattern Recognition, vol. 1, p. I–134. IEEE (2003)
3. Yan, J., El Ahmad, A.S.: A low-cost attack on a microsoft captcha. In: Proceedings of the 15th ACM Conference on Computer and Communications Security, pp. 543–554. ACM (2008)

4. Tam, J., Simsa, J., Hyde, S., Von Ahn, L.: Breaking audio CAPTCHAs. Advances in Neural Information Processing Systems 1(4) (2008)
5. Bursztein, E., Beauxis, R., Paskov, H., Perito, D., Fabry, C., Mitchell, J.: The failure of noise-based non-continuous audio CAPTCHAs. In: IEEE Symposium on Security and Privacy, pp. 19–31. IEEE (2011)
6. Bursztein, E., Bethard, S., Fabry, C., Mitchell, J.C., Jurafsky, D.: How good are humans at solving CAPTCHAs? A large scale evaluation. In: IEEE Symposium on Security and Privacy, pp. 399–413. IEEE (2010)
7. Chellapilla, K., Larson, K., Simard, P., Czerwinski, M.: Computers beat humans at single character recognition in reading based human interaction proofs (HIPs). In: Proceedings of the Second Conference on Email and Anti-Spam, pp. 21–22 (2005)
8. Von Ahn, L., Maurer, B., McMillen, C., Abraham, D., Blum, M.: reCAPTCHA: Human-based character recognition via web security measures. Science 321(5895), 1465–1468 (2008)
9. Rabiner, L.R.: A tutorial on hidden markov models and selected applications in speech recognition. Proceedings of IEEE 77(2), 257–286 (1989)
10. Young, S.J., Young, S.: The HTK hidden Markov model toolkit: Design and philosophy. Citeseer (1993)
11. Tiwari, V.: MFCC and its applications in speaker recognition. International Journal on Emerging Technologies 1(1), 19–22 (2010)
12. Umesh, S., Cohen, L., Nelson, D.: Frequency warping and the mel scale. IEEE Signal Processing Letters 9(3), 104–107 (2002)
13. Furui, S.: Speaker-independent isolated word recognition using dynamic features of speech spectrum. IEEE Transactions on Acoustics, Speech and Signal Processing 34(1), 52–59 (1986)
14. Welch, L.R.: Hidden Markov Models and the Baum-Welch Algorithm. IEEE Information Theory Society Newsletter 53(4) (2003)
15. Lee, K.F., Hon, H.W.: Large-vocabulary speaker-independent continuous speech recognition using hmm. In: Proceedings of International Conference on Acoustics, Speech, and Signal Processing (ICASSP), pp. 123–126. IEEE (1988)
16. Nakagawa, S., Hanai, K., Yamamoto, K., Minematsu, N.: Comparison of syllable-based hmms and triphone-based hmms in japanese speech recognition. In: Proceedings of International Workshop on Automatic Speech Recognition and Understanding, pp. 393–396 (1999)
17. Halley, J.M., Kunin, W.E.: Extinction risk and the 1/f family of noise models. Theoretical Population Biology 56(3), 215–230 (1999)
18. Maekawa, K.: Corpus of spontaneous japanese: Its design and evaluation. In: ISCA & IEEE Workshop on Spontaneous Speech Processing and Recognition (2003)
19. Chellapilla, K., Larson, K., Simard, P.Y., Czerwinski, M.: Building segmentation based human-friendly human interaction proofs (HIPs). In: Baird, H.S., Lopresti, D.P. (eds.) HIP 2005. LNCS, vol. 3517, pp. 1–26. Springer, Heidelberg (2005)
20. Bronkhorst, A.W.: The cocktail party phenomenon: A review of research on speech intelligibility in multiple-talker conditions. Acta Acustica United with Acustica 86(1), 117–128 (2000)

Constructions of Almost Secure Frameproof Codes Based on Small-Bias Probability Spaces

José Moreira[1], Marcel Fernández[1], and Grigory Kabatiansky[2]

[1] Department of Telematics Engineering
Universitat Politècnica de Catalunya, 08034 Barcelona, Spain
{jose.moreira,marcel}@entel.upc.edu
[2] Institute for Information Transmission Problems
Russian Academy of Sciences, 127994 Moscow, Russia
kaba@iitp.ru

Abstract. Secure frameproof code is the name given to a separating code when studied in relation to fingerprinting schemes. Separating codes are combinatorial objects that have found to be useful in many areas such as technical diagnosis and the protection of distribution rights. A relaxed definition of the properties of separation and frameproofness, in the sense of only requiring the properties to hold with high probability, shows that for the relaxed definitions these notions are different. In this paper we address the construction of almost secure frameproof codes based on small-bias probability spaces.

Keywords: Secure frameproof code, separating code, traitor tracing, fingerprinting.

1 Introduction

Fingerprinting codes are used to deter the illegitimate redistribution of protected contents. A distributor that wishes to protect some content delivers marked copies to the users. Each marked copy identifies a particular user. This fact discourages naive redistribution of individual copies. However, several users (traitors) can participate in a collusion attack generating a pirated copy that is a combination of their copies. Therefore, the goal of the distributor is to obtain a set of marks (fingerprinting code) such that it is possible to identify a traitor in the presence of a collusion attack.

This paper discusses de construction of almost secure frameproof codes and their application to explicitly construct fingerprinting codes. The term "secure frameproof code" [1–4] was the name given to a separating code [5–11] when they were rediscovered for its application in traitor-tracing and fingerprinting schemes. A relaxed version of these families of codes, coined as almost secure frameproof codes, was introduced in [12]. There, they were proved to be useful in fingerprinting schemes. For instance, almost secure frameproof codes are useful to construct a family of fingerprinting codes, in the style of [13], improving the lower bound on the asymptotical rate of the previous constructions. In practical terms, this means that replacing ordinary separating codes (that is, secure

K. Sakiyama and M. Terada (Eds.): IWSEC 2013, LNCS 8231, pp. 53–67, 2013.

frameproof codes) by almost secure frameproof codes in these constructions allows the distributor to use shorter fingerprints, reducing the cost of embedding the fingerprints into the content and the cost of traitor identification [12, 18].

Let C be a code. Informally, we can say that two disjoint subsets $U, V \subseteq C$ are *separated* if there is a position in which the set of entries of the codewords in U is disjoint with the set of entries of the codewords in V in this same position. The code C is called a (c, c)-*separating code* [5–11] if every pair of disjoint subsets $U, V \subseteq C$ of c codewords each are separated.

Suppose now that, given a subset $U \subseteq C$ of at most c codewords, we generate a new word in which the entry in each position is one of the entries of the codewords in the subset U. A word generated in this way is called a *descendant* of the subset U. Since the codewords correspond to the marks that identify the users, the descendant models the word embedded in the pirated copy. A descendant of U is called *uniquely decodable* if it is not a descendant of any other subset of at most c codewords of C that is disjoint with U. A code C in which every descendant of a subset of at most c codewords is uniquely decodable is called a c-*secure frameproof code* [1–4]. Note that this is equivalent to the condition that for every pair of disjoint subsets of at most c elements their respective sets of descendants are also disjoint. It is easy to see that this is the same as a (c, c)-separating code.

Now we relax both definitions in the sense of not requiring neither strict separation nor strict frameproofness. This brings two different notions, as shown in [12]. An *almost (c, c)-separating code* is a code in which a subset of at most c codewords is separated from all other disjoint subsets of at most c codewords with high probability. On the other hand an *almost c-secure frameproof code* is a code in which every descendant is uniquely decodable with high probability.

In this paper we will connect the concepts defined above with the concept of small-bias probability spaces [14, 15]. A *small-bias probability space* defined on M binary random variables is a probability space in which the parity of every subset of variables is either zero or one with "almost" equal probability.

A small-bias probability space is readily seen to also be almost independent. Let S be a subset of size at most t, of the binary random variables of a probability space. An ε-*biased from t-wise independence space* is a probability space in which the distribution of every subset S is "close" to the uniform distribution.

Since, an ε-biased from t-wise independence is close to the uniform distribution then the resulting space has the following property: for a small enough value of ε every possible configuration $\{0, 1\}^t$ appears in every subset of binary random variables of size at most t. A space with this property is called a (M, t)-*universal set*. This observation will prove very useful for our purposes, since a (M, t)-universal set is a c-secure frameproof code for $t = 2c$.

From the definitions it is easy to see that a code is a (c, c)-separating code if and only if it is a c-secure frameproof code. However, when the definitions of separation and frameproofness are relaxed then both notions are different. Intuitively it seems clear that almost separation is a more strict requirement than almost secure frameproofness. More precisely, it has been proved that there exist almost secure frameproof codes with a much higher rate than almost separating codes [12]. The strategy used to establish the existing lower bounds in the code rates of almost separating and almost secure frameproof codes is to use a standard probabilistic argument. It can be shown that there exist codes that achieve such rates, within an ensemble of codes in which every codeword $(u_1, \ldots, u_n)$ has been chosen at random with $\Pr\{u_i = 0\} = \Pr\{u_i = 1\} = 1/2$ for each position $i = 1, \ldots, n$.

We are now in the position to underline the structure of the paper. In Section 2 we provide the necessary definitions and a brief overview of previous results. Our contributions are discussed in Section 3. We begin by proving that the above choice of probabilities $\Pr\{u_i = 0\} = \Pr\{u_i = 1\} = 1/2$ is in fact the appropriate one to use. With this in mind we move into spaces where this choice of probabilities is slightly biased. By adjusting the bias of the probability space we provide explicit constructions for "almost" universal sets. Finally we show that these constructions can be used to explicitly construct almost secure c-frameproof codes, yielding explicit constructions of fingerprinting codes.

2   Definitions and Previous Results

We start by introducing some definitions. Given an alphabet Q of size $|Q| = q$, we denote by Q^n the set of all q-ary n-tuples (vectors) over Q. We denote the vectors in boldface, e.g. $\mathbf{u} = (u_1, \ldots, u_n) \in Q^n$. A subset $C \subseteq Q^n$ of M elements is called a q-ary (n, M)-code. The elements of a code C are called codewords. If the code alphabet Q is the finite field of q elements, denoted $\mathbb{F}_q$, then a code $C \subseteq \mathbb{F}_q^n$ is called a (linear) $[n, k]$-code if it forms a vector subspace of dimension k. The minimum distance of a code C, denoted $d(C)$, is the smallest Hamming distance between any two of its codewords. The rate of a q-ary (n, M)-code C is defined as

$$R(C) \overset{\text{def}}{=} n^{-1} \log_q M.$$

2.1   Almost Separating and Almost Secure Frameproof Codes

For an (n, M)-code C, a subset $U = \{\mathbf{u}^1, \ldots, \mathbf{u}^c\} \subseteq C$ of size c is called a c-coalition. Let $P_i(U)$ denote the projection of U on the ith position, i.e., the set of elements of the code alphabet in the ith position,

$$P_i(U) \overset{\text{def}}{=} \{u_i^1, \ldots, u_i^c\}.$$

Given two c-coalitions $U, V \subseteq C$, we say that U and V are separated if there is a position i that "separates" them, that is $P_i(U) \cap P_i(V) = \emptyset$. We call such a position i a separating position. Also, we say that the c-coalition U is a separated c-coalition if it is separated from any other disjoint c-coalition $V \subseteq C$.

Definition 1. *A code C is (c, c)-separating if every pair of disjoint c-coalitions $U, V \subseteq C$ have a separating position. Equivalently, all c-coalitions $U \subseteq C$ are separated c-coalitions.*

Separating codes were introduced in [5] by Graham et al. more than 40 years ago. A separating code is a natural combinatorial object that have countless applications. We mention for example the application to hash function, testing of combinatorial circuits and automata synthesis. Separating codes have subsequently been investigated by many authors, e.g. in [6–11]. Nontrivial lower and upper bounds have been derived and relationships with similar notions have been established. See for instance the overviews [6] and [10].

Recently, separating codes have draw more attention in connection with fingerprinting settings. Let $U = \{\mathbf{u}^1, \ldots, \mathbf{u}^c\} \subseteq C$ be a c-coalition from an (n, M)-code C over Q. In a collusion attack, the *marking assumption* [16,17] states that the positions i such that all codewords from U have the same symbol must remain unchanged in the pirated word $\mathbf{z}$ that they generate. Moreover, under the *narrow-sense envelope model* [13], for every position i we have that $z_i \in P_i(U)$. Hence, the set of all pirated words that coalition U can generate under the narrow-sense envelope model, denoted desc(U), is defined as

$$\text{desc}(U) \overset{\text{def}}{=} \{\mathbf{z} = (z_1, \ldots, z_n) \in Q^n : z_i \in P_i(U), \ 1 \leq i \leq n\}.$$

Often, the codewords in U are called *parents* and the words in desc(U) are called *descendants*. Also, the *c-descendant code* of C, denoted desc$_c(C)$, is defined as

$$\text{desc}_c(C) \overset{\text{def}}{=} \bigcup_{U \subseteq C, |U| \leq c} \text{desc}(U).$$

A descendant $\mathbf{z} \in \text{desc}_c(C)$ is called *c-uniquely decodable* if $\mathbf{z} \in \text{desc}(U)$ for some c-coalition $U \subseteq C$ and $\mathbf{z} \notin \text{desc}(V)$ for any c-coalition $V \subseteq C$ such that $U \cap V = \emptyset$.

Definition 2. *A code C is c-secure frameproof if for any $U, V \subseteq C$ such that $|U| \leq c$, $|V| \leq c$ and $U \cap V = \emptyset$, then $\text{desc}(U) \cap \text{desc}(V) = \emptyset$. Equivalently, all $\mathbf{z} \in \text{desc}_c(C)$ are c-uniquely decodable.*

The concepts of frameproof and secure frameproof codes were introduced in [1,2,16,17]. It is not difficult to see that the definition of a c-secure frameproof code coincides with the definition of a (c, c)-separating code. Moreover, in the fingerprinting literature, $(c, 1)$-separating codes are also known as *c-frameproof codes*.

Let $R_q(n, c)$ denote the rate of an optimal (i.e., maximal) (c, c)-separating code of length n over a q-ary alphabet Q, i.e.,

$$R_q(n, c) \overset{\text{def}}{=} \max_{\substack{C \subseteq Q^n \text{ s.t. } C \text{ is} \\ (c, c)\text{-separating}}} R(C).$$

Also, consider the corresponding asymptotical rates

$$\underline{R}_q(c) \overset{\text{def}}{=} \liminf_{n \to \infty} R_q(n, c), \qquad \overline{R}_q(c) \overset{\text{def}}{=} \limsup_{n \to \infty} R_q(n, c).$$

Lower bounds on $(2,2)$-separating codes were studied in [5, 7]. Some important, well-known results for binary separating codes are, for example, $\underline{R}_2(2,2) \geq 1 - \log_2(7/8) = 0.0642$, from [6,7], which also holds for linear codes [7]. Also, for general codes, it was shown in [13] that

$$\underline{R}_2(c) \geq -\frac{\log_2(1 - 2^{-2c+1})}{2c - 1}.$$

Regarding the upper bounds, in [6, 9] it was shown that $\overline{R}_2(2) < 0.2835$ for arbitrary codes, and in [6] that $\overline{R}_2(2) < 0.108$ for linear codes.

Note that the existence bounds for separating codes shown above give codes of low rate. In order to obtain codes with better rates, in [12] two relaxed versions of separating codes were presented.

Definition 3. *A code $C \subseteq Q^n$ is ε-almost c-separating if the ratio of separated c-coalitions (among all c-coalitions) is at least $1 - \varepsilon$.*

A sequence of codes $\{C_i\}_{i \geq 1}$ of growing length n_i is an asymptotically almost (c,c)-separating family *if every code C_i is ε_i-almost (c,c)-separating and $\lim_{i \to \infty} \varepsilon_i = 0$.*

Definition 4. *A code $C \subseteq Q^n$ is an ε-almost c-secure frameproof code if the ratio of c-uniquely decodable vectors (among all vectors in $\mathrm{desc}_c(C)$) is at least $1 - \varepsilon$.*

A sequence of codes $\{C_i\}_{i \geq 1}$ of growing length n_i is an asymptotically almost c-secure frameproof family *if every code C_i is an ε_i-almost c-secure frameproof code and $\lim_{i \to \infty} \varepsilon_i = 0$.*

It is worth noting that the previous definitions allow to separate the concepts of separation and frameproofness, which coincide when we consider ordinary separation. Moreover, the new notions introduced allow to obtain codes with better rates when the separating or frameproof properties are only required with high probability.

For a family of codes $\{C_i\}_{i \geq 1}$, we define its asymptotical rate as

$$R(\{C_i\}) = \liminf_{i \to \infty} R(C_i).$$

Hence, we are interested in estimating the maximal possible asymptotical rate among all asymptotically almost c-separating families, denoted $R_q^*(c)$, and among all c-secure frameproof families, denoted $R_q^{(f)*}(c)$.

For instance, for binary codes and coalitions of size $c = 2$ the best existence bounds are $R_2^*(2) \geq 0.1142$, from [18], and $R_2^{(f)*}(2) \geq 0.2075$, from [12].

2.2 Small-Bias Probability Spaces

In this section we present the concepts about small-bias probability spaces that will be used in our constructions below. We will concentrate on the binary case, since our goal is to construct almost secure frameproof binary codes. For a more detailed exposition, we refer the reader to [14, 15, 19].

Consider the binary alphabet $Q = \mathbb{F}_2 = \{0, 1\}$. That is, Q is the set $\{0, 1\}$ with all operations reduced modulo 2. A *binary (n, M)-array* A is an $n \times M$ matrix, where the entries of A are elements from $\mathbb{F}_2$.

For a binary (n, M)-array A and a subset of indices $S \subseteq \{1, \ldots, M\}$ of size s, let us denote $N_S(\mathbf{a}; A)$ the frequency of rows of A, whose projection onto S equals the s-tuple $\mathbf{a} \in \mathbb{F}_2^s$. We will omit the subindex S whenever $s = M$, i.e., when we are considering the whole rows of array A. In particular, for a binary n-tuple $\mathbf{u}$, viewed as a binary $(n, 1)$-array, $N(0; \mathbf{u})$ and $N(1; \mathbf{u})$ denote the frequency, in $\mathbf{u}$, of symbols 0 and 1 respectively.

Definition 5. *Let* $\mathbf{u} = (u_1, \ldots, u_n) \in \mathbb{F}_2^n$. *The* bias *of vector* $\mathbf{u}$ *is defined as*

$$n^{-1} |N(0; \mathbf{u}) - N(1; \mathbf{u})|.$$

That is, a vector $\mathbf{u}$ which has approximately the same number of 0's and 1's will have small bias.

Definition 6. *Let* $0 \leq \varepsilon < 1$. *A binary (n, M)-array is ε-biased if every non-trivial linear combination of its columns has bias $\leq \varepsilon$.*

In other words, the bias of array A is the bias of the binary linear code C generated by its columns. By definition, the bias of A is low if the bias of every nonzero codeword from C is low. Explicit constructions of ε-biased (n, M)-arrays exist, with $n = 2^{O(\log M + \log \varepsilon^{-1})}$ [14].

The previous definition can be restricted by allowing a maximum number of columns in the linear combination.

Definition 7. *Let* $0 \leq \varepsilon < 1$. *A binary (n, M)-array is t-wise ε-biased if every nontrivial linear combination of at most t columns has bias $\leq \varepsilon$.*

We will also need the concepts of ε-dependent and ε-away from t-wise independence arrays.

Definition 8. *Let* $0 \leq \varepsilon < 1$. *A binary (n, M)-array A is t-wise ε-dependent if for every subset $S \subseteq \{1, \ldots, M\}$ of $s \leq t$ columns and every vector $\mathbf{a} \in \mathbb{F}_2^s$, the frequency $N_S(\mathbf{a}; A)$ satisfies*

$$|n^{-1} N_S(\mathbf{a}; A) - 2^{-s}| \leq \varepsilon.$$

Definition 9. *Let* $0 \leq \varepsilon < 1$. *A binary (n, M)-array A is ε-away from t-wise independence if for every subset $S \subseteq \{1, \ldots, M\}$ of $s \leq t$ columns we have*

$$\sum_{\mathbf{a} \in \mathbb{F}_2^s} |n^{-1} N_S(\mathbf{a}; A) - 2^{-s}| \leq \varepsilon.$$

Observe that if an array A is t-wise ε-dependent, then it is $2^M \varepsilon$-away from t-wise independence, and if A is ε-away from t-wise independence, then it is t-wise ε-dependent.

The definitions above have an interpretation as a small-bias probability space. Consider a set of M binary random variables $X_1, \ldots X_M$ that take the corresponding values of a row, chosen uniformly at random, from an (n, M)-array A. If the array A is ε-away from t-wise independence, then any t of the random variables are "almost independent," provided that ε is small. Hence, one would like to obtain such arrays A with n (the size of the probability space) as small as possible.

For our purposes, the most important concept will be that of (M, t)-universal set. We have the following definition.

Definition 10. *An (M, t)-universal set B is a subset of $\mathbb{F}_2^M$ such that for every subset $S \subseteq \{1, \ldots, M\}$ of t positions the set of projections of the elements of B on the indices of S contains every configuration $\mathbf{a} \in \mathbb{F}_2^t$.*

Let A be a binary (n, M)-array. Note that if for every subset $S \subseteq \{1, \ldots, M\}$ of t columns and every vector $\mathbf{a} \in \mathbb{F}_2^t$ we have $N_S(\mathbf{a}; A) > 0$, then the rows of A form an (M, t)-universal set. We are interested in universal sets of as small size as possible.

In [14] the relationship between this concept and ε-away from t-wise independence arrays was shown.

Proposition 1. *Let A be a binary (n, M)-array A. For $\varepsilon \leq 2^{-t}$, if A is ε-away from t-wise independence, then the rows of A yield an (M, t)-universal set of size n.*

Moreover, the following result [14, 20, 21] also relates these concepts with the concept of ε-biased arrays.

Corollary 1. *Let A be a binary (n, M)-array A. If A is ε-biased, then A is $2^{t/2} \varepsilon$-away from t-wise independence.*

Hence, the construction of universal sets is reduced to the construction of ε-away from t-wise independence arrays by Proposition 1, which is reduced to the construction of ε-biased arrays by Corollary 1.

We will have occasion to use Corollary 1 in the next section, where an even more convenient method to construct ε-away from t-wise independence arrays will be discussed.

3   Constructions

In this section we present our constructions for almost secure frameproof codes. Before dwelling into explicit details we give an intuitive reasoning of our discussion.

First, we will show that the expected value of the ratio of separated c-coalitions in a random binary (n, M)-code is maximized when the codewords are generated according to a probability vector $\mathbf{p} = (p_1, \ldots, p_n)$ with $p_1 = \cdots = p_n = 1/2$. That is, we generate M random codewords $(u_1, \ldots, u_n)$ such that $\Pr\{u_i = 1\} = p_i = 1/2$. But since we are interested in *almost* secure frameproof codes, we will be able to allow a small bias on these probabilities and therefore consider small-bias probability spaces.

By using the definitions and results from the previous section it can be seen that from ε-away from t-wise independence arrays we can obtain (M, t)-universal sets of size $O(2^t \log M)$. If we arrange the vectors of this universal set as the rows of a matrix, the columns of that matrix form a c-secure frameproof code for $t = 2c$. This code has size M, length $O(2^{2c} \log M)$ and rate $O(2^{-2c})$. The main idea is to impose a value of 0 to the probability of a given number of configurations in the universal set, yielding "almost" universal sets. We finally prove that "almost" universal sets can be used to generate ε-almost c-secure frameproof code with ε a function of the number of configurations with probability 0.

3.1   Separation in Random Codes

We start by making some observations about random codes. Let us assume that C is an (n, M)-random code generated according to a probability vector $\mathbf{p} = (p_1, \ldots, p_n)$, where $\mathbf{p}$ is chosen according to pmf $f_{\mathbf{p}}$. That is, we first generate a probability vector $\mathbf{p}$ of length n, distributed according to $f_{\mathbf{p}}$, and then we randomly generate M binary vectors $\mathbf{u} = (u_1, \ldots, u_n)$ such that $\Pr\{u_i = 1\} = p_i$. We would like to know which probability distribution $f_{\mathbf{p}}$ maximizes the ratio (probability) of separated c-coalitions in a code generated in that way.

Lemma 1. *Let C be an (n, M)-random code, whose codewords are generated according to the probability vector $\mathbf{p} = (p_1, \ldots, p_n)$. If the entries of $\mathbf{p}$ are iid r.v.'s, then the expected value of the ratio (probability) of separated c-coalitions is maximized by taking $p_1 = \cdots = p_n = 1/2$.*

Proof. Note that for a given $\mathbf{p}$ the probability that two c-coalitions U, V are not separated is $\prod_{i=1}^{n}(1 - 2p_i^c(1 - p_i)^c)$.

Let us denote ε the probability that a c-coalition U is not a separated c-coalition, which in general is a random variable. Hence, the expectation of the probability that U is a separated c-coalition, averaged over all the possible choices of $\mathbf{p}$, can be expressed as

$$E_{f_{\mathbf{p}}}[1 - \varepsilon] = E_{f_{\mathbf{p}}}\left[\left(1 - \prod_{i=1}^{n}(1 - 2p_i^c(1 - p_i)^c)\right)^{\binom{M-c}{c}}\right].$$

Observe that this expectation is maximized simply by considering a pmf that takes 1 on the maximum of the argument of the expectation and 0 otherwise. Therefore, this can be translated into finding the pmf that maximizes the following expectation

$$E_{f_{\mathbf{p}}}\left[1 - \prod_{i=1}^{n}(1 - 2p_i^c(1 - p_i)^c)\right] = 1 - (1 - 2E_{f_p}[p^c(1 - p)^c])^n,$$

which follows after assuming that the components of $\mathbf{p}$ are i.i.d. random variables distributed according to f_p. It is easy to see that $p^c(1 - p)^c$ is symmetric around $1/2$, and the expected value is maximized simply by taking $p = 1/2$ with probability 1.

The previous lemma suggests that codes with approximately the same number of 0's and 1's in each row of the codebook are good candidates to be (c, c)-separating and c-secure frameproof codes. Equivalently, for each set of $2c$ rows of the codebook, one would expect that all the 2^{2c} possible $2c$-tuples exhibit a uniform distribution approximately. In fact, there exist constructions of (c, c)-separating codes which are based on this observation [22].

3.2   Universal and Almost Universal Sets

Universal sets have been described in Definition 10. Moreover, it has been shown that the construction of universal sets can be reduced to the construction of ε-biased arrays.

It is easy to see that an $(M, 2c)$-universal set of size n also yields a (c, c)-separating (n, M)-code [22]. To see this, let A be an (n, M)-array whose rows form an $(M, 2c)$-universal set. Now, regard the columns of A as the codewords of a code C. Consider two disjoint c-subsets $U, V \subseteq C$, i.e., $2c$ columns of A. Since the rows of A are a $(M, 2c)$-universal set, this means that for the selected $2c$ columns all $\mathbb{F}_2^{2c}$ possible configurations appear. In particular, there is a row i where all the columns corresponding to U contain symbol 0 and all the columns corresponding to V contain symbol 1 in that particular row. Hence i is a separating position for coalitions U, V, i.e., $P_i(U) \cap P_i(V) = \emptyset$, as desired. Recall again that this is the same as a c-secure frameproof code when we are talking about absolute separation.

Efficient constructions of $(M, 2c)$-universal sets using ε-biased from $2c$-wise independence arrays are presented in [14], by virtue of Proposition 1 and Corollary 1. These constructions yield a (c, c)-separating code of length $2^{O(c)} \log M$. Using this idea, we aim to relax the constraint that the $(M, 2c)$-universality imposes to obtain a shorter array, i.e., a code with a better rate. In fact, we do not need that every possible $\mathbb{F}_2^{2c}$-tuple appears in the code. Hence, we propose to relax Definition 10 by allowing a given number of vectors $\mathbf{a} \in \mathbb{F}_2^{2c}$, say z, not to appear in the projection of a subset $S \subseteq \{1, \ldots, M\}$ of $2c$ positions. This is formalized in the following definition.

Definition 11. *An (M, t, z)-universal set B is a subset of $\mathbb{F}_2^M$ such that for every subset $S \subseteq \{1, \ldots, M\}$ of t positions the set of projections of the elements of B on the indices of S contains every configuration $\mathbf{a} \in \mathbb{F}_2^t$ except, at most z.*

Again, if A is a binary (n, M)-array, the rows of A generate an (M, t, z)-universal set provided that there are at least $2^t - z$ vectors $\mathbf{a} \in \mathbb{F}_2^t$ such that $N_S(\mathbf{a}; A) > 0$, for every subset $S \subseteq \{1, \ldots, M\}$ of t columns.

Similarly as Proposition 1, the following result shows the connection between (M, t, z)-universal sets and ε-away from t-wise independence arrays.

Proposition 2. *Let A be a binary (n, M)-array A. For $\varepsilon \leq (z + 1)2^{-t}$, if A is ε-away from t-wise independence, then the rows of A yield an (M, t, z)-universal set of size n.*

Proof. Assume by contradiction that the rows of A do not yield an (M, t, z)-universal set. In other words, there is a subset $S \subseteq \{1, \ldots, M\}$ of t columns such that there are strictly more than z vectors $\mathbf{a} \in \mathbb{F}_2^t$ such that $N_S(\mathbf{a}; A) = 0$. For this particular subset S we have that

$$\sum_{\mathbf{a} \in \mathbb{F}_2^t} |n^{-1} N_S(\mathbf{a}; A) - 2^{-t}|$$

$$\geq (z+1)2^{-t} + \sum_{\substack{\mathbf{a} \in \mathbb{F}_2^t \text{ s.t.} \\ N_S(\mathbf{a}; A) > 0}} |n^{-1} N_S(\mathbf{a}; A) - 2^{-t}| > (z+1)2^{-t+1} > \varepsilon,$$

which contradicts the fact that array A is ε-away from t-wise independence.

3.3 Construction of (M, t, z)-Universal Sets

As Proposition 2 states, the construction of (M, t, z)-universal sets reduces to constructing an $(z + 1)2^{-t}$-away from t-wise independence array, and by Corollary 1, it reduces to the construction of an ε-biased array. Moreover, it is easy that the array A from Corollary 1 can be regarded as a t-wise ε-biased array, which is a less restrictive condition than a ε-biased array.

A standard construction of t-wise ε-biased binary arrays is also presented in [14].

Theorem 1. *Let A be an ε-biased binary (n, M')-array, and let H be the parity-check matrix of a binary $[M, M - M']$-code of minimum distance $t + 1$. Then, the matrix product $A \times H$ is a t-wise ε-biased (n, M)-array.*

Usually, the matrix H used in Theorem 1 above is the parity-check matrix of a binary $[M, M - M']$-BCH code of minimum distance $t + 1$. In this case, the matrix H has M columns and $M' = t \log M$ rows. It is shown in [14] that, by using Theorem 1 in Corollary 1, the number of rows of an (n, M)-array ε-away from t-wise independence can be reduced from $n = 2^{O(t + \log M + \log \varepsilon^{-1})}$ to $n = 2^{O(t + \log \log M + \log \varepsilon^{-1})}$.

The problem now reduces to obtain binary ε-biased (n, M')-arrays with n as small as possible. From [19], one can see that explicit constructions exist for such arrays with

$$n \leq 2^{2(\log_2 M' + \log_2 \varepsilon^{-1})}.$$

However in [15], better explicit construction of ε-biased arrays are given, when the parameters satisfy some required conditions. The best construction shown there is based in Suzuki codes. Below we rewrite [15, Theorem 10] in our notation.

Theorem 2. *If* $\log M' > 3 \log \varepsilon^{-1}$*, then there exists an explicit construction of a binary* (n, M')*-array* A *that is* ε*-biased, with* $n = 2^{3/2 \, (\log_2 M' + \log_2 \varepsilon^{-1}) + 2}$*.*

Hence, to construct an (M, t, z)-universal set we can proceed as follows.

1. Take $\varepsilon = (z + 1)2^{-3t/2}$.
2. Construct an (n, M')-array A' that is ε-biased, where $M' = t \log M$.
3. Construct the parity-check matrix H of a BCH code of length M, codimension $M' = t \log M$ and minimum distance $t + 1$.
4. The matrix product $A = A' \times H$ generates a t-wise ε-biased (n, M)-array.
5. The array A is also ε'-away from t-wise independence, with $\varepsilon' = 2^{t/2} \varepsilon = (z + 1)2^{-t}$. Hence, the rows of A generate an (M, t, z)-universal set.

Observe that the conditions of Theorem 2 apply, when $\log_2 M' > 3 \log_2 \varepsilon^{-1}$, that is

$$\log_2 t + \log_2 \log_2 M > 9 \, t/2 - 3 \log_2(z + 1),$$

The resulting (M, t, z)-universal set from the construction above has size

$$n = 2^{3/2(3 \, t/2 + \log_2 t + \log_2 \log_2 M - \log_2(z+1)) + 2}.$$

We remark that the condition above, even though analitically meaningful, it is satisfied for impractically large values of M. That is, it will lead to codes with an excessively large number of codewords. For practical scenarios, using the constructions for ε-biased (n, M')-arrays given in [19], the resulting (M, t, z)-universal sets have size

$$n = 2^{2(3 \, t/2 + \log_2 t + \log_2 \log_2 M - \log_2(z+1))}.$$

3.4 ε-Almost c-Secure Frameproof Codes

Recall that an $(M, 2c)$-universal set of size n generates a c-secure frameproof (n, M)-code. Now, take an (n, M)-array A whose rows generate an $(M, 2c, z)$-universal set B, and regard its columns as the codewords of an ε-almost c-secure frameproof code C. Observe that for $z < 2^c$ the $(M, 2c, z)$-universal set B is in fact an (M, c)-universal set. To see this, note that if a configuration from $\mathbb{F}_2^c$ does not appear in B, it would mean that there are, at least, 2^c missing configurations from $\mathbb{F}_2^{2c}$, which contradicts the definition of an $(M, 2c, z)$-universal set with $z < 2^c$.

Given a code C constructed using an $(M, 2c, z)$-universal set as above, in order to ease the analysis, we will assume that for each c-coalition $U \subseteq C$, each possible configuration from $\mathbb{F}_2^c$ appears approximately with uniform probability.

Note that we are regarding the columns of an (n, M)-array A as the codewords of an almost secure frameproof code, which means that the resulting code has rate $R = \log M / n$. The following corollary formalizes the relationship between almost secure frameproof codes and (M, t, z)-universal sets.

Corollary 2. *Let $M > 0$, $c \geq 2$, $z < 2^c$, and $\varepsilon \geq p(M, c, z)$, where*

$$p(M, c, z) \overset{\text{def}}{=} M^c(1 - 2^{-c})^n.$$

Then, an $(M, 2c, z)$-universal set of size n yields an ε-almost c-secure frameproof code of rate $R = \log M/n$.

Proof. Consider a code C generated from an $(M, 2c, z)$-universal set, as stated above. Let $\mathbf{z}$ be a descendant generated by some c-coalition of the code, $\mathbf{z} \subseteq \text{desc}_c(C)$. By the assumptions stated above, the probability that $\mathbf{z}$ belongs to another c-coalition V is $(1 - 2^{-c})^n$. Hence, using the union bound, we can bound the probability that $\mathbf{z}$ is generated by some other coalition of the code as

$$p(M, c, z) = M^c(1 - 2^{-c})^n.$$

The ratio (probability) of not uniquely decodable descendants in $\text{desc}_c(C)$ is therefore $\leq p(M, c, z)$, which means that C is an ε-almost c-secure frameproof code. ∎

3.5 Results for Some Coalition Sizes

In the Table 1 we show the derived rates for the case of coalitions of size $c = 2$ and 3. The maximum number of missing $\{0, 1\}^{2c}$ configurations is denoted by z, and the probability that a descendant is not uniquely decodable is denoted by ε. Observe that when $z = 0$ the code is (c, c)-separating, that is $\varepsilon = 0$. The value of ε provided corresponds to the worst-case for the given row. The code rates have been computed for code sizes of $M = 10^3$, 10^4, 10^5, 10^6 and 10^7 users, using the constructions of $(M, 2c, z)$-universal sets derived from constructions of ε-biased (n, M')-arrays given in [19].

Table 1. Some attainable code rates for explicit constructions of ε-almost c-secure frameproof codes of size between 10^3 and 10^7

			Code rates				
c	z	$\log_2 \varepsilon$	$M = 10^3$	$M = 10^4$	$M = 10^5$	$M = 10^6$	$M = 10^7$
2	0	n/a	1.531×10^{-6}	1.148×10^{-6}	9.187×10^{-7}	7.656×10^{-7}	6.562×10^{-7}
2	1	-1.201×10^6	6.124×10^{-6}	4.593×10^{-6}	3.675×10^{-6}	3.062×10^{-6}	2.625×10^{-6}
2	2	-5.336×10^5	1.378×10^{-5}	1.034×10^{-5}	8.268×10^{-6}	6.890×10^{-6}	5.906×10^{-6}
2	3	-3.001×10^5	2.450×10^{-5}	1.837×10^{-5}	1.470×10^{-5}	1.225×10^{-5}	1.050×10^{-5}
3	0	n/a	1.063×10^{-8}	7.975×10^{-9}	6.380×10^{-9}	5.316×10^{-9}	4.557×10^{-9}
3	1	-8.025×10^7	4.253×10^{-8}	3.190×10^{-8}	2.552×10^{-8}	2.127×10^{-8}	1.823×10^{-8}
3	3	-2.006×10^7	1.701×10^{-7}	1.276×10^{-7}	1.021×10^{-7}	8.506×10^{-8}	7.291×10^{-8}
3	5	-8.917×10^6	3.828×10^{-7}	2.871×10^{-7}	2.297×10^{-7}	1.914×10^{-7}	1.640×10^{-7}
3	7	-5.016×10^6	6.805×10^{-7}	5.104×10^{-7}	4.083×10^{-7}	3.402×10^{-7}	2.916×10^{-7}

4   Constructions of Fingerprinting Codes

In this section we show how binary ε-almost c-secure frameproof codes can be used to construct a family of binary fingerprinting codes with an efficient decoding algorithm.

For a fingerprinting scheme to achieve a small error probability a single code is not sufficient, but a *family of codes* $\{C_j\}_{j \in T}$ is needed, where T is some finite set. The family $\{C_j\}_{j \in T}$ is publicly known. The distributor chooses secretly a code C_j with probability $\pi(j)$. This choice is kept secret. Then, codewords are assigned correspondingly.

In [18] existence conditions for a family of concatenated fingerprinting codes is proposed, using an almost separating code as inner code. We remark that the almost separating code can be replaced by an almost secure frameproof code, yielding an explicit construction of a fingerprinting code. Hence, combining [18, Corollary 1] with our results we have the following result.

Corollary 3. *Let C_{out} be an extended Reed-Solomon $[n, k]$-code over $\mathbb{F}_q$ of rate*

$$R_o = R(C_{\text{out}}) < \frac{1 - \sigma}{c(c + 1)},$$

and let C_{in} be an ε-almost c-secure frameproof (l, q)-code of rate $R_i = R(C_{\text{in}})$, with $\varepsilon < \sigma$. Then, there exists an explicit construction of a binary c-secure family of fingerprinting codes $\{C_j\}_{j \in T}$ with outer code C_{out} and inner code C_{in}, with polynomial-time tracing algorithm, rate $R = R_i R_o$ and probability of error decreasing exponentially as

$$p_e \leq 2^{-n l \left(\frac{1-\sigma}{c} R_i - (c+1)R + o(1)\right)} + 2^{-n D(\sigma \| \varepsilon)}.$$

Finally, it is worth noting here that, as shown in [12, 18], the use of almost secure frameproof codes instead of ordinary secure frameproof codes introduces an additional error term in the identification process, as stated in Corollary 3. Note, however, that this error term decreases exponentially with the outer code length.

5   Conclusion

Almost separating and almost secure frameproof codes are two relaxed versions of separating codes. In this paper, we have presented the first explicit constructions of almost secure frameproof codes.

Our work has departed from the study of the connection between small-bias probability spaces and universal sets, and the subsequent connection between universal sets and separating codes.

Starting with this idea, we have introduced a relaxation in the definition of a universal set. We show that an almost universal set can be used to construct an almost secure frameproof code. This observation has lead us to the explicit constructions of almost secure frameproof codes. We have proposed a construction based on Suzuki codes, which provide one of the best constructions known for small-bias probability spaces. For practical uses, however, we have to switch to the constructions of small-bias probability spaces proposed by Alon et al.

We remark that, as expected, the explicit constructions presented are somewhat far from the theoretical existence bounds shown in earlier works. For example, probabilistic arguments show the existence of asymptotically almost 2-secure frameproof families of codes of rate $R = 0.2075$, whereas the explicit constructions that we have presented above provide codes of rate below this figure. Nevertheless, the main point of our work is to present the first explicit and practical-use constructions for such families of codes.

We have also shown how the proposed constructions can be used to explicitly construct a family of concatenated fingerprinting codes. The construction presented is based on the theoretical existence results of a previous work, which assumed the existence of almost secure frameproof codes. Hence, one of the main contributions of the present work has been to provide a "real" implementation of such a theoretical existence result for a fingerprinting scheme. As discussed also in earlier works, replacing ordinary secure frameproof codes by almost secure frameproof codes introduces an additional error term in the identification of guilty users that, fortunately, decreases exponentially with the outer code length.

Finally, we would like to note that even though a universal set is a separating code, the relationship between an almost universal set and an almost separating code is by no means evident and will we the subject of future research.

Acknowledgement. We would like to thank the IWSEC2013 anonymous Referees for their useful comments. J. Moreira and M. Fernández have been supported by the Spanish Government through projects Consolider Ingenio 2010 CSD2007-00004 "ARES" and TEC2011-26491 "COPPI", and by the Catalan Government through grant 2009 SGR-1362. G. Kabatiansky has been supported by the Russian Foundation for Basic Research through grants RFBR 13-07-00978 and RFBR 12-01-00905.

References

1. Stinson, D.R., van Trung, T., Wei, R.: Secure frameproof codes, key distribution patterns, group testing algorithms and related structures. J. Stat. Plan. Infer. 86(2), 595–617 (2000)
2. Staddon, J.N., Stinson, D.R., Wei, R.: Combinatorial properties of frameproof and traceability codes. IEEE Trans. Inf. Theory 47(3), 1042–1049 (2001)
3. Tonien, D., Safavi-Naini, R.: Explicit construction of secure frameproof codes. Int. J. Pure Appl. Math. 6(3), 343–360 (2003)
4. Stinson, D.R., Zaverucha, G.M.: Some improved bounds for secure frameproof codes and related separating hash families. IEEE Trans. Inf. Theory 54(6), 2508–2514 (2008)
5. Friedman, A.D., Graham, R.L., Ullman, J.D.: Universal single transition time asynchronous state assignments. IEEE Trans. Comput. C-18(6), 541–547 (1969)

6. Sagalovich, Y.L.: Separating systems. Probl. Inform. Transm. 30(2), 105–123 (1994)
7. Pinsker, M.S., Sagalovich, Y.L.: Lower bound on the cardinality of code of automata's states. Probl. Inform. Transm. 8(3), 59–66 (1972)
8. Sagalovich, Y.L.: Completely separating systems. Probl. Inform. Transm. 18(2), 140–146 (1982)
9. Körner, J., Simonyi, G.: Separating partition systems and locally different sequences. SIAM J. Discr. Math. (SIDMA) 1(3), 355–359 (1988)
10. Cohen, G.D., Schaathun, H.G.: Asymptotic overview on separating codes. Department of Informatics, University of Bergen, Norway, Tech. Rep. 248 (August 2003)
11. Cohen, G.D., Schaathun, H.G.: Upper bounds on separating codes. IEEE Trans. Inf. Theory 50(6), 1291–1294 (2004)
12. Fernández, M., Kabatiansky, G., Moreira, J.: Almost separating and almost secure frameproof codes. In: Proc. IEEE Int. Symp. Inform. Theory (ISIT), Saint Petersburg, Russia, pp. 2696–2700 (August 2011)
13. Barg, A., Blakley, G.R., Kabatiansky, G.: Digital fingerprinting codes: Problem statements, constructions, identification of traitors. IEEE Trans. Inf. Theory 49(4), 852–865 (2003)
14. Naor, J., Naor, M.: Small-bias probability spaces: Efficient constructions and applications. SIAM J. Comput (SICOMP) 22(4), 838–856 (1993)
15. Bierbrauer, J., Schellwat, H.: Almost independent and weakly biased arrays: Efficient constructions and cryptologic applications. In: Bellare, M. (ed.) CRYPTO 2000. LNCS, vol. 1880, pp. 533–544. Springer, Heidelberg (2000)
16. Boneh, D., Shaw, J.: Collusion-secure fingerprinting for digital data. In: Coppersmith, D. (ed.) CRYPTO 1995. LNCS, vol. 963, pp. 452–465. Springer, Heidelberg (1995)
17. Boneh, D., Shaw, J.: Collusion-secure fingerprinting for digital data. IEEE Trans. Inf. Theory 44(5), 1897–1905 (1998)
18. Moreira, J., Kabatiansky, G., Fernández, M.: Lower bounds on almost-separating binary codes. In: Proc. IEEE Int. Workshop Inform. Forensics, Security (WIFS), Foz do Iguaçu, Brazil, pp. 1–6 (November 2011)
19. Alon, N., Goldreich, O., Håstad, J., Peralta, R.: Simple constructions of almost k-wise independent random variables. Random Struct. Alg. 3(3), 289–304 (1992)
20. Vazirani, U.V.: Randomness, adversaries and computation. Ph.D. dissertation, Dept. Elect. Eng. Comp. Sci., Univ. California, Berkeley (1986)
21. Diaconis, P.: Group Representations in Probability and Statistics. Inst. Math. Stat., Beachwood (1988)
22. Alon, N., Guruswami, V., Kaufman, T., Sudan, M.: Guessing secrets efficiently via list decoding. ACM Trans. Alg. 3(4), 1–16 (2007)

Differential Power Analysis
of MAC-Keccak at Any Key-Length

Mostafa Taha and Patrick Schaumont*

Secure Embedded Systems
Center for Embedded Systems for Critical Applications
Bradley Department of ECE
Virginia Tech, Blacksburg, VA 24061, USA

Abstract. Keccak is a new hash function selected by NIST as the next
SHA-3 standard. Keccak supports the generation of Message Authentication Codes (MACs) by hashing the direct concatenation of a variable-
length key and the input message. As a result, changing the key-length
directly changes the set of internal operations that need to be targeted
with Differential Power Analysis. The proper selection of these target operations becomes a new challenge for MAC-Keccak, in particular when
some key bytes are hidden under a hierarchical dependency structure.
In this paper, we propose a complete Differential Power Analysis of
MAC-Keccak under any key-length using a systematic approach to identify the required target operations. The attack is validated by successfully breaking several, practically difficult, case studies of MAC-Keccak,
implemented with the reference software code on a 32-bit Microblaze
processor.

1 Introduction

The recent SHA-3 hashing competition, organized by the National Institute of
Standards and Technology (NIST), has recently concluded with Keccak as the
winner [6]. Keccak was particularly selected for being built with a new algorithmic construction, the *Sponge* construction [5], which is entirely different from
the previous hashing standards. This new construction opens new questions,
challenges and opportunities in Side-Channel Analysis (SCA).

The idea of the *Sponge* construction is to have an internal state that is big-
ger than the input and output block sizes. This feature prevents the input block
from directly affecting the entire state and protects the internal state from being
fully exposed at the output. This construction allowed Keccak to securely create Message Authentication Codes (MACs) by hashing the direct concatenation
between the key and the message in a cryptographic mode called MAC-Keccak
[5]. Although previous MACs required that the key is hashed in a separate input
block, MAC-Keccak allowed the secret key and the message to share one input

---

* This research was supported in part by the VT-MENA program of Egypt, and by
the National Science Foundation Grant no. 1115839.

K. Sakiyama and M. Terada (Eds.): IWSEC 2013, LNCS 8231, pp. 68–82, 2013.

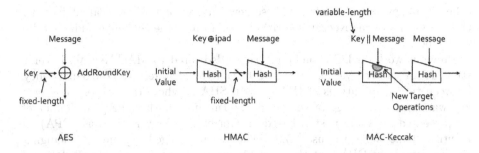

Fig. 1. The difference between the DPA of AES, HMAC, and MAC-Keccak

block. Furthermore, the designers allowed the use of a variable-length key, which opens a new dimension in the analysis. In this paper, we will study how to mount a Differential Power Analysis (DPA) on MAC-Keccak under any key-length.

Differential Power Analysis (DPA) is an implementation attack that aims at recovering the secret key of a cryptographic module by monitoring the instantaneous power consumption [12]. With DPA, the attacker searches for the key that maps the change in the power consumption from trace to trace, to the corresponding change in the input message. Only the correct secret key will make this mapping meaningful. Typically, only one point in the power trace is enough to mount an attack, the point where the key and the message get mixed in an internal operation. We call this internal operation 'the target operation' and we assume that the input message can be monitored.

The difference between the DPA of MAC-Keccak and that of the typical cryptographic algorithms (e.g. AES and HMAC) is highlighted in Fig. 1. The length of the secret key in typical cryptographic algorithms is fixed. Hence, the algorithm can be analyzed easily to select the set of target operations. The AES encryption algorithm, for example, uses a fixed length key of 128, 192 or 256 bits and the target operation is the AddRoundKey [11]. Similarly, DPA on HMAC aims at recovering the internal state after hashing the key (XORed with the ipad) [4]. In this case, the size of the required unknown is also fixed and equals to the size of the internal state, where the target operation is the input to the next hashing cycle. On the contrary, the secret key in MAC-Keccak has a variable length, and it shares the message in one input block. As a result, changing the key-length (and consequently the message-length) within the input block, changes the set of target operations that need to be monitored for a successful DPA. Also, the target operations are located deep inside the algorithm itself instead of being only at the input. For instance, the key-length can be selected near the input block size to shrink the proportion of the input message. Here, a single target operation will not be sufficient due to the difference in size between the key and the known message. In such case, several consecutively dependent operations will be targeted to reach the point in the algorithm where every key-byte is affected by at least one message-byte. Hence, the DPA of MAC-Keccak requires new methodologies, which were not needed in the analysis of other cryptographic algorithms. These new features of MAC-Keccak along with the complexity of the Keccak algorithm itself formed the challenge addressed by this paper.

Previous work on DPA, including those dedicated to MAC-Keccak, haven't studied DPA under a variable key-length. McEvoy *et al.* resolved the dependency between target operations of HMAC with SHA-2, where the size of the secret key was fixed to the size of the secret intermediate hash value [13]. Zohner *et al.* presented an analysis step based on Correlation Power Analysis (CPA) to identify the key-length in use [15,9] . They acknowledged the effect of changing the key-length on DPA, but they did not study the problem in detail. Relying on their method of identify the key-length, we will assume that the key-length of MAC-Keccak is known upfront. Taha *et al.* studied the effect of changing the key-length on the difficulty of DPA, but still with an ad-hoc selection of the target operations [14]. The Keccak developers proposed a side-channel countermeasure for MAC-Keccak based on secret sharing (masking) [10]. They presented results for attacking both protected and unprotected implementations of MAC-Keccak using simulated traces [7]. In their analysis, they studied the effect of changing the state-size assuming that the key-length is fixed and equals to the input block size (i.e. similar to previous MAC constructions). If the key-length changes, they will have to apply high-order DPA following the same attack methodology of this paper.

In this paper, we will show that the complete recovery of the secret key requires mounting DPA against several consecutively dependent operations. We will use a systematic approach to analyze MAC-Keccak under any key-length, to extract the required target operations, in the proper order of dependency. We will present complete case studies of MAC-Keccak under several practically difficult key-lengths, with all the necessary details. Finally, the analysis will be validated by successfully breaking these case studies implemented with the reference software code on a 32-bit Microblaze processor [2].

The paper is organized as follows: Section 2 introduces the required background on the *Sponge* construction, the Keccak-function and MAC-Keccak. Section 3 presents a detailed analysis of some examples of MAC-Keccak. This analysis is intended to show the consecutive dependency of target operations, and the systematic approach used to resolve that dependency under any key-length. The complete case studies are presented in Section 4. Section 5 validates the analysis by presenting the results of a practical attack mounted against the considered cases. Section 6 presents the conclusion and future work.

2   Background

Keccak uses a new *Sponge* construction chaining mode with a fixed permutation function called the Keccak-function [5]. The size of the internal state of the *Sponge* is $b = 25 * 2^l$ bits and $l = [0 : 6]$ arranged in a $5 * 5 * 2^l$ array. The size of the input block is called the Rate r, which is strictly less than the state size. The difference between the Rate and the state-size is called the Capacity c, which is

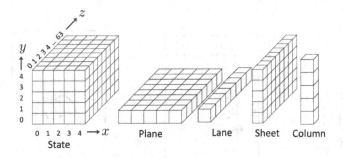

Fig. 2. Terminology used in Keccak

the number of state bits that are not directly affected by the current input. The Rate and Capacity are used as design parameters to trade security strength for throughput. Their sum equals to the state-size ($b = r + c$). To match the NIST output length requirements, Keccak designers proposed $b = 1600$ bits ($l = 6$), and $r = 1152, 1088, 832$ and 576 bits for output length of 224, 256, 384 and 512 bits respectively [8].

The internal state of Keccak is arranged in a 3-D array as shown in Fig. 2. Each state-bit is addressed with three coordinates, written as $S(X, Y, Z)$. The Keccak state also defines a *plane, lane, sheet* and *column*. These are defined as follows: a plane $\mathbb{P}(y)$ contains all state-bits $S(X, Y, Z)$ for which $Y=y$; a lane $\mathbb{L}(x, y)$ contains all state-bits for which $X=x$ and $Y=y$; a sheet $\mathbb{S}(x)$ contains all state-bits for which $X=x$; a column $\mathbb{C}(x, z)$ contains all state-bits for which $X=x$ and $Z=z$. The state is filled with r new message bits starting from $S(0, 0, 0)$ and filling in the Z direction, followed by the X direction, followed by the Y direction. The remaining part of the state (the Capacity) is kept unchanged; it is filled with zeros in the first hashing operation. This filling sequence puts the new input bits in the lower planes (from $Y = 0$) leaving the zero bits at the upper planes.

The Keccak-function consists of 24 rounds of five sequential steps. The steps are briefly discussed here. Further details can be found in the Keccak reference [6]. The output of each round is:

$$Output = \iota \circ \chi \circ \pi \circ \rho \circ \theta(Input) \tag{1}$$

Throughout the following; operations on X and Y are done modulo 5, and operations on Z are done modulo 64.

– θ is responsible for diffusion. It is a binary XOR operation with 11 inputs and a single output, as shown in Fig. 3. Every bit of the output state is the result of XOR between itself, and the bits of two neighbor columns:

$$S(X, Y, Z) = S(X, Y, Z) \oplus \left(\oplus_{i=0}^{4} S(X - 1, i, Z)\right) \oplus \left(\oplus_{i=0}^{4} S(X + 1, i, Z - 1)\right) \tag{2}$$

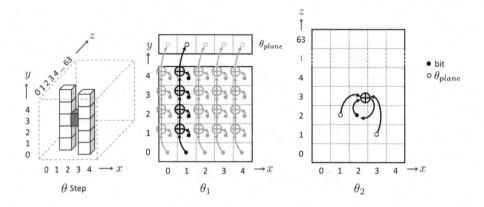

Fig. 3. θ step, θ_1 and θ_2

The θ operation is done over two successive steps. The first step θ_1 calculates the parity of each column, which is called θ_{plane}:

$$\theta_{plane}(X, Z) = \oplus_{i=0}^{4} S(X, i, Z) \tag{3}$$

In software implementations, θ_1 is implemented in incremental steps starting from $Y = 0$. We name each XOR operation by the first input, e.g. $\theta_1(X, Y, Z)$ is an XOR operation with inputs: $S(X, Y, Z)$ and $S(X, Y + 1, Z)$.

The second step θ_2 computes the XOR between every bit of the state and two parity bits of θ_{plane}.

$$S(X, Y, Z) = S(X, Y, Z) \oplus \theta_{plane}(X - 1, Z) \oplus \theta_{plane}(X + 1, Z - 1) \tag{4}$$

We name these operations by the name of the output bit, e.g. $\theta_2(X, Y, Z)$ is an XOR operation with three inputs: the state-bit $S(X, Y, Z)$, the parity bit $\theta_{plane}(X - 1, Z)$ and the parity bit $\theta_{plane}(X + 1, Z - 1)$.

- ρ is a binary rotation over each lane of the state.
- π is a binary permutation over lanes of the state as shown in Fig. 4. Every lane is replaced with another lane. In other words, π shuffles every row of lanes to a corresponding column.
- χ is responsible for the non-linearity. It flips a state-bit if its two adjacent bits along X are 0 and 1:

$$S(X, Y, Z) = S(X, Y, Z) \oplus \left(\overline{S(X + 1, Y, Z)} \cdot S(X + 2, Y, Z) \right) \tag{5}$$

We name the χ operation by its output bit, e.g. $\chi(X, Y, Z)$ takes three state-bits: $S(X, Y, Z)$, $S(X + 1, Y, Z)$ and $S(X + 2, Y, Z)$.

- ι is a binary XOR with a round constant.

It is clear that θ and χ are the only operations that can be targeted with DPA, as they involve mixing between different bits of the state. However, ρ and

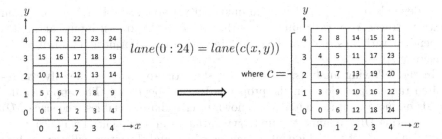

Fig. 4. Lane indices before and after the π step

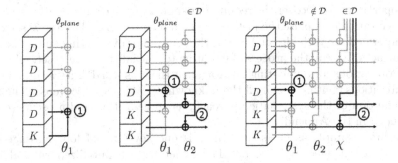

Fig. 5. DPA applied to some examples of MAC-Keccak

π operations cannot be ignored; they are needed to track the location of different bits in the χ operation.

As mentioned earlier, Keccak recommends the direct MAC construction, which is secured depending on the characteristics of the *Sponge* [5].

$$MAC(M, K) = H(K\|M) \tag{6}$$

This construction features an arbitrary-length key, which raised the new challenge addressed by this paper. The presented analysis can be applied directly to other MAC constructions (e.g. HMAC) by setting the key-length to the Rate.

3 Analysis of MAC-Keccak Examples

This section is mandatory to understand how the key-length affects DPA. We will present detailed analysis of some examples of MAC-Keccak and how the consecutive dependency shows up. We will also show the DPA steps required to resolve that dependency in different cases.

The studied examples of MAC-Keccak are shown in Fig. 5. We focus on one column of the state and study the propagation through θ_1, θ_2 and χ. The effect of ρ and π is neglected in the figure for clarity.

We define a 'data-dependent variable' as any intermediate variable that depends on the input message, and 'unknown variable' as any intermediate variable that depends only on the key. The unknown variable should be constant from trace to trace. Also, we define '$\mathcal{D}$' as the set of all the known inputs and data-dependent variables that can be calculated using the information known to the adversary.

In the following analysis, we used a systematic approach to identify the required target operations, in the proper order of dependency. The approach depends on increasing the number of known intermediate variables (the size of $\mathcal{D}$) by mounting DPA against the unknown variables in a sequential way. First, we initialize the $\mathcal{D}$ set to include all the message bits, and the intermediate variables that directly depend on them. Then, we select all the internal operations that process one unknown variable with one element of $\mathcal{D}$. We mount DPA against these operations, to recover the required unknown variables. If there are still any unknown variables, we use the just recovered unknowns to update the $\mathcal{D}$ set with more intermediate variables and repeat the same steps of selecting and mounting DPA attacks. We call the process of updating (or initializing) the $\mathcal{D}$ set, selecting target operations and mounting DPA against them as a 'DPA iteration'. These DPA iterations continue until all the unknown variables are recovered. The formal definition of the systematic approach, as an algorithm and a pseudo-code, is included in the Appendix.

The first example assumes that there is one key-bit and four message-bits in every column (see Fig. 5, left). The $\mathcal{D}$ set will be initialized with all the input message-bits. In this case, we will select and mount DPA against the first θ_1 operation, which involves one unknown variable (the secret key), and one element of $\mathcal{D}$ (the message-bits). The complete secret key should be recovered after this DPA iteration.

The second example assumes that there are two key-bits in every column (see Fig. 5, middle). Similarly, the $\mathcal{D}$ will be initialized with all the input message-bits. However, the second θ_1 operation will be selected in this case, as the first operation has two unknown inputs. By mounting DPA against the selected operation, the involved unknown variable should be recovered, which is the output of XORing the two key-bits. Unfortunately, this recovered unknown is not enough to uncover the original secret key, nor it can be used to forge a MAC digest because each key-bit will have its individual effect in later operations. In this case, we will have to go for another DPA iteration. We will use the information available from the just recovered unknown to calculate the θ_{plane}, and add it to $\mathcal{D}$. In this iteration, we will select the first two operations of θ_2, for having unknown variables (secret key-bits) and elements of $\mathcal{D}$ (θ_{plane}-bits) at their inputs. By mounting DPA against the selected operations, the complete secret key should be recovered. Note that, we cannot skip the θ_1 operation and directly attack θ_2, as every θ_2 operation involves two columns of the state, which greatly increases the search space.

The third example is yet more complicated (see Fig. 5, right). Similar to the previous example, we assume that there are two key-bits in every column. However, we assume that the θ_{plane}-bits required in the second DPA iteration are also unknown. Although this scenario is only possible if the key and message bits are interleaved in the input block, it shows that certain key-configurations may require targeting operations that are deep in the algorithm. In this case, the first DPA iteration will be similar to the previous example, and the θ_{plane} should be partially recovered. The partially recovered θ_{plane} will be added to $\mathcal{D}$. We will trace the Keccak-function passing through ρ and π, where the location of bits within the state will be mixed. Finally, the χ operation will be selected, targeted and the unknown should be recovered. The recovered unknown can be used to trace-back the Keccak-function to calculate the original key.

The analysis of these MAC-Keccak examples shows several interesting findings:

- A cryptographic algorithm can be designed in such a way that builds a hierarchical dependency structure between different key-bytes. Attacking one key-byte depends on the successful recovery of another key-byte.
- The order of attacking the internal operations should respect the consecutive dependency between different key-bytes.
- The probability of achieving a successful attack is affected by the number of dependant DPA iterations.
- The order of attacking internal operations and the number of DPA iterations in the attack depends on the key-length and the location of the key-bytes (key configurations) within the input block.
- Key configurations can be selected to maximize the effort required to mount a successful attack, which will be the focus of our future work.

4 Case Studies

The selected target operations depend heavily on the location of key-bits within the state. Hence, it becomes important to visualize the relative locations of key-bits and message-bits within the state. We assume a state-size of $b = 1600$ bit, arranged in a $5 * 5 * 64$ array, and an input block size of $r = 1088$ bit. As the state fills bottom-up, the new input block will first fill the lower three planes ($\mathbb{P}([0:2])$ in Fig. 2), followed by the first two lanes of the fourth plane ($\mathbb{L}([0:1], 3)$). We assume a key-length less than the Rate; i.e. the first input block will contain key-bits prepended on the message-bits. The key-bits will be in the lower planes of the state while the message-bits will be in the middle planes. Since this is the first hash block of the chain, the upper planes will contain zeros.

While there is no current standard for the key-length of MAC-Keccak, we present detailed analysis of three cases with different key-lengths; starting from 768 bits and adding 128 bits (or two Keccak lanes) in each case. The results of attacking these cases are highlighted in the following section. The results of two other cases, namely (Key-length = 832 and 960), are also presented without detailed analysis as they follow the same attack methodology.

The reasoning behind our choice of long key-lengths is that it gives the MAC-Keccak implementation higher resistance against DPA. Typically, the attack complexity increases linearly with the key-length where a separate attack is required for every key-byte. However, the complexity of attacking MAC-Keccak

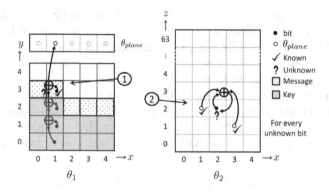

Fig. 6. The DPA of Key-length = 768 bits

increases faster than linear with the key-length due to the consecutive dependency between different key-bytes. This behavior is validated by the results presented in the following section. Also, long key-lengths are not uncommon in the cryptographic community. They have been practically used in the applications of RSA [3].

The analysis of shorter key-lengths (128, 256 and up to 320 bits) is trivial and can be solved by only one DPA iteration (the easiest case in Section 3). The studied key-lengths are chosen to highlight the dependency of attack complexity on key-length, where every case requires a new DPA iteration. The proportional reduction of the amount of message-bits in each case will make the DPA increasingly harder. We considered cases where the key fills complete lanes, hence the Z index is always $Z = [0 : 63]$ however, the analysis can be applied to any other key-length.

4.1   Key-Length = 768 Bits

The key-length is chosen so that the input message-length = 320 bits, the size of one plane of the state. There will be a single message bit in every column. Fig. 6 shows the position of the key-bits, message-bits and zeros using shaded, dotted and white squares respectively.

This case study is similar to the second example discussed in the previous section (Fig. 5, middle). The attack involves the following steps:

- Add all the input message-bits to the set $\mathcal{D}$.
- Select and target the output of θ_1 XOR operations between the key-bits and message-bits: $\theta_1([0 : 1], 2, Z)$ and $\theta_1([2 : 4], 1, Z)$.
- Recover the parity of the key-bits in every column.
- Calculate the θ_{plane}, and add it to $\mathcal{D}$.
- Select and target the output of θ_2 XOR operations between every key-bit and the corresponding θ_{plane}-bits: $\theta_2([0 : 1], [0 : 2], Z)$ and $\theta_2([2 : 4], [0 : 1], Z)$.
- This should recover the required key-bits.

This attack required two DPA iterations.

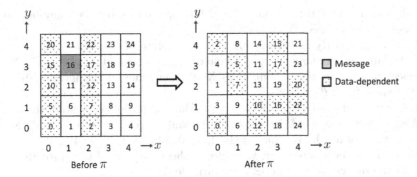

Fig. 7. Data-dependent bits of key-length=1024 bits, before and after π step

4.2 Key-Length = 896 Bits

The key-length is chosen such that the input message-length = 192 bits. Here, both sheets $\mathbb{S}(2)$ and $\mathbb{S}(3)$ have no message-bits. This case study is more difficult than the previous one. The all-unknown sheets will lead to all unknown lanes in the θ_{plane}. These unknown lanes of θ_{plane} will require more effort in the attack.

Our attack will follow the following steps:

- Add all the input message-bits to the set $\mathcal{D}$.
- Select and target the output of θ_1 XOR operations between key-bits and message-bits in sheets $\mathbb{S}([0,1,4])$: $\theta_1([0:1],2,Z)$ and $\theta_1(4,1,Z)$.
- Recover the parity of the key-bits in each column of those sheets.
- Calculate the partially recovered θ_{plane}, and add it to $\mathcal{D}$.
- Select and target the output of θ_2 XOR operations of the message-bits of the neighboring sheets $\mathbb{S}([1,4])$: $\theta_2(1,3,Z)$ and $\theta_2(4,2,Z)$.
- Recover the two missing lanes of θ_{plane} ($\mathbb{L}([2,3],0)$).
- Add the recovered lanes to $\mathcal{D}$.
- Select and target the output of θ_2 XOR operations between every key-bit and the corresponding θ_{plane}-bits: $\theta_2([0:3],[0:2],Z)$ and $\theta_2(4,[0:1],Z)$.
- This should recover the required key-bits.

The attack in this case required three DPA iterations.

4.3 Key-Length = 1024 Bits

The length of the input message is 64 bits, only one lane ($\mathbb{L}(1,3)$) with 4 all-unknown sheets $\mathbb{S}([0,2,3,4])$. The attack in this case will logically be in these steps: recover the Keccak state after θ step, apply the mapping of ρ and π steps as shown in Fig. 7, and recover the key using the χ step.

The exact attack will be as follows:

- Add the input message-bits to the set $\mathcal{D}$.

- Select and target the output of θ_1 XOR operations between key-bits and message-bits in sheet $\mathbb{S}(1)$: $\theta_1(1, 2, Z)$.
- Recover the parity of the key-bits in each column of that sheet.
- Calculate the partially recovered lane of θ_{plane}, and add it to $\mathcal{D}$.
- Select and target the output of θ_2 XOR operations of the message-bits of that sheet $\mathbb{S}(1)$: $\theta_2(1, 3, Z)$.
- Recover the data-dependent variables of lane $\mathbb{L}(1, 3)$ of the state after θ.
- Also in the same DPA iteration, select and target the output of θ_2 XOR operations of all the key-bits of sheets $\mathbb{S}([0, 2])$: $\theta_2([0, 2], [0 : 4], Z)$.
- Recover the data-dependent variables of sheets $\mathbb{S}([0, 2])$ of the state after θ.
- Add the recovered data-dependent variables to $\mathcal{D}$.
- Select and target the output of χ operations between an unknown variable and two elements of $\mathcal{D}$: $\chi(0, 0, Z)$, $\chi(1, 1, Z)$, $\chi(3, 1, Z)$, $\chi(4, 2, Z)$, $\chi(1, 3, Z)$, $\chi(3, 4, Z)$.
- Recover the targeted unknown variables, and add them to $\mathcal{D}$.
- Select and target the output of χ operations between an unknown variable, one elements of $\mathcal{D}$, and one just recovered unknown: $\chi(1, 0, Z)$, $\chi(4, 0, Z)$, $\chi(0, 2, Z)$, $\chi(3, 2, Z)$, $\chi(0, 3, Z)$, $\chi(2, 3, Z)$, $\chi(2, 4, Z)$, $\chi(4, 4, Z)$
- This should recover all the rest of unknown variables.

The recovered state can be used to trace-back the Keccak-function to retrieve the original key, or it can be used directly to forge a MAC digest by inserting the recovered state directly in the χ step. The attack in this case required four DPA iterations.

5   Practical Results

The experimental evaluation of our analysis was conducted on the reference software code of Keccak [1] running on 32-bit Microblaze processor [2] built on top of a Xilinx Spartan-3e FPGA. We used a Tektronix MIDO4104-3 oscilloscope with a CT-2 current probe to capture the instantaneous current of the FPGA core as an indication of the power consumption. To reduce the measurement noise, the processor was programmed to execute the same hashing operation 16 times, while the oscilloscope calculates the average of them. Figure 8 shows one recorded trace indicating all the steps of the Keccak-function.

The attack was conducted using the Correlation Power Analysis [9] with Hamming Weight power model. The power model was built using an 8-bit key guess at a time. This choice was motivated to reduce the algorithmic noise (because it is a 32-bit processor, where the processing of the remaining 24 bits will be considered noise) at a practical search space (256 different key guesses).

The results of attacking MAC-Keccak with key-lengths = (768, 896 and 1024), as discussed in the previous section, are shown in Fig. 9. The figure also shows the results of attacking two other cases; key-lengths = (832 and 960), that were analyzed using the same approach. The figure shows the success rate of each case study as a function of the number of traces used in the analysis, where the success rate is the percentage of key-bytes that have been recovered successfully.

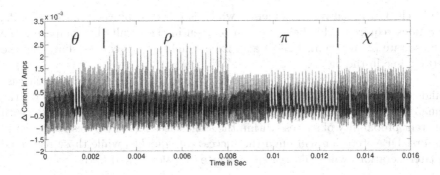

Fig. 8. A power trace of MAC-Keccak

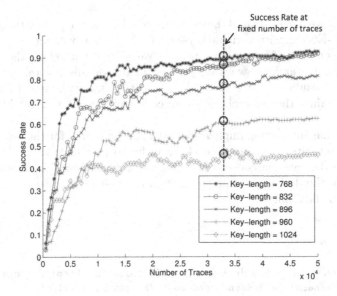

Fig. 9. Success rate of different case studies

The consecutive dependency between different key-bytes in MAC-Keccak affects the success rate of the attack as follows. Assuming that a single DPA is successful with probability p. The key-bytes recovered in the first DPA iteration will have probability of success p. Key-bytes of the second DPA iteration will be successful only if all the involved key-bytes of the previous iteration were recovered successfully and the second iterations itself was successful. Assuming that the number of previously recovered and involved key-bytes is n, the probability of success in the second DPA iteration will be $p^{(n+1)}$. Similarly, the probability of success of the third and forth DPA iterations will drop very quickly depending on the number of previous, involved key-bytes. As a result, the complexity of a complete DPA attack on MAC-Keccak increases faster than linear by increasing the levels of consecutive dependency, which is reflected by the number of DPA iterations required. This behavior can be seen in the results by comparing the success rate of the studied cases at a fixed number of traces, as shown by the vertical line in the figure.

There is another remark in the figure, where the success rate of each case builds up very quickly in the first 10,000 traces, then in a slower way through the remaining traces. The reason of this behavior is that key-bytes are recovered with different probability of success within the same case. The key-bytes recovered in the first DPA iteration will build the success rate quickly, while those recovered in later iterations will build the success rate in a slower and flatter way.

6   Conclusion and Future Work

In this paper, we demonstrated the challenge of selecting the proper target operation in MAC-Keccak with variable key-length. We used a systematic approach to increase the number of known intermediate variables by attacking the unknown variables in a hierarchical way. We studied in full details several, practically-difficult, case studies of MAC-Keccak. The attack was validated by a practical attack against the reference software code running on a 32-bit Microblaze processor.

Our work has interesting implications. We plan to search for the best possible location of key-bits in an input block, in order to maximize the effort for the attacker. We believe the concept of hierarchical dependency of unknowns also has value for the design and implementation of other cryptographic algorithms that offer this flexibility.

References

1. Keccak reference code submission to NIST (round 3), http://csrc.nist.gov/groups/ST/hash/sha-3/Round3/documents/Keccak_FinalRnd.zip
2. Xilinx microblaze soft processor core, http://www.xilinx.com/tools/microblaze.htm
3. RSA cryptography standard PKCS# 1 v2.2. RSA Laboratories, p. 63 (2012)
4. Bellare, M., Canetti, R., Krawczyk, H.: Keying hash functions for message authentication. In: Koblitz, N. (ed.) CRYPTO 1996. LNCS, vol. 1109, pp. 1–15. Springer, Heidelberg (1996)
5. Bertoni, G., Daemen, J., Peeters, M., Van Assche, G.: Cryptographic sponge functions 0.1 (2011), http://sponge.noekeon.org/CSF-0.1.pdf
6. Bertoni, G., Daemen, J., Peeters, M., Van Assche, G.: The keccak reference. Submission to NIST (Round 3) 3.0 (2011), http://keccak.noekeon.org/Keccak-reference-3.0.pdf
7. Bertoni, G., Daemen, J., Debande, N., Le, T.H., Peeters, M., Assche, G.V.: Power Analysis of Hardware Implementations Protected with Secret Sharing (2013), published: Cryptology ePrint Archive, Report 2013/067, http://eprint.iacr.org/
8. Bertoni, G., Daemen, J., Peeters, M., Van Assche, G.: The keccak SHA-3 submission. Submission to NIST (Round 3) (2011)

9. Brier, E., Clavier, C., Olivier, F.: Correlation power analysis with a leakage model. In: Joye, M., Quisquater, J.-J. (eds.) CHES 2004. LNCS, vol. 3156, pp. 16–29. Springer, Heidelberg (2004)

10. Daemen, J., Bertoni, G., Peeters, M., Van Assche, G., Van Keer, R.: Keccak implementation overview. Technical report, NIST (2012)

11. Daemen, J., Rijmen, V.: The Design of Rijndael. Springer-Verlag New York, Inc., Secaucus (2002)

12. Kocher, P., Jaffe, J., Jun, B.: Differential power analysis. In: Wiener, M. (ed.) CRYPTO 1999. LNCS, vol. 1666, pp. 388–789. Springer, Heidelberg (1999)

13. McEvoy, R., Tunstall, M., Murphy, C.C., Marnane, W.P.: Differential power analysis of HMAC based on SHA-2, and countermeasures. In: Kim, S., Yung, M., Lee, H.-W. (eds.) WISA 2007. LNCS, vol. 4867, pp. 317–332. Springer, Heidelberg (2008)

14. Taha, M., Schaumont, P.: Side-channel analysis of MAC-Keccak. In: 2013 IEEE International Symposium on Hardware-Oriented Security and Trust (HOST) (June 2013)

15. Zohner, M., Kasper, M., Stöttinger, M., Huss, S.: Side channel analysis of the SHA-3 finalists. In: Design, Automation Test in Europe Conference Exhibition (DATE), pp. 1012–1017 (March 2012)

A   Systematic Analysis for DPA of MAC-Keccak

The systematic approach used in the paper to identify the required target operations is composed of the following steps:

1. Add all the message bytes to the set $\mathcal{D}$.
2. Calculate all the data-dependent variables that depend on the available information.
3. Add these new calculated variables to the set $\mathcal{D}$.
4. Select the operations that process an element of $\mathcal{D}$ and a *constant* unknown.
5. Target the output of these operations with DPA to recover the unknown.
6. If the recovered unknown is enough to recover the required key, finish.
   Else,
   - Repeat from Step 2 using the information that became available from the just recovered unknown.

   MAC-Keccak can be viewed as a Directed Acyclic Graph, where vertices (V) represent the internal operations and edges (E) represent inputs and intermediate variables. The vertices are numbered similar to the order of executing the operations within the algorithm. The pseudo-code of the systematic approach is shown as follows, where the output is the correct value of all the edges, including the required secret key.

   We used the following data objects:

   - Flag[e]: C for constant edges, D for unknown data-dependent edges, SetD for known data-dependent edges.
   - Init[e]: The initial values of Flag[e]; C for key-bytes, SetD for message bytes, D for all the rest.

- `Required[v]`: The set of operations that need to be targeted to recover the full key.
- `e.Eval()`: A function to evaluate the value of edge `e`.
- `v.DPA()`: A function to apply DPA on the output of vertex `v`.

The code consists of three phases. The initialization phase flags the initial values of the edges. The exploration phase flags the constant edges with C and marks the set of target internal operations. The attack phase performs a greedy search for new target operations and mounts DPA attacks against them.

Algorithm 1. Systematic Analysis for DPA of MAC-Keccak

Require: Graph G(V,E)
Require: Init[E]
                                                          ▷ Initialization Phase: step 1
   Flag[E] = Init[E];
                                                          ▷ Exploration Phase
   for each vertex v in V **do**
      if all Flag[v.input] == C **then**
         Flag[v.output] = C;
      end if
      if any Flag[v.input] == C & (any Flag[v.input] == D or SetD) **then**
         Required[v] = 1;
      else
         Required[v] = 0;
      end if
   end for
                                                          ▷ Attack Phase:
   while any Required == 1 **do**
      for each vertex v in V **do**
         if all Flag[v.input] == SetD **then**
            v.output.Eval();   ▷ step 2
            Flag[v.output] = SetD;   ▷ step 3
         end if
      end for
      for each vertex v in V **do**
         if any Flag[v.input] == C & any Flag[v.input] == SetD) **then**   ▷ step 4
            v.DPA();   ▷ step 5
            Flag[v.input] == SetD;
            Required[v] = 0;   ▷ step 6
         end if
      end for
   end while
      return Value of all E

Generic State-Recovery and Forgery Attacks on ChopMD-MAC and on NMAC/HMAC

Yusuke Naito[1], Yu Sasaki[2], Lei Wang[3], and Kan Yasuda[2]

[1] Mitsubishi Electric Corporation, Japan
[2] NTT Secure Platform Laboratories, Japan
[3] Nanyang Technological University, Singapore
Naito.Yusuke@ce.MitsubishiElectric.co.jp,
{sasaki.yu,yasuda.kan}@lab.ntt.co.jp, Wang.Lei@ntu.edu.sg

Abstract. This paper presents new attacks on message authentication codes (MACs). Our attacks are generic and applicable to (secret-prefix) ChopMD-MAC and to NMAC/HMAC, all of which are based on a Merkle-Damgård hash function. We show that an internal state value of these MACs can be recovered with time/queries less than $O(2^n)$— roughly, with an $O(2^n/n)$ complexity, where ChopMD has $2n$-bit state and NMAC/HMAC n-bit. We also show that state-recovery can be extended to MAC-security compromise, such as almost universal forgeries and distinguishing-H attacks. While our results remain to be of theoretical interest due to the high attack complexity, they lead to profound consequences. Namely, our analyses provide us with *proper* understanding of these MAC constructions, for in the literature the complexity has been implicitly and explicitly assumed to be $O(2^n)$. Since the complexity is very close to 2^n, we make a precise calculation of attack complexities and of success probabilities in order to show that the total complexity is indeed less than 2^n. Moreover, we perform an experiment by computer simulation to demonstrate that our calculation is correct.

Keywords: Generic attack, internal state recovery, multi-collision, 2^n security, almost universal forgery, distinguishing-H.

1 Introduction

A message authentication code (MAC) is a secret-key primitive that ensures integrity of data. A MAC is a function which takes as input a secret-key K and a variable-length message M, and outputs a fixed-length tag σ. The secret-key K must be confidentially shared between two parties prior to communication. The sender produces a tag σ of a message M by using the secret-key K, and sends the pair (M, σ) to the receiver. The receiver re-produces a tag σ^* of the message M by using its own secret-key K, and checks if $\sigma = \sigma^*$. If they are equal, then it means that the received message M is indeed sent correctly from the other party. Otherwise, it means that the message M or the tag σ or both must have been modified or forged during transmission.

K. Sakiyama and M. Terada (Eds.): IWSEC 2013, LNCS 8231, pp. 83–98, 2013.

An effective way to construct an efficient MAC scheme is to key a hash function, where the key is kept secret. Many of the existing hash functions (e.g. [1,10,14,25]) are based on an iterative structure called the Merkle-Damgård construction [8,19]. As a result, there exist a number of MAC schemes that are realized by keying the Merkle-Damgård construction, examples being secret-prefix LPMAC [30] and ChopMD-MAC [7], as well as the widely standardized NMAC/HMAC [3].

State-Recovery, Forgery and Distinguishing Attacks. Most devastating to MACs is the recovery of secret keys, as in [6,13,16,31]. However, to iterative MACs, almost equally destructive is the recovery of an internal state value. The internal state values are produced as a result of iteration, and usually all of them need to be kept secret in order for the MAC scheme to be secure.[1] An internal state recovery would allow an adversary to manipulate succeeding values. Such a state recovery can lead to serious compromise of MAC security, such as almost universal forgeries (the notion introduced by Dunkelman et al. in [12]) and distinguishing-H attacks (the notion introduced by Kim et al. in [15]; a distinguishing-H attack is to differentiate a MAC scheme calling a specific compression function from the one calling a random compression function).

The security of hash-based MAC schemes against state recovery is not as well-understood as the pseudo-randomness of these MACs. LPMAC is known to be pseudo-random up to $O(2^{n/2})$ queries [4],[2] NMAC/HMAC also $O(2^{n/2})$ [2], and ChopMD-MAC $O(2^n/n)$ [5,23] (when the state size is $2n$ bits and the tag size n bits, using Th. 3.1 of [23]). These results only tell us that the security against state recovery should lie somewhere between these figures and $O(2^n)$. Moreover, exactly what an adversary can do after a recovery to extend it to forgery/distinguishing attacks heavily depends on which particular MAC construction is in question.

Previous Work. Sasaki [26] has shown that, for narrow-pipe (meaning the internal state size equal to the tag length, both n bits) LPMAC, by using collisions, one can recover an internal state with just $O(2^{n/2})$ work. Sasaki's method is generic, treating the underlying compression function as a black-box. In the same work, Sasaki has utilized the recovered state value for mounting $O(2^{n/2})$ almost universal forgery and distinguishing-H attacks on narrow-pipe LPMAC, where the forgery is created without prior knowledge about the length of the target message.

Peyrin et al. [20] have presented a generic state recovery attack on HMAC with a complexity $O(2^{n/2})$. Their attack exploits a related key. Also, in the same work, Peyrin et al. have presented a distinguishing-H attack, where the complexity is $O(2^{n/2})$. Their attacks are not applicable to NMAC.

---

[1] Only a limited number of MAC constructions (e.g. [11]) are resistant to the leakage of internal state values.

[2] The notation $O(\cdots)$ means that we are omitting the exact constants.

Table 1. Generic state-recovery attacks: comparison of our results with previous work

type	setting	scheme	complexity	optimal?	source
Secret-prefix	narrow	LPMAC	$O(2^{n/2})$	✓	[26]
	wide	**ChopMD-MAC**	$O(2^n/n)$	✓	**this work**
Envelope	related-key	HMAC	$O(2^{n/2})$	✓	[20]
	single-key	**NMAC/HMAC**	$O(2^n/n)$		**this work**

Problem Statement. Sasaki's method of internal state recovery on LPMAC is not effective in attacking ChopMD-MAC. In the case of ChopMD-MAC, its wide size of internal state (which we assume is $2n$ bits) makes it intangible to cause internal state collisions. Also, Peyrin et al.'s generic state recovery crucially relies on the adversary's access to a related key, which makes it infeasible to use the same idea and mount a similar attack without a related key. Therefore, we pose the following questions:

1. *Is it possible to recover an internal state of ChopMD-MAC with a complexity less than $O(2^n)$?*
2. *Without using a related key, can we recover an internal state of HMAC with less than $O(2^n)$ complexity?*

Our Contributions. We answer the two questions affirmative. Table 1 summarizes our generic state-recovery attacks. Moreover, we make the following contributions:

- We show that our state-recovery attack on secret-prefix ChopMD-MAC can be extended to almost universal forgery and distinguishing-H attacks.
- Similarly, our state-recovery attack on HMAC can be extended to a distinguishing-H attack. Our attack on HMAC is also applicable to NMAC.
- Our techniques exploit a multi-collision of internal state values, and the complexities of the recovery become roughly $O(2^n/n)$.
- Since the attack complexities are very close to 2^n, we shall perform a precise calculation of complexities and of success probabilities, verifying that the total complexity is actually less than 2^n.
- Moreover, we shall conduct an experiment by computer simulation to demonstrate that our calculation is correct.

Implications. While our attacks remain largely theoretical due to their high complexities, our results make significant contributions to the *proper* understanding about MAC security; by our results, we are now forced to abandon part of the popular belief in the $O(2^n)$ security of these MAC schemes.

Specifically, for the ChopMD construction, there exists previous work [9] that has claimed $O(2^n)$ security with respect to the notion called indifferentiability [7, 18]. Our generic (state-recovery, forgery and distinguishing-H) attacks on secret-prefix ChopMD-MAC indicate that, together with Th. 3.1 of [23], such an $O(2^n)$

claim cannot be made and the security proof in [9] is not entirely correct.[3] Also, by [5] and Th. 3.1 of [23], we know that our attack on ChopMD-MAC is essentially optimal, making the bound $O(2^n/n)$ basically tight.

For NMAC/HMAC, several pieces of previous work [15, 21, 22, 27, 32] have implicitly or explicitly assumed $O(2^n)$ security against distinguishing-H attacks. The previous work [21] contains a dedicated attack that has a complexity higher than "$O(2^n/n)$" (when the constants are evaluated exactly), and such an attack now becomes of little consequence due to our generic attack.

At the same time with this paper, Leurent et al. also published generic state-recovery and distinguishing-H attacks on NMAC/HMAC [17]. Differently from our attacks based on the multicollision technique, their attacks are based on cycle-detection technique. On one hand, their attacks must query extremely long messages, e.g. up to $2^{n/2}$ blocks, which is impractical, while our attacks use short messages. On the other hand, their attacks achieved a lower complexity $O(2^{n/2})$.

2   Background and Related Work

2.1   Hash Function

Merkle-Damgård Hash Function. Given an input message $M \in \{0,1\}^*$, a Merkle-Damgård hash function [8,19] first pads it with a value *pad* so that the length of $M\|pad$ is multiple block long, then divides the padded message to blocks $M[0]\|M[1]\|\cdots\|M[t]$, and processes these blocks sequentially from $M[0]$ to $M[t]$:

$$v[i+1] \leftarrow h(v[i], M[i]), \ 0 \leq i \leq t,$$

where h is compression function with fixed-length inputs, and $v[0]$ is a public constant denoted as initial vector (IV). Finally Merkle-Damgård hash function applies a finalization function g to $v[t+1]$, and outputs $v = g(v[t+1])$ as the hash digest. If the bit length of v is equal to that of $v[i]$ ($0 \leq i \leq t+1$), the hash function is called *narrow-pipe*. And if the bit length of v is less than that of $v[i]$ ($0 \leq i \leq t+1$), the hash function is called *wide-pipe*.

ChopMD [7] is a wide-pipe Merkle-Damgård hash function. The finalization g is chopping several bits of $v[t+1]$, and outputting the other bits as the hash digest.

r-Multi-collision on $g(h(\cdot,\cdot))$. Our attacks on ChopMD MAC in Section 3 use an r-multi-collision on the last compression function call and the finalization function $g(h(v[t], M[t]))$. More precisely, we fix the value of $v[t]$ as a constant, and find a set of r distinct values of $M[t]$, denoted as $M[t]_i$ ($0 \leq i \leq r-1$), satisfying the following relation

$$g(h(v[t], M[t]_0)) = g(h(v[t], M[t]_1)) = \cdots = g(h(v[t], M[t]_{r-1})).$$

---

[3] Our result does not necessarily invalidate the whole framework of [9]; it seems that the proof in [9] misses the case of our attack, which should be treated as a bad event.

Note that such an r-multi-collision is generated by online interaction with ChopMD MAC. Recall that $v[t]$ is confidential in ChopMD MAC and the outputs of $g(h(\cdot, \cdot))$ are public as tags. We keep the message blocks $M[0]\|M[1]\|\cdots\|M[t-1]$ as constants, which makes $v[t]$ a constant, then vary the value of the last message block $M[t]$, and finally derive an r-multi-collision by observing the tag values.

r-Multi-collision on $h(\cdot, \cdot)$. Our attacks on NMAC/HMAC in Section 4 use an r-multi-collision on a compression function call $v[i+1] = h(v[i], M[i])$. More precisely, we fix the value of $M[i]$ as a constant, and find a set of r distinct values of $v[i]$, denoted as $v[i]_j$ $(0 \le j \le r-1)$, satisfying the following relation

$$h(v[i]_0, M[i]) = h(v[i]_1, M[i]) = \cdots = h(v[i]_{r-1}, M[i]).$$

Note that such an r-multi-collision is generated by offline computations. We set $M[i]$ to a constant value, choose random values as $v[i]$ and compute $h(v[i], M[i])$ to search an r-multi-collision.

2.2   Definitions of Hash-Based MACs

Secret-Prefix ChopMD MAC. There are two common methods to build a secret-prefix MAC based on a ChopMD hash function $H(IV, \cdot)$. The first one replaces IV by a secret key K, and computes the tag of a message M by $H(K, M)$. The second one prepends a secret key K of a single block size to a message M, and computes the tag by $H(IV, K\|M)$.

NMAC and HMAC. These two schemes [3] are MACs based on a Merkle-Damgård hash function $H(IV, \cdot)$. NMAC keys a Merkle-Damgård hash function by replacing IV with a secret key K. For a message M, the tag of NMAC is derived with two secret keys, an inner key K_1 and an outer key K_2, as below:

$$\text{NMAC}(K_1, K_2, M) = H(K_2, H(K_1, M)).$$

HMAC is a variant of NMAC and uses a secret key K. For a message M, the tag of HMAC is derived as below:

$$\text{HMAC}(K, M) = H(IV, (K \oplus \text{opad})\|H(IV, (K \oplus \text{ipad})\|M)),$$

where ipad and opad are two distinct constants.

2.3   Security of Hash-Based MAC

As a cryptographic primitive, a hash-based MAC should receive continuous security evaluation. We briefly describe several attacks on hash-based MACs, which are related to this paper.

Distinguishing-H Attack. It was introduced by [15] for hash-based MAC constructions. Let $C[h]$ be a hash-based MAC. A distinguishing-H adversary A then tries to distinguish the real oracle $C[h](\cdot)$ from an oracle $C[f](\cdot)$, where f denotes a random compression function. That is, the oracle $C[f](\cdot)$ is just like the real oracle $C[h](\cdot)$ except that its component h is now replaced with a random compression function f. The advantage measure of an adversary A is defined as

$$\text{Adv} := \Pr\left[A^{C[h](\cdot)} = 1\right] - \Pr\left[A^{C[f](\cdot)} = 1\right].$$

State-Recovery Attack. As briefly stated in Section 1, the internal states $v[i]$s of hash-based MACs should also be kept confidential. A state-recovery adversary then tries to recover the value of some internal state of a (chosen) message. Let $H(K, \cdot)$ be the target hash-based MAC. The adversary is allowed to interact with $H(K, \cdot)$ by sending chosen messages to receive the corresponding tags. In the end, the adversary produces a pair (M, v), where M can be one of previous queried messages. If v is equal to some internal state $v[i]$ of $H(K, M)$, the adversary wins, namely succeeding in state-recovery.

2.4   Previous Work on MACs with a Specific Hash Function

Several pieces of previous work have presented distinguishing-H attacks on MACs using specific hash functions such as MD4, MD5, SHA-0, and reduced SHA-1. Kim et al. [15] showed distinguishing-H attacks on HMACs based on MD4, 33-step MD5, SHA-0 and 43-step SHA-1, where the complexities are $2^{121.5}$, $2^{126.1}$, 2^{109} and $2^{154.9}$, respectively. Also, Rechberger and Rijmen [21] proposed distinguishing-H attacks on HMAC based on 50-step SHA-1, where the complexity is $2^{153.5}$. Wang et al. [32] presented distinguishing-H attacks on HMAC based on MD5, where the complexity is 2^{97}. Very recently, Sasaki and Wang [27] reduced the complexity of their attacks to 2^{89}.

3   New Generic Attacks on ChopMD-MAC

In this section, we describe several attacks on ChopMD-MAC having a $2n$-bit state size and an n-bit tag size. All of the attacks are based on an observation that, for any secret-key, we can succeed in an internal state recovery attack by utilizing an n-bit multi-collision generated with a complexity roughly $O(2^n/n)$. The attack can be converted to a distinguishing-H attack and an almost universal forgery.

3.1   Internal State Recovery Attack

Overview. Fig. 1 illustrates the sketch of the internal state recovery attack. This case deals with g which chops the first half bits of an input and outputs remaining bits. To simply the explanation, this sketch omits the padding value. First of all, the adversary performs on-line queries. She chooses $O(2^n/n)$ distinct

Try $O(2^n/n)$ values online

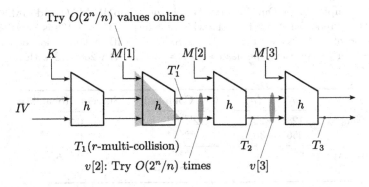

Fig. 1. Attack on ChopMD-MAC

first blocks, $M[1]$ to obtain an r-multi-collision of ChopMD-MAC, where T_1 is the r-multi-collision tag. Now, it knows that the half bits of the internal state after processing the first block is T_1. However, it cannot directly observe the r-chopped values T_1'. To solve this problem, it fixes the second block $M[2]$ and chooses $2^n/r$ distinct internal states $v[2]$ such that the last n-bit value is T_1 to specify one of the r-values T_1'. For each $v[2]$, it computes $v[3] \leftarrow h(v[2], M[2])$. For each $M[1]$, it makes a query $M[1]\|M[2]$ to obtain the tag T_2. Then one can expect that one of r-internal state values $T_1'\|T_1$ connects with one of $2^n/r$ values $v[2]$. This connection can be observed by the collision of $g(v[3])$ and T_2 and by a collision of third block. If $T_2 = g(v[3])$, it chooses third block messages $M[3]$, makes queries $M[1]\|M[2]\|M[3]$ to obtains the tag T_3, and checks whether $T_3 = g(h(v[3], M[3]))$. If the collision occurs, it finds the internal state $v[2]$ after processing the first block.

Attack Procedure. In this attack, we shall construct an r-multi-collision of ChopMD-MAC. We postpone determining the exact value of r till we finish describing our attack. We set $N := \lceil (r!)^{1/r} \cdot 2^{(r-1)n/r} \rceil$.

1. Choose first message block $M[1]$ so that the message with the padding value pad_1, denoted by $M[1]\|pad_1$, fits in the first block, and make queries $M[1]$ to obtain the corresponding tag. Iterate this until an r-multi-collision of the tag is generated (about N times). Let $M[1]_i$, where $i \in \{1, 2, \ldots, r\}$, be the r values forming an r-multi-collision, and let T_1 be the corresponding tag. If no r-multi-collision is found, abort.

2. Fix a value of the second message block $M[2]$ so that the message with the padding value pad_2, denoted by $M[2]\|pad_2$, fits in the second block.

3. Choose $2^n/r$ distinct internal state values $v[2]_j$ so that $g(v[2]_j) = T_1$, and compute its next state values $v[3]_j \leftarrow h(v[2]_j, M[2]\|pad_2)$, where we have $j \in \{1, 2, \ldots, 2^n/r\}$. Store the pairs $(v[2]_j, g(v[3]_j))$, where $j = \{1, 2, \ldots, 2^n/r\}$.

4. For each first message block $M[1]_i\|pad_1$, where $i \in \{1, 2, \ldots, r\}$, make queries $M[1]_i\|pad_1\|M[2]$ to obtain the corresponding tags $T_{2,i}$. Check the match of $T_{2,i}$ and $g(v[3]_j)$ for $j = 1, 2, \ldots, N$. If the match is found, choose

Table 2. Complexity of Our Generic Attacks on ChopMD-MAC and HMAC with Several Parameters. An example of the wide-pipe hash with $n = 128$ is SHA-256 with a truncation of half bits with $n = 256$ is SHA-512 with a truncation of half bits. An example of the narrow-pipe hash with $n = 256$ is SHA-256 and with $n = 512$ is SHA-512.

Type	n	$r^*(n)$	Attack Complexity	Success Probability
ChopMD	128	22	$2^{125.36}$	0.316
	256	35	$2^{252.48}$	0.316
HMAC	256	38	$2^{253.34}$	0.316
	512	63	$2^{508.61}$	0.316

a value of the third message block $M[3]$ so that the message with the padding value pad_3, denoted by $M[3]\|pad_3$, fits in the third block, and make a query $M[1]_i\|pad_1\|M[2]\|pad_2\|M[3]$ to obtain the tag T_3. If we have $T_3 = g(h(v[3]_j, M[3]\|pad_3))$, then $v[2]_j$ becomes the internal state after processing $M[1]_i\|pad_1$.

Complexity and Success Probability. Step 1 requires to make N queries to obtain the r-multi-collision with a probability $\approx 1/2$ [28, 29]. The memory requirement is N for the internal state values on finding the r-multi-collision, and r for storing the pairs $(v[2], v[3])$. Step 2 is negligible. Step 3 requires $\lceil 2^n/r \rceil$ offline computations of h. Step 4 requires to make r 2-block queries, which is $2r$ queries, and if $T_i = g(v[3]_j)$, requires an offline computation of h and requires to make a 3-block query. Finally, we can conclude that the query complexity is $N + 5r$ for Steps 1 and 4, the time complexity is $\lceil 2^n/r \rceil + r$ for Steps 3 and 4, and the memory complexity is $N + \lceil 2^n/r \rceil$ for Steps 1 and 3.

The success probability of Step 1 is roughly $1/2$ [28,29] and the success probability of Step 4 is roughly $1 - 1/e$. Finally, the success probability of the entire attack is $1/2 \cdot (1 - 1/e) \approx 0.316$.

Determining r for Typical Parameters. The attack complexity, being the maximum of query, time, and memory complexities, lies somewhere between $2^n/n$ and 2^n. It can be minimized by appropriately choosing a value of r. The total query complexity is $N + 5r$ and the time complexity is $\lceil 2^n/r \rceil + r$. Now let $r^* = r^*(n)$ be the integral value of r that minimizes the difference between $N + 5r$ and $\lceil 2^n/r \rceil + r$. Table 2 gives us values of r^* for typical choices of n where we choose $r^* = 22$ and 35 for 128- and 256-bit functions, respectively.

Machine Experiment. We carried out a small experiment to verify the evaluation of the multi-collision. We modifed the compression function of SHA-256 so that the output is truncated to 16 bits. For $n = 16$, according to the evaluation, the value of r^* is $r^* = 7$, which achieves the time and memory complexities of

$2^{15.47}$ and the query complexity of $2^{15.78}$. We then checked the probability that a 7-multi-collision is generated with $2^{15.78}$ queries.

We chose 1000 groups of $2^{15.78}$ randomly chosen queries, and counted how many groups generated a 7-multi-collision. As a result, 507 out of 1000 groups generated a 7-multi-collision. This matches the evaluation of [28, 29], where a 7-multi-collision is generated with probability about $1/2$.

3.2 Distinguishing-H Attack

The internal state recovery attack in Section 3.1 is immediately converted to a distinguishing-H attack with the same complexity. Since the internal state recovery attack is based on the simulation of h, the above attack fails with a non-negligible probability if ChopMD-MAC uses a random function f, while the above attack is succeeded with a non-negligible probability if ChopMD-MAC uses a compression function h. A distinguishing-H attack can be obtained by modifying Step 4 and by adding Step 5.

4. For each first message block $M[1]_i \| pad_1$, where $i \in \{1, 2, \ldots, r\}$, make queries $M[1]_i \| pad_1 \| M[2]$ to obtain the corresponding tags $T_{2,i}$. Check the match of $T_{2,i}$ and $g(v[3]_j)$ for $j = 1, 2, \ldots, N$. If the match is found, choose a value of the third message block $M[3]$ so that the message with the padding value pad_3, denoted by $M[3] \| pad_3$, fits in the third block, and make a query $M[1]_i \| pad_1 \| M[2] \| pad_2 \| M[3]$ to obtain the tag T_3. If $T_3 = g(h(v[3]_j, M[3] \| pad_3))$, output 1.
5. Output 0.

Evaluation of the Advantage. Here, we evaluate the advantage of the attack. Let $\Pr_h$ be the probability that the adversary outputs 1 when he interacts with the oracle instantiating h. Also let $\Pr_f$ be the probability that the adversary outputs 1 when it interacts with the oracle instantiating a randomly chosen compression function f with the same range and domain as h. We calculate the advantage $\text{Adv} := |\Pr_h - \Pr_f|$.

First, we evaluate the probability $\Pr_h$. The success probability of Step 1 is roughly $1/2$ [28, 29] and the success probability of Step 4 is roughly $1 - 1/e$. Finally, $\Pr_h \approx 1/2 \cdot (1 - 1/e)$.

On the other hand, suppose that the compression function is f. The adversary outputs 1 only if an event (i) occurs and then an event (ii) occurs.

(i) An r-multi-collision is generated at Step 1.
(ii) $T_{2,i} = g(v[3]_j)$ at Step 4 and $T_3 = g(h(v[3]_i, M[3] \| pad_3))$.

The probability of the event (i) is about $1/2$. The probability of the event (ii) is about $\sum_{j=1}^{\lceil 2^n/r \rceil}((1/2^n) \cdot (r/2^n)) = 1/2^n$. Hence, $\Pr_f = (1/2^n) \cdot (1/2)$.

Finally, we can compute the advantage $\text{Adv} = |\Pr_h - \Pr_f| = (1/2) \cdot (1 - 1/e - 1/2^n) \approx 0.316$, which is big enough to be a valid distinguisher.

3.3   Existential and Almost Universal Forgery Attacks

If the value of the internal state $v[1]$ after processing the first message block $M[1]$ can be recovered, one can generate the valid tag of a message $M[1]\|M$ for any M. Therefore, an almost universal forgery attack (and also an existential forgery attack) can be performed with the same complexity as the internal state recovery attack.

3.4   Observations

Optimality of Our Attack. Our generic forgeries (and attacks) immediately convert to the differentiable attack on ChopMD. This is ensured by Theorem 3.1 of [24]: there is the following relation between the advantages of a (existential or almost universal) forgery and of indifferentiability from a random oracle ([24]): $\mathrm{Adv}^{\mathrm{forge}}_{\mathrm{chopMD}} \leq \mathrm{Adv}^{\mathrm{forge}}_{\mathrm{RO}} + \mathrm{Adv}^{\mathrm{indiff}}_{\mathrm{chopMD}}$. $\mathrm{Adv}^{\mathrm{forge}}_{\mathrm{H}}$ is the advantage of the forgery on a secret prefix MAC based on a hash function H, and $\mathrm{Adv}^{\mathrm{indiff}}_{\mathrm{H}}$ is the indifferentiable advantage on H.

The complexity for ChopMD to be differentiable from a random oracle is at least $O(2^n/n)$ [5]. Therefore the above relation offers the lower bounds of the complexities of the forgeries on ChopMD-MAC which is at least $O(2^n/n)$, while our forgeries guarantee that the complexities are at most $O(2^n/n)$. Therefore, the complexity of our forgery is *optimal*.

Though Daubignard *et al.* [9] claimed that the complexity to be differentiable from a random oracle is at least $O(2^n)$, our result proved that their result is *incorrect*.

Generalized Finalization Function g. In the above discussions, we focused on g being a chop function, while the attacks can also be performed on other g when it has the following properties.

1. For any $y \in \{0,1\}^n$, there are exactly 2^n values which are mapped onto the value y.
2. The 2^n values are efficiently computable.

The properties 1 and 2 are required to recover the values $v[3]_i$ in Step 1. The finalization function g which has these property is for example $g(x_1\|x_2) = x_1 \oplus x_2$ where $x_1, x_2 \in \{0,1\}^n$.

4   Generic Attacks on NMAC/HMAC

In this section, we describe our state-recovery and distinguishing-H attacks on NMAC/HMAC that instantiates a generic narrow-pipe Merkle-Damgård hash function. We focus on HMAC, as HMAC is much more widely used in practice. Essentially the same attack can be applied to NMAC. All of our attacks are based on an observation that, for any key, we can recover an internal state value by utilizing a multi-collision generated with a complexity of roughly $O(2^n/n)$.

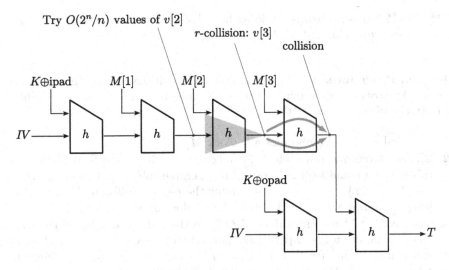

Fig. 2. Attack on HMAC

4.1   State-Recovery Attack

We start with the state-recovery attack. Fig. 2 illustrates the attack on HMAC, where the padding value is omitted.

Overview. The adversary first performs offline computations; it fixes the second block $M[2]$ of a message M and simulates the output of the compression-function computation $h(v[2], M[2])$ for $O(2^n/n)$ distinct values of chaining variable $v[2]$ to obtain an r-multi-collision of the output. The value of r is chosen between 2 and n, depending on the size of the chaining variable. Let $v[3]^*$ be the value of the r-multi-collision obtained. Now, the adversary knows that $v[3]^*$ should occur more frequently than other values of $v[3]$ (roughly r times) as long as the second message block is fixed to $M[2]$. The adversary would not be able to observe directly the output value of the second block. However, the adversary can detect that the value is equal to $v[3]^*$ as follows: it searches for a pair of third message blocks $(M[3], M[3]')$ that produce a collision when the chaining variable is $v[3]^*$, so that

$$h(v[3]^*, M[3]) = h(v[3]^*, M[3]').$$

Hence, at the end of this precomputation phase, the adversary has found r choices of $v[2]$, together with the blocks $M[2]$, $M[3]$ and $M[3]'$.

   After the offline computation, the adversary chooses $2^n/r$ distinct first message blocks $M[1]$ and makes queries $M[1] \parallel M[2] \parallel M[3]$ and $M[1] \parallel M[2] \parallel M[3]'$. If a collision occurs between two messages, then the adversary knows that the internal state $v[2]$ takes only r possibilities and $v[3]$ is uniquely determined as $v[3]^*$. Then, the real value of $v[2]$ can be easily identified from r possibilities with a few more queries. The attack complexity, which is max{query, time, memory}, is between

$6 \cdot 2^n/n$ and 2^n. In other words, the lower bound of the attack complexity is $6 \cdot 2^n/n$. The success probability is $1 - 1/e$.

Detailed Procedure. We shall construct an r-multi-collision of internal state values. We postpone determining the exact value of r till we finish describing our attack. We set $N := \lceil (r!)^{1/r} \cdot 2^{(r-1)n/r} \rceil$.

1. Fix a value of the second message block $M[2]$.
2. Choose an internal state value $v[2]$ and compute $v[3] \leftarrow h(v[2], M[2])$. Iterate this until an r-multi-collision of $v[3]$ is generated (about N times). Let $v[2]_i$, $i \in \{1, 2, \ldots, r\}$, be the r values forming the r-multi-collision. Also, let $v[3]^*$ be the colliding value of $v[3]$. Store the r values of $v[2]_i$ and $v[3]^*$.
3. Choose $2^{(n/2)+1}$ distinct values of $M[3]$ so that $M[3]\|pad_3$ fits in the third block. Compute $v[4] \leftarrow h(v[3]^*, M[3]\|pad_3)$ and obtain a pair of $M[3]$ which makes a collision of $v[4]$. Let $(M[3], M[3]')$ be two messages satisfying the relation $h(v[3]^*, M[3]\|pad_3) = h(v[3]^*, M[3]'\|pad_3)$.
4. Choose $\lceil 2^n/r \rceil$ distinct $M[1]_j$ for $1 \le j \le \lceil 2^n/r \rceil$, and for each j, make queries $M[1]_j\|M[2]\|M[3]$ and $M[1]_j\|M[2]\|M[3]'$ to obtain the corresponding tags T_j and T_j'. Check the match of T_j and T_j'. If the match is found, let $M[1]^*$ be the corresponding $M[1]_j$.
5. For each value of $v[2]_i$ where $i \in \{1, 2, \ldots, r\}$, choose $2^{(n/2)+1}$ distinct values of $M[2]$ so that $M[2]\|pad_2$ fits in the second block. Then, compute the corresponding $v[3] \leftarrow h(v[2]_i, M[2]\|pad_2)$ and find a pair of message $M[2]_i$ and $M[2]_i'$ that make a collision of $v[3]$.
6. For each $i \in \{1, 2, \ldots, r\}$, make queries $M[1]^*\|M[2]_i$ and $M[1]^*\|M[2]_i'$ and check the match of two tags. If they match, the corresponding $v[2]_i$ is the internal state after processing $M[1]^*$.

Evaluation of the Attack Complexity. Step 1 is negligible. After N computations at Step 2, we expect to find an r-multi-collision of $v[3]$ with a probability $\approx 1/2$ [28,29]. Therefore, Step 2 requires N offline computations of h to obtain the r-multi-collision. The memory requirement is N for storing tag values on finding the r-multi-collision, r for storing $M_j[1]$. Step 3 requires $2^{(n/2)+1}$ offline computations. Step 4 requires to make $\lceil 2^n/r \rceil$ 3-block paired queries, which is $6 \cdot \lceil 2^n/r \rceil$ queries. Step 5 requires $r \cdot 2^{(n/2)+1}$ offline computations. Step 6 requires r 2-block paired queries, which is $4r$ queries. Because r is much smaller than N, the complexities for Steps 3, 5, and 6 are negligible. Finally, we can conclude that the query complexity is $6 \cdot \lceil 2^n/r \rceil$ for Step 4 and both of the time and memory complexities are N for Step 2, where $N = \lceil (r!)^{1/r} \cdot 2^{(r-1)n/r} \rceil$.

The success probability of Step 2 is roughly $1/2$ [28,29] and the success probability of Step 4 is roughly $1 - 1/e$. Note that the success probabilities of other steps can increase to almost 1 with trying more message values because the complexities of those steps are much smaller than the dominant parts. Finally, the success probability of the entire attack is evaluated as $1/2 \cdot (1 - 1/e) \approx 0.316$.

Determining r for Typical Parameters. The attack complexity, being the maximum of query, time, and memory complexities, lies somewhere between $2^n/n$ and 2^n. It can be minimized by appropriately choosing a value of r. The total query complexity is $6 \cdot \lceil 2^n/r \rceil$ blocks arising from Step 4. We see that Step 2 needs the most running time; $\lceil (r!)^{1/r} \cdot 2^{(r-1)n/r} \rceil$ computations of h. The memory requirement is dominated by Step 2, which is $\lceil (r!)^{1/r} \cdot 2^{(r-1)n/r} \rceil + r$.

Now let $r^* = r^*(n)$ be the integral value of r that minimizes the difference between $6 \cdot \lceil 2^n/r \rceil$ and $\lceil (r!)^{1/r} \cdot 2^{(r-1)n/r} \rceil$. Table 2 gives us values of r^* for typical choices of n. By setting $r = r^*$ our attack complexity becomes optimal, which is given by $\max\{\lceil (r^*!)^{1/r^*} \cdot 2^{(r^*-1)n/r^*} \rceil + r^*, 6 \cdot \lceil 2^n/r^* \rceil\}$. Table 2 gives us values of r^* for typical choices of n where we choose $r^* = 38$ and 63 for 256- and 512-bit functions, respectively.

4.2   Distinguishing-H Attack

The internal state recovery attack in Sect. 4.1 is immediately converted to a distinguishing-H attack with the same complexity. At Step 4, only if the compression function is a target algorithm h, a collision can be observed about r times faster than other cases, thus at Step 4, a collision between T_j and T'_j is obtained with $\lceil 2^n/r \rceil$ choices of $M[1]_j$. If the compression function is not h, reaching one of $v[2]_i$ does not help to generate a collision of the tag. This is because $M[3]$ and $M[3]'$ are generated under the assumption that the compression function is h, a collision cannot be observed at Step 4. Finally, we can distinguish whether the compression function is h or not.

We modify Steps 2 and 4 of the attack procedure for the internal state recovery attack as shown below. Moreover, Steps 5 and 6 are removed.

2. Abort the procedure and output 0 if no r-multi-collision is found with N different values of $v[2]$.
4. Check the match of T_j and T'_j. If the match is found, output 1. Otherwise, output 0.

Evaluation of the Advantage. We evaluate the advantage of the attack. Let Pr_h and Pr_f be the probabilities that the adversary outputs 1 when she interacts with the oracle instantiating h and with the oracle instantiating a randomly chosen compression function f, respectively. We calculate the advantage $\mathrm{Adv} := |\mathrm{Pr}_h - \mathrm{Pr}_f|$.

Consider the case where the adversary interacts with the oracle instantiating h. The adversary outputs 1 only if an event (i) occurs and then either of events (ii) or (iii) occurs.

(i): an r-multi-collision is generated at Step 2.
(ii): $M[1]_j$ reaches one of $v[2]_i$, then the match of T_j and T'_j occurs with probability 1.
(iii): event (ii) does not occur, but collision occurs at $v[4]$ or tag.

The probability of the event (i) is about $1/2$. The probability of the event (ii) is $1 - 1/e$. For the event (iii), the probability that one pair messages do not reach one of $v[2]_i$ but causes a collision at $v[4]$ is $1 - (1/e)^{1/r}$. The same is applied for obtaining a collision at the tag. Hence, the probability of the event (iii) is $2 \cdot 1/e \cdot (1 - (1/e)^{1/r})$. Therefore,

$$\Pr_h \approx \frac{1}{2} \cdot \left\{ \left(1 - \frac{1}{e}\right) + \frac{2}{e} \cdot \left(1 - \left(\frac{1}{e}\right)^{\frac{1}{r}}\right) \right\} = \frac{1}{2} \cdot \left(1 + \frac{1}{e}\right) - \left(\frac{1}{e}\right)^{\frac{1}{r}+1}.$$

On the other hand, suppose that the oracle is instantiating f. The adversary outputs 1 only if

(I) an r-multi-collision is generated at Step 2, and then
(II) a collision occurs at $v[4]$ or tag.

The probability of the event (I) is about $1/2$. The probability of the event (II) is $2 \cdot (1 - (1/e)^{1/r})$. Hence,

$$\Pr_f = \frac{1}{2} \cdot 2 \cdot \left(1 - \left(\frac{1}{e}\right)^{\frac{1}{r}}\right) = 1 - \left(\frac{1}{e}\right)^{\frac{1}{r}}.$$

Finally, the advantage is computed as follows:

$$\frac{1}{2} \cdot \left(-1 + \frac{1}{e}\right) + \left(\frac{1}{e}\right)^{\frac{1}{r}} \cdot \left(1 - \frac{1}{e}\right) = \left(-\frac{1}{2} + \left(\frac{1}{e}\right)^{\frac{1}{r}}\right) \cdot \left(1 - \frac{1}{e}\right).$$

The advantage is big enough to be a valid distinguisher. For example when choosing $r = 38$ and 63 for 256- and 512-bit functions, respectively, the advantage becomes 0.300 for $r = 38$ and 0.306 for $r = 63$.

Acknowledgments. The authors would like to thank the anonymous reviewers for their helpful comments.

References

1. Barreto, P.S.L.M., Rijmen, V.: The Whirlpool hashing function. NESSIE (2003)
2. Bellare, M.: New proofs for NMAC and HMAC: Security without collision-resistance. In: Dwork, C. (ed.) CRYPTO 2006. LNCS, vol. 4117, pp. 602–619. Springer, Heidelberg (2006)
3. Bellare, M., Canetti, R., Krawczyk, H.: Keying hash functions for message authentication. In: Koblitz, N. (ed.) CRYPTO 1996. LNCS, vol. 1109, pp. 1–15. Springer, Heidelberg (1996)
4. Bellare, M., Canetti, R., Krawczyk, H.: Pseudorandom functions revisited: The cascade construction and its concrete security. In: FOCS 1996, pp. 514–523. IEEE Computer Society (1996)
5. Chang, D., Nandi, M.: Improved indifferentiability security analysis of chopMD hash function. In: Nyberg, K. (ed.) FSE 2008. LNCS, vol. 5086, pp. 429–443. Springer, Heidelberg (2008)

6. Contini, S., Yin, Y.L.: Forgery and partial key-recovery attacks on HMAC and NMAC using hash collisions. In: Lai, X., Chen, K. (eds.) ASIACRYPT 2006. LNCS, vol. 4284, pp. 37–53. Springer, Heidelberg (2006)
7. Coron, J.-S., Dodis, Y., Malinaud, C., Puniya, P.: Merkle-Damgård revisited: How to construct a hash function. In: Shoup, V. (ed.) CRYPTO 2005. LNCS, vol. 3621, pp. 430–448. Springer, Heidelberg (2005)
8. Damgård, I.B.: A design principle for hash functions. In: Brassard, G. (ed.) CRYPTO 1989. LNCS, vol. 435, pp. 416–427. Springer, Heidelberg (1990)
9. Daubignard, M., Fouque, P.-A., Lakhnech, Y.: Generic indifferentiability proofs of hash designs. In: Chong, S. (ed.) CSF 2012, pp. 340–353. IEEE (2012)
10. Dobbertin, H., Bosselaers, A., Preneel, B.: RIPEMD-160: A strengthened version of RIPEMD. In: Gollmann, D. (ed.) FSE 1996. LNCS, vol. 1039, pp. 71–82. Springer, Heidelberg (1996)
11. Dodis, Y., Steinberger, J.: Message authentication codes from unpredictable block ciphers. In: Halevi, S. (ed.) CRYPTO 2009. LNCS, vol. 5677, pp. 267–285. Springer, Heidelberg (2009)
12. Dunkelman, O., Keller, N., Shamir, A.: ALRED blues: New attacks on AES-based MAC's. Cryptology ePrint Archive, Report 2011/095 (2011)
13. Fouque, P.-A., Leurent, G., Nguyen, P.Q.: Full key-recovery attacks on HMAC/NMAC-MD4 and NMAC-MD5. In: Menezes, A. (ed.) CRYPTO 2007. LNCS, vol. 4622, pp. 13–30. Springer, Heidelberg (2007)
14. Gallagher, P.: Secure hash standard (SHS). FIPS PUB 180-3, NIST (2008)
15. Kim, J., Biryukov, A., Preneel, B., Hong, S.: On the security of HMAC and NMAC based on HAVAL, MD4, MD5, SHA-0 and SHA-1 (extended abstract). In: De Prisco, R., Yung, M. (eds.) SCN 2006. LNCS, vol. 4116, pp. 242–256. Springer, Heidelberg (2006)
16. Lee, E., Chang, D., Kim, J., Sung, J., Hong, S.: Second preimage attack on 3-Pass HAVAL and partial key-recovery attacks on HMAC/NMAC-3-Pass HAVAL. In: Nyberg, K. (ed.) FSE 2008. LNCS, vol. 5086, pp. 189–206. Springer, Heidelberg (2008)
17. Leurent, G., Peyrin, T., Wang, L.: New Generic Attacks Against Hash-based MACs. In: ASIACRYPT 2013 (2013)
18. Maurer, U., Renner, R., Holenstein, C.: Indifferentiability, impossibility results on reductions, and applications to the random oracle methodology. In: Naor, M. (ed.) TCC 2004. LNCS, vol. 2951, pp. 21–39. Springer, Heidelberg (2004)
19. Merkle, R.C.: One way hash functions and DES. In: Brassard, G. (ed.) CRYPTO 1989. LNCS, vol. 435, pp. 428–446. Springer, Heidelberg (1990)
20. Peyrin, T., Sasaki, Y., Wang, L.: Generic related-key attacks for HMAC. In: Wang, X., Sako, K. (eds.) ASIACRYPT 2012. LNCS, vol. 7658, pp. 580–597. Springer, Heidelberg (2012)
21. Rechberger, C., Rijmen, V.: New results on NMAC/HMAC when instantiated with popular hash functions. J. UCS 14(3), 347–376 (2008)
22. Rechberger, C., Rijmen, V.: On authentication with HMAC and non-random properties. In: Dietrich, S., Dhamija, R. (eds.) FC 2007 and USEC 2007. LNCS, vol. 4886, pp. 119–133. Springer, Heidelberg (2007)
23. Ristenpart, T., Shacham, H., Shrimpton, T.: Careful with composition: Limitations of indifferentiability and universal composability. Cryptology ePrint Archive, Report 2011/339 (2011)
24. Ristenpart, T., Shacham, H., Shrimpton, T.: Careful with composition: Limitations of the indifferentiability framework. In: Paterson, K.G. (ed.) EUROCRYPT 2011. LNCS, vol. 6632, pp. 487–506. Springer, Heidelberg (2011)

25. Rivest, R.L.: The MD5 message-digest algorithm. RFC 1321, IETF (1992)
26. Sasaki, Y.: Cryptanalyses on a Merkle-Damgård based MAC—almost universal forgery and distinguishing-H attacks. In: Pointcheval, D., Johansson, T. (eds.) EUROCRYPT 2012. LNCS, vol. 7237, pp. 411–427. Springer, Heidelberg (2012)
27. Sasaki, Y., Wang, L.: Improved Single-Key Distinguisher on HMAC-MD5 and Key Recovery Attacks on Sandwich-MAC-MD5. In: Selected Areas in Cryptography (2013)
28. Suzuki, K., Tonien, D., Kurosawa, K., Toyota, K.: Birthday paradox for multi-collisions. In: Rhee, M.S., Lee, B. (eds.) ICISC 2006. LNCS, vol. 4296, pp. 29–40. Springer, Heidelberg (2006)
29. Suzuki, K., Tonien, D., Kurosawa, K., Toyota, K.: Birthday paradox for multi-collisions. IEICE Transactions 91-A(1), 39–45 (2008)
30. Tsudik, G.: Message authentication with one-way hash functions. In: INFOCOM 1992, vol. 3, pp. 2055–2059. IEEE (1992)
31. Wang, L., Ohta, K., Kunihiro, N.: New key-recovery attacks on HMAC/NMAC-MD4 and NMAC-MD5. In: Smart, N.P. (ed.) EUROCRYPT 2008. LNCS, vol. 4965, pp. 237–253. Springer, Heidelberg (2008)
32. Wang, X., Yu, H., Wang, W., Zhang, H., Zhan, T.: Cryptanalysis on HMAC/NMAC-MD5 and MD5-MAC. In: Joux, A. (ed.) EUROCRYPT 2009. LNCS, vol. 5479, pp. 121–133. Springer, Heidelberg (2009)

New Property of Diffusion Switching Mechanism on CLEFIA and Its Application to DFA

Yosuke Todo and Yu Sasaki

NTT Secure Platform Laboratories
{todo.yosuke,sasaki.yu}@lab.ntt.co.jp

Abstract. In this paper, we show a new property for the diffusion switching mechanism (DSM) which was proposed by Shirai and Shibutani in 2006, and propose new differential fault attacks (DFAs) on CLEFIA. The DSM is an effective mechanism to design Feistel ciphers, and Feistel ciphers using the DSM are more secure against the differential and the linear cryptanalysis. By applying the DSM to the generalized Feistel network, Shirai et al. proposed a 128-bit block cipher CLEFIA which was adopted as an ISO standard. Shirai and Shibutani proposed two types DSMs; one is using two matrices and the other is using three matrices. It was considered that the security difference between two types DSMs was quite small. In this paper, we propose a new property for the DSM. Our property can be applied to two types DSMs, in particular, it can be applied to the one using two matrices efficiently. We show a small security advantage of the DSM using three matrices, and our results contribute to the comprehension of the DSM. Moreover we can improve DFAs on CLEFIA by using our property. Existing DFAs can not execute without exploiting several faults induced after the 14-th round, but our new DFAs can execute by exploiting several faults induced after the 12-th round. The position where several faults are induced of new DFAs is improved, and it is two rounds earlier than that of existing works.

Keywords: Block cipher, Feistel cipher, CLEFIA, Diffusion switching mechanism, Cryptanalysis, Differential fault attack.

1 Introduction

There are two powerful cryptanalyses on the block cipher: one is the differential cryptanalysis [4] and the other is the linear cryptanalysis [11,10]. The block ciphers must guarantee the security against these cryptanalyses. In 2006, Shirai and Shibutani proposed an effective design method for KSP-type Feistel ciphers, and they called this method the diffusion switching mechanism (DSM) [15,14,16]. KSP-type Feistel ciphers are practical Feistel ciphers whose round function is calculated as follows: First an input is XORed with a round key. Next each byte value is substituted by using one or several S-boxes. Finally several substituted values are mixed by multiplying a diffusion matrix. By using the DSM, we can design KSP-type Feistel ciphers which are more secure against the differential and the linear cryptanalysis. KSP-type Feistel ciphers adopting the DSM use

K. Sakiyama and M. Terada (Eds.): IWSEC 2013, LNCS 8231, pp. 99–114, 2013.

Table 1. Fault models of related DFAs and our DFAs

Reference	Fault models
Common	1. Attackers can induce random byte faults to target branches. 2. The fault values are unknown.
[5]	3. The byte position where attackers induce faults is unknown. 4. The fault values does not collide when several faults are induced to the same byte position.
[18]	3. The byte position where attackers induce faults is unknown.
[2,3]	3. The byte position where attackers induce faults is known.
Our	3. The byte position where attackers induce faults is known (Attackers can induce random byte fault to the target byte position.). 4. The fault values does not collide when several faults are induced to the same byte position.

several diffusion matrices for different rounds. In [16], Shirai et al. proposed two types DSMs; one is using two matrices and the other is using three matrices. However the security difference between two types DSMs is regarded as small. Shirai et al showed that the DSM with three matrices provides stronger security than with two matrices only when the number of rounds is nine. There does not exist any security difference for other number of rounds.

By applying the DSM using two matrices to the 4 branch generalized Feistel network, Shirai et al. proposed a 128-bit block cipher CLEFIA [17] which was adopted as an ISO standard. CLEFIA has 18, 22 and 26 rounds for 128-bit, 192-bit and 256-bit key lengths, respectively. It has two 32-bit round functions F_0 and F_1. In [1], the security of CLEFIA against several well-known attacks is reported. Several cryptanalyses of CLEFIA are reported in [20,9,8].

In this paper, we pay attention to differential fault attacks (DFAs) on Feistel ciphers using the DSM. In DFAs, attackers induce several faults at positions which they choose, and get the correct ciphertext and several faulty ciphertexts. Attackers recover the secret key by comparing these ciphertexts. Next we explain the countermeasure against DFAs. Major countermeasures against fault attacks involve error checking and recalculation. However they are costly especially in software-based implementation. Thus implementers are motivated to reduce the number of error checking and recalculation while maintaining the security. Therefore, it is important to investigate the exploitable rounds. Several DFAs exploiting faults induced at an earlier round have been discussed, and several results are given for AES [12,7,6,13].

For CLEFIA with 128-bit key length, several DFAs have been proposed [5,18,19,21]. We show fault models of existing DFAs in Table 1, and we show the complexity in Table 2. In [5], they proposed the DFA which exploits random byte faults induced at the 15-th round. They can recover the secret key by using 18 faults. This indicated that the last four rounds should be protected against the fault injection. In [18], they improved the number of necessary faults, and they showed the DFA which uses only 2 faults. The complexity is $2^{19.02}$. In [2,3], they improved the condition of the fault injection position, and proposed the DFA

Table 2. The complexity of related DFAs and our DFAs

Reference	Fault Location	#Faults	Complexity
[5]	15 round	18	1
[18]	15 round	2	$2^{19.02}$
[2,3]	14 round	2	$2^{25.507}$
Our	13 round	4	† 6×2^{16}
Our	12 round	8	† 10×2^{40}

†:It shows the complexity which is necessary after the fault induction. Note that the complexity before the fault induction is 2^{24} which is feasible to implement.

which exploits random byte faults induced at the 14-th round. The complexity is about $2^{25.507}$. They argued that protecting the last four rounds of CLEFIA is not enough against DFAs and protecting the last five rounds is necessary.

Our Contribution. Shirai et al. evaluated the security of two types DSMs against the classical differential/linear cryptanalyses. They concluded that the security difference between two types DSMs was small. In this paper, we propose a new property for the DSM and evaluate the security against other cryptanalyses. Our property can be applied to two types DSMs, in particular, it can be applied to the one using two matrices efficiently. It shows that there exists another small security advantage of the DSM using three matrices. Our results contribute to the comprehension of the DSM.

By applying our new property to CLEFIA, we propose new practical DFAs on CLEFIA. Table 1 shows fault models of existing DFAs and our DFAs. In our DFAs, attackers can induce random byte faults to the target byte position, because they must induce several random byte faults to the same byte position. Moreover the fault values must be different. If attackers can not induce different faults, attackers must prepare a new plaintext-ciphertext pair and induce faults again. We show the complexity of our DFAs in Table 2. Our attack can exploit faults induced at an earlier round, which is the 13-th round and is one round earlier than that of existing DFAs. The complexity for recovering the secret key is 6×2^{16}. Moreover our attack can exploit faults induced at the 12-th round and it is two rounds earlier than that of existing DFAs. The complexity for recovering the secret key is 10×2^{40}.

This paper is organized as follows. Section 2 gives several preliminaries. Section 3 gives new properties for the DSM using two matrices, and Sect. 4 gives new DFAs on CLEFIA. We conclude this paper in Sect. 5.

2 Preliminaries

2.1 KSP-Type Feistel Ciphers

Definition 1 ((n, ℓ)-KSP-type round functions)
Let n be the number of S-boxes in a round function, and ℓ be the size of the S-boxes. Let $k \in \{0,1\}^{n\ell}$ be a round key, and k is divided into $k = k[0]\| \cdots \|k[n-1]$

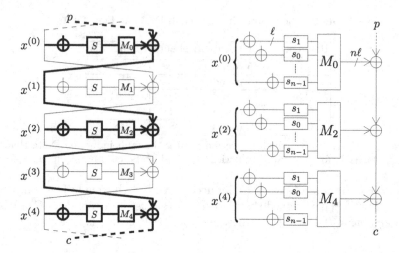

Fig. 1. The left is KSP-type Feistel ciphers. The right is the structure which is picked out the bold lines from KSP-type Feistel ciphers.

where $k_i \in \{0,1\}^\ell$ $(i = 0, \ldots, n-1)$. Let s_i $(i = 0, \ldots, n-1) : \{0,1\}^\ell \to \{0,1\}^\ell$ be the i-th S-box. Let M be $(GF(2^\ell))^{n \times n}$ diffusion matrices. Let $x \in \{0,1\}^{n\ell}$ be an input of a round function and $y \in \{0,1\}^{n\ell}$ be an output of a round function, and x and y are divided into $x = x[0]\|\cdots\|x[n-1]$ and $y = y[0]\|\cdots\|y[n-1]$ where $x[i] \in \{0,1\}^\ell$ $(i = 0, \ldots, n-1)$ and $y[i] \in \{0,1\}^\ell$ $(i = 0, \ldots, n-1)$, respectively. In this time, output y is calculated as follows:

$$y = M \cdot (s_0(x[0] \oplus k[0]), \ldots, s_{n-1}(x[n-1] \oplus k[n-1]))^{\mathrm{T}},$$

and we express it as $y = M \cdot S(x \oplus k)$ for simplicity.

In this paper, we define KSP-type Feistel ciphers as Feistel ciphers which have KSP-type round functions.

2.2 Diffusion Switching Mechanism

Shirai and Shibutani proposed an effective design method called Diffusion Switching Mechanism (DSM) for KSP-type Feistel ciphers in 2006. By using the DSM, we can guarantee a high minimal number of active S-boxes, and it means that we can design Feistel ciphers which are more secure against the differential and the linear cryptanalyses[1]. We show KSP-type Feistel ciphers in the left of Fig. 1, and the right of Fig. 1 is the structure which is picked out the bold lines from KSP-type Feistel ciphers in the left of Fig. 1. In this picked structure, the DSM uses different diffusions matrices alternately. For example, we show two types DSMs proposed in [16]; one is using two matrices, and the other is using three matrices. For the former case, two different diffusion matrices A and

---

[1] The proof based on the number of active S-boxes only gives the security against the classical differential/linear cryptanalysis.

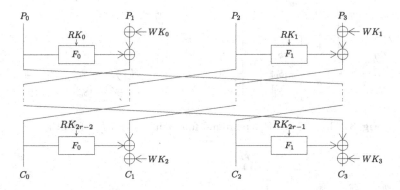

Fig. 2. The network of CLEFIA

B are used, and $A = M_0, M_3, M_4, M_7, \ldots$ and $B = M_1, M_2, M_5, M_6, \ldots$ are satisfied. Namely $M_i = M_{i+4}$ and $M_i \neq M_{i+2}$ are always satisfied for any i. For the latter case, three different diffusion matrices A, B and C are used, and $A = M_0, M_5, M_6, M_{11}, \ldots$, $B = M_1, M_4, M_7, M_{10}, \ldots$ and $C = M_2, M_3, M_8, M_9, \ldots$ are satisfied. Namely $M_i = M_{i+6}$ and $M_i \neq M_{i+2} \neq M_{i+4}$ are always satisfied for any i. However, the difference between the former and the latter is only the number of lower bounds of active S-boxes on the 9-th round and the 18-th round differential cryptanalyses. Accordingly, they argued that the DSM using three matrices should be taken into consideration only if the 9-th round immunity against the differential cryptanalysis is important.

2.3   CLEFIA

CLEFIA developed by Shirai et al. in 2007 is a 128-bit block cipher. CLEFIA has the 4-branch generalized Feistel network and has 18, 22 and 26 rounds for 128-bit, 192-bit and 256-bit key lengths, respectively. See Fig. 2 in which symbols are defined as follows. Let $P, C \in \{0,1\}^{128}$ be a plaintext and a ciphertext, and we divide P and C into $P = P_0 \| P_1 \| P_2 \| P_3$ and $C = C_0 \| C_1 \| C_2 \| C_3$, where $P_i, C_i \in \{0,1\}^{32}$ for $i = 0, \ldots, 3$. Let $P_i[j]$ and $C_i[j]$ be the j-th byte of P_i and C_i, respectively. Let $X^{(i)}$ be the i-th round input data, and we divide $X^{(i)}$ into $X^{(i)} = X_0^{(i)} \| X_1^{(i)} \| X_2^{(i)} \| X_3^{(i)}$, where $X_j^{(i)} \in \{0,1\}^{32}$ for $j = 0, \ldots, 3$. Let $X_j^{(i)}[h]$ be the h-th byte of $X_j^{(i)}$. Let $WK_0, \ldots, WK_3 \in \{0,1\}^{32}$ be whitening keys and $RK_0, \ldots, RK_{2r-1} \in \{0,1\}^{32}$ be round keys, where r is the number of rounds. Let $WK_i[j]$ and $RK_i[j]$ be the j-th byte of WK_i and RK_i, respectively. These keys are generated by the key scheduling algorithm. The key scheduling algorithm of CLEFIA generates an intermediate key L by applying a 12-round generalized Feistel network which takes twenty-four 32-bit constant values as round keys and K as an input. Next it generates WK_i and RK_i as follows:

$$WK_0 \| WK_1 \| WK_2 \| WK_3 = K,$$

$$RK_{4i+0} \| RK_{4i+1} \| RK_{4i+2} \| RK_{4i+3} = \Sigma^i(L) \oplus CON_i \qquad (i = 0, 2, 4, 6),$$

$$RK_{4i+0} \| RK_{4i+1} \| RK_{4i+2} \| RK_{4i+3} = \Sigma^i(L) \oplus K \oplus CON_i \quad (i = 1, 3, 5, 7),$$

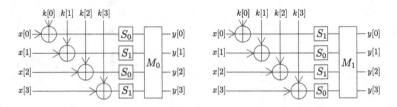

Fig. 3. The (4,8)-KSP-type round functions F_0 and F_1 of CLEFIA

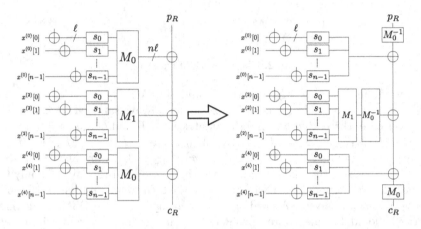

Fig. 4. The new property for the DSM

where CON_i is a 128-bit constant values and Σ is defined as $Y = \Sigma(X) = X[7-63]\|X[121-127]\|X[0-6]\|X[64-120]$.

In this paper, we only use the property that attackers can derive the secret key from RK_{30}, RK_{31}, RK'_{32}, RK'_{33}, RK_{34} and RK_{35}, where RK'_{32} and RK'_{33} denote $RK_{32} \oplus WK_3$ and $RK_{33} \oplus WK_2$, respectively.

CLEFIA uses the DSM using two matrices. CLEFIA has two different (4,8)-KSP-type round functions F_0 and F_1 (see Fig. 3), where M_0 and M_1 denote the diffusion matrices. In this paper, M_0^{-1} and M_1^{-1} denote the inverse matrices of M_0 and M_1, respectively, and $M_0^{-1} = M_0$ and $M_1 = M_1^{-1}$ hold for CLEFIA.

3   New Property of the DSM on CLEFIA

3.1   Our Property of the DSM Using Two Matrices

We show a new property of the DSM, and we can improve the DFA on CLEFIA by applying this property.

As mentioned in Sect. 2.2, the right half of ciphertexts c_R is calculated by XORing the right half of plaintexts p_R with three values passed KSP-type round functions for 5 rounds KSP-type Feistel ciphers. The left of Fig. 4 shows the structure which is picked out from 5-round KSP-type Feistel ciphers with the DSM using two matrices, where $x^{(0)}, x^{(2)}$ and $x^{(4)}$ are $n\ell$-bit values which depend on the plaintext (p_L, p_R) and round keys $(k^{(0)}, \ldots, k^{(4)})$. In this case, the ciphertext $c_R \in \{0,1\}^{n\ell}$ is calculated from the plaintext $p_R \in \{0,1\}^{n\ell}$ as follows:

$$c_R = p_R \oplus M_0 S(x^{(0)} \oplus k^{(0)}) \oplus M_1 S(x^{(2)} \oplus k^{(2)}) \oplus M_0 S(x^{(4)} \oplus k^{(4)}), \quad (1)$$

Let us discuss the difference when two different plaintexts $(p, \bar{p})$ are processed. They derive two values $(x^{(0)}, \bar{x}^{(0)})$, $(x^{(2)}, \bar{x}^{(2)})$ and $(x^{(4)}, \bar{x}^{(4)})$ and two ciphertexts $(c_R, \bar{c}_R)$, respectively. For simplicity, let Δx and $\Delta s(x \oplus k)$ be $x \oplus \bar{x}$ and $s(x \oplus k) \oplus s(\bar{x} \oplus k)$, respectively. In the following, we denote any byte position by ap, where $0 \leq ap \leq n - 1$. Note that δ denotes $\Delta s_{ap}(x^{(2)}[ap] \oplus k^{(2)}[ap])$ and it is an unknown value.

Property 1 (New property of the DSM using two matrices)
δ *can be calculated by guessing at most two bytes, if the following three conditions is satisfied.*

1. *Attackers know the difference $\Delta p_R \oplus \Delta c_R$.*
2. *Attackers know $(x^{(0)}[i], \bar{x}^{(0)}[i])$ or $\Delta s_i(x^{(0)}[i] \oplus k^{(0)}[i])$, and $(x^{(4)}[i], \bar{x}^{(4)}[i])$ or $\Delta s_i(x^{(4)}[i] \oplus k^{(4)}[i])$ for any i, where $0 \leq i \leq n - 1$.*
3. *Attackers know that ap-th value of $x^{(2)}$ is active and the others of $x^{(2)}$ are passive.*

Proof. From Eq. 1, we can get the following equation

$$M_0^{-1} M_1 S(x^{(2)} \oplus k^{(2)}) = M_0^{-1}(p_R \oplus c_R) \oplus S(x^{(0)} \oplus k^{(0)}) \oplus S(x^{(4)} \oplus k^{(4)}).$$

By calculating the difference between two values, we get the following equation

$$M_0^{-1} M_1 \Delta S(x^{(2)} \oplus k^{(2)}) = M_0^{-1}(\Delta p_R \oplus \Delta c_R) \oplus \Delta S(x^{(0)} \oplus k^{(0)}) \oplus \Delta S(x^{(4)} \oplus k^{(4)}).$$

Now, the ap-th value of $\Delta S(x^{(2)} \oplus k^{(2)})$ is δ and the others are 0. Then, we can describe $M_0^{-1} M_1 \Delta S(x^{(2)} \oplus k^{(2)})$ as $(m_0 \delta, m_1 \delta, \ldots, m_{n-1} \delta)$, where $(m_0, m_1, \ldots, m_{n-1})$ are public because M_0 and M_1 are public and the conditions 3 is satisfied. Then we calculate δ by the following equation

$$\delta = m_i^{-1}(M_0^{-1}(\Delta p \oplus \Delta c)[i] \oplus \Delta s_i(x^{(0)}[i] \oplus k^{(0)}[i]) \oplus \Delta s_i(x^{(4)}[i] \oplus k^{(4)}[i])). \quad (2)$$

We can calculate this equation by guessing at most two bytes $(k^{(0)}[i]$ and $k'^{(4)}[i])$, if the conditions 1 and 2 are satisfied. $\square$

By using our property, we construct differential attacks. For any i and any j satisfying the condition 2, we calculate each δ by guessing $(k^{(0)}[i], k^{(4)}[i])$ and $(k^{(0)}[j], k^{(4)}[j])$ in parallel. When each δ has the different value, we can know that the key $(k^{(0)}[i], k^{(0)}[j], k^{(4)}[i], k^{(4)}[j])$ is a wrong key. By using one pair, the remaining key space is reduced from $2^{4\ell}$ to $2^{3\ell}$. Moreover the complexity to calculate δ is order of $2^{2\ell}$. By using 4 pairs, the remaining key space becomes small enough, and we can recover the correct key by exhaustive search.

3.2 Our Property of the DSM Using Three Matrices

For the DSM using three matrices, Eq. 2 are replaced with

$$\delta = m_i^{-1} \cdot (M_0^{-1}(\Delta p \oplus \Delta c)[i] \oplus \Delta s_i(x^{(0)}[i] \oplus k^{(0)}[i]) \oplus M_0^{-1} M_2(\Delta S(x^{(4)} \oplus k^{(4)}))[i]).$$

In this time, a byte-wise guess of $k^{(4)}$ does not provide any useful information, and we need to guess $(k^{(0)}[i], k^{(4)})$ and $(k[j], k^{(4)})$ in parallel for any i and j in order to calculate δ. When each δ has the different value, we can know that the key $(k^{(0)}[i], k^{(0)}[j], k^{(4)})$ is a wrong key. The complexity to calculate δ is order of $2^{\ell+n\ell}$, and it is $2^{n\ell-\ell}$ times as much as that of the DSM using two matrices. We argue that the complexity difference is a security advantage of the DSM using three matrices.

4   Applications to Differential Fault Attack

Until today, CLEFIA is the only Feistel cipher applying the DSM. Consequently, we show applications of our property by using a cryptanalysis on CLEFIA instead of a cryptanalysis on general Feistel ciphers with the DSM. For simplicity, we first show a new DFA which exploits faults induced at the 13-th round and simulation results. Next we show security of modified CLEFIA which has the DSM using three matrices. Finally we show how to extend our attack to a new DFA which exploits faults induced at the 12-th round.

4.1   DFA Exploiting Faults at the 13-th Round

Outline. Our attack consists of five steps. In the 1-st step, attackers induce several faults at the 13-th round. In the 2-nd step, attackers reduce each key space of RK_{34} and RK_{35} to about 2^8 by using our property, where each complexity is 6×2^8. In the 3-rd step, attackers recover (RK'_{32}, RK_{35}) and (RK'_{33}, RK_{34}), where each complexity is 6×2^{16}. In the 4-th step, attackers recover $(RK_{30}, \ldots, RK_{35})$. Finally attackers recover the secret key in the final step. Each complexity of the 4-th step and the final step is negligible.

Precomputation. In each step, we use two precomputation tables; s_0-table and s_1-table. We prepare the s_0-table that the input is x, $\bar{x}$ and y and the output is k satisfying $y = s_0(x \oplus k) \oplus s_0(\bar{x} \oplus k)$. Similarly, we prepare the s_1-table. The complexity to prepare these table is at most 2×2^{23} which is feasible to implement. Moreover we can prepare two precomputation tables before inducing faults.

1-st Step: Induction of Faults. In our DFA, attackers induce several byte-oriented faults as follows. Attackers first get a correct ciphertext corresponding to a plaintext. Next attackers induce two random byte faults to $X_1^{(13)}$ which is calculated from the same plaintext and secret key, and get two faulty ciphertexts. The position where several faults are induced can be any byte position, but attackers can know the position and two faults must be induced at the same byte position. We denote these faulty ciphertexts by left faulty texts. Similarly, attackers induce two faults at the data $X_3^{(13)}$ and get two faulty ciphertexts. We denote these faulty ciphertexts by right faulty texts. In this paper, for simplicity, we assume that the faults are induced to $X_1^{(13)}[0]$ and $X_3^{(13)}[0]$.

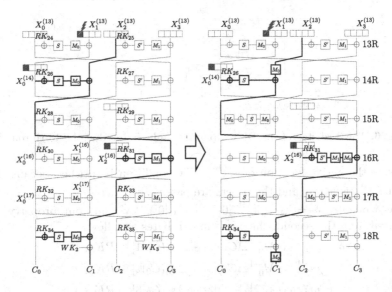

Fig. 5. The 2-nd step of the DFA exploiting faults induced at the 13 round

2-nd Step: Recovering RK_{34} and RK_{35}. In this step, we reduce each key space of RK_{34} and RK_{35} to about 2^8 from 2^{32}. We know that C_1 and C_3 of ciphertext are calculated as follows:

$$C_1 = X_2^{(13)} \oplus M_0 S(X_0^{(14)} \oplus RK_{26}) \oplus M_1 S'(X_2^{(16)} \oplus RK_{31})$$
$$\oplus M_0 S(X_0^{(18)} \oplus RK_{34}) \oplus WK_2,$$

$$C_3 = X_0^{(13)} \oplus M_1 S'(X_2^{(14)} \oplus RK_{27}) \oplus M_0 S(X_0^{(16)} \oplus RK_{30})$$
$$\oplus M_1 S'(X_2^{(18)} \oplus RK_{35}) \oplus WK_3.$$

We define the C_i-paths as these paths. For example, we show the C_1-path in Fig. 5.

We show how to reduce the key space of RK_{34} by using left faulty texts and the correct ciphertext. We pay attention to the C_1-path and apply our property. Now we consider whether the C_1-path satisfies three conditions of our property when faults are induced at the $X_1^{(13)}[0]$ (see Fig. 5). The condition 1 of our property is expressed that attackers can calculate $\Delta X_2^{(13)} \oplus \Delta C_1$. The condition is satisfied because $\Delta X_2^{(13)} = 0$ holds and attackers can know C_1 and $\bar{C}_1$. Here the whitening key WK_2 does not affect the condition 1 because the difference is 0. Next we consider whether the condition 2 of our property is satisfied. The condition 2 of our property is expressed that attackers know $(X_0^{(14)}[i], \bar{X}_0^{(14)}[i])$ or $\Delta S(X_0^{(14)} \oplus RK_{26})[i]$, and $(C_0[i], \bar{C}_0[i])$ or $\Delta S(C_0 \oplus RK_{34})[i]$ for any i. Now attackers know C_0 and $\bar{C}_0$. Moreover it satisfies $\Delta S(X_0^{(14)} \oplus RK_{26}) = (\delta', 0, 0, 0)$ because each byte of $X_0^{(14)}$ is passive except the first byte. Thus the condition 2 is satisfied in the second, third and fourth bytes. Finally the condition 3 of our property is expressed that attackers know the only one byte of $X_2^{(16)}$ is active

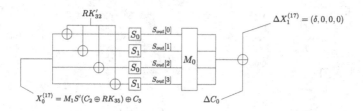

Fig. 6. The 3-rd step of the DFA exploiting faults induced at the 13-th round

and know the active location of $X_2^{(16)}$. The condition is satisfied because first byte of $X_2^{(16)}$ is active and the others are passive.

Now δ denotes $\delta = \Delta s_1(X_2^{(16)}[0] \oplus RK_{31}[0])$. By using our property, δ is calculated from the second, third and fourth bytes as follows:

$$\delta = m_1^{-1} \cdot ((M_0^{-1}\Delta C_1)[1] \oplus \Delta s_1(C_0[1] \oplus RK_{34}[1]),$$
$$\delta = m_2^{-1} \cdot ((M_0^{-1}\Delta C_1)[2] \oplus \Delta s_0(C_0[2] \oplus RK_{34}[2]),$$
$$\delta = m_3^{-1} \cdot ((M_0^{-1}\Delta C_1)[3] \oplus \Delta s_1(C_0[3] \oplus RK_{34}[3]),$$

where m_1^{-1}, m_2^{-1} and m_3^{-1} is given 0x33, 0x39 and 0x70, respectively. If we use the correct $(RK_{34}[1], RK_{34}[2], RK_{34}[3])$, all δ have the same value. For wrong $(RK_{34}[1], RK_{34}[2], RK_{34}[3])$, the probability that all δ have the same value is 2^{-16}. When we use two faults, the probability that wrong keys remain in round key candidates is $2^{24} \times (2^{-16})^2 = 2^{-8}$. Then 2^8 round key RK_{34} remain in the candidates, where $(RK_{34}[1], RK_{34}[2], RK_{34}[3])$ are correct values and $RK_{34}[0]$ is any value from 0 to 255. We recover $RK_{34}[0]$ in the 3-rd step.

We show the attack procedure. Attackers first guess $RK_{34}[1]$ and calculate two differences δ by using two left faulty texts and the correct ciphertext. The complexity is 3×2^8 and the complexity is dominant complexity. Next, attackers get $RK_{34}[2]$ by using the s_0-table. Here the input of the s_0-table is $x = C_0[2]$, $\bar{x} = \bar{C}_0[2]$ and $y = m_2\delta \oplus (M_0^{-1}\Delta C_1)[2]$. Similarly, attackers get $RK_{34}[3]$ by using the s_1-table. Finally, keys $(*, RK_{34}[1], RK_{34}[2], RK_{34}[3])$ remain as candidates of the round key RK_{34}, where $*$ is any value from 0 to 255.

Similarly, we reduce key space of RK_{35} by using right faulty texts and the correct ciphertext. In this time, we pay attention to the C_3-path and apply our property. The complexity is about 3×2^8, thus the total complexity in the 2-nd step is 6×2^8.

3-rd Step: Recovering RK_{32}' and RK_{33}'. We show the method to recover RK_{32}' by using left faulty texts and the correct ciphertext. In this time, we recover RK_{35} at the same time by searching remaining candidates of RK_{35} after the 2-nd step. We pay attention to that it satisfies $\Delta X_1^{(17)} = (\delta, 0, 0, 0)$ where δ is an unknown value for attackers (see Fig. 6). By using $\Delta X_1^{(17)}$ and ΔC_0, each byte of ΔS_{out} is expressed as a linear function of δ. On the other hand, By using each byte of RK_{32}' and using 2^8 remaining candidates of RK_{35}, each byte of ΔS_{out} can be calculated. Thus we can calculate δ independently from

$RK'_{32}[i]$ and RK_{35} for any i. If we use the correct RK_{35} and RK'_{32}, all δ have the same value. For wrong RK'_{32} and RK_{35}, the probability that all δ have the same value is 2^{-24}. When we use two faults, the probability that wrong RK'_{32} and RK_{35} remain in round key candidates is $2^8 \times 2^{32} \times (2^{-24})^2 = 2^{-8}$. Then we can recover RK_{35} and RK'_{32}.

We show the attack procedure. Attackers first guess RK_{35} whose key space is reduced to 2^8 in the 2-nd step, and calculate three $X_0^{(17)}$ by using two left faulty texts and the correct ciphertext. Next attackers guess $RK'_{32}[0]$ and calculate two differences δ. The complexity is 3×2^{16} and this complexity is dominant complexity to recover RK_{35} and RK'_{32}. Next, attackers get $RK'_{32}[1]$, $RK'_{32}[2]$ and $RK'_{32}[3]$ by using the s_0-table and the s_1-table.

Similarly, we recover RK_{34} and RK'_{33} by using right faulty texts and the correct ciphertext. In this time, we pay attention to that it satisfies $X_3^{(17)} = (\delta, 0, 0, 0)$. The complexity to recover RK_{34} and RK'_{33} is 3×2^{16}, and the total complexity in the 3-rd step is 6×2^{16}.

4-th Step: Recovering RK_{30} and RK_{31}. We show the method to recover RK_{30} by using left faulty texts and the correct ciphertext. Now, we know that $X_1^{(16)}$ are the same value for left faulty texts and the correct ciphertext. Then we calculate differences of $X_1^{(16)}$ and check whether the differences have 0. Since $\Delta X_1^{(16)} = 0$, it satisfies the following equations

$$0 = M_0 \cdot \Delta S(X_0^{(16)} \oplus RK_{30}) \oplus \Delta X_0^{(17)}, \tag{3}$$

where $X_0^{(16)}$ and $\Delta X_0^{(17)}$ are calculated as follows:

$$X_0^{(16)} = M_1 S'(M_0 S(C_0 \oplus RK_{34}) \oplus C_1 \oplus RK'_{33}) \oplus C_2,$$

$$\Delta X_0^{(17)} = M_1 \Delta S'(C_2 \oplus RK_{35}) \oplus \Delta C_3.$$

By using RK_{35}, RK_{34} and RK'_{33} which remain in key candidates after the 2-nd and the 3-rd step, we can calculate $X_0^{(16)}$ and $\Delta X_0^{(17)}$, respectively. Since we can evaluate whether Eq. 3 is satisfied by using the s_0-table and the s_1-table, we can recover each byte of $RK_{30}[i]$ independently. For wrong $RK_{30}[i]$, the probability satisfying Eq. 3 is 2^{-8}. When we use two faults, the probability that wrong $RK_{30}[i]$ remain in round key candidates is $1 \times 2^8 \times (2^{-8})^2 = 2^{-8}$. Then we can recover $RK_{30}[i]$ for any i. Similarly, by using right faulty texts and the correct ciphertext, we can recover RK_{31}. The complexity of the 4-th step is negligible compared with that of the entire attack.

Final Step: Recovering the Secret Key. After the 4-th step, a few wrong keys sometimes remain in the candidates. In this time, we first calculate round keys $RK_{24}, \ldots RK_{29}$ and whitening keys WK_2 and WK_3 from the candidates of $RK_{30}, \ldots, RK_{35}$. Next we calculate three $X^{(13)}$ from the left faulty texts, the right faulty texts and the correct ciphertext, and evaluate the validity. If we use the correct round and whitening keys, three $X_0^{(13)}$ (and $X_2^{(13)}$) must correspond for all texts. Since the probability satisfying this condition is negligible for wrong

Table 3. Simulation results of our attack

Target	RK_{35}	RK_{34}	RK'_{33}	RK'_{32}	RK_{31}	RK_{30}	Complexity	Simulation time
Precomputation step	-	-	-	-	-	-	2×2^{23}	4.91 sec
2-nd step	294.349	273.587	-	-	-	-	6×2^{8}	0.1 ms
3-rd step	1.043	1.070	1.209	1.151	-	-	6×2^{16}	17.8 ms
4-th step	1.000	1.000	1.000	1.000	1.106	1.113	negligible	0.02 ms
Final step	1.000	1.000	1.000	1.000	1.000	1.000	negligible	0.01 ms

We omit the result of 1-st step because we can not evaluate this step by simulation.

keys, we can recover the secret key. The complexity of the final step is negligible compared with that of the entire attack.

4.2 Simulation Results and Discussions

To confirm the feasibility, we implemented the simulation of our attack. This simulation was written in C++ programming and executed on a single core in an Intel Core-i7 3770 3.4GHz desktop machine. In each experiment, we used random 128-bit keys and induced four random byte faults at the 13-th round. The average number of the result is shown in Table 3 for 10,000 samples. In Table 3, each number of $RK_{30}, \ldots, RK_{35}$ denotes the average number which remains as candidates of each round key after execution of each step. Complexity denotes the complexity in each step. Simulation time denotes the execution time in each step.

The complexity to prepare two precomputation tables is about 2×2^{23}, and it is dominant complexity in our attacks. However we can calculate it before the fault injection. After the fault injection, the dominant complexity is about 6×2^{16}.

4.3 Security on Modified CLEFIA

We define modified CLEFIA as CLEFIA which uses the DSM using three matrices. The left of Fig. 7 shows the last 6 rounds of modified CLEFIA.

The attack procedure in the 1-st step is the same as that against original CLEFIA. In the 2-nd step, we recover RK_{34} and RK_{35}. We first show how to recover RK_{34} by using left faulty texts and the correct ciphertext. We pay attention to the C_1-path and apply our property (see Fig. 7). We pay attention to $Y_0^{(14)}$ which is defined as $Y_0^{(14)}$ in right of Fig. 7. For original CLEFIA, it satisfies $\Delta Y_0^{(14)} = \Delta S(X_0^{(14)} \oplus RK_{26}) = (\delta', 0, 0, 0)$. However, for modified CLEFIA, it satisfies $\Delta Y_0^{(14)} = M_0 M_2 \Delta S(X_0^{(14)} \oplus RK_{26}) = M_0 M_2(\delta', 0, 0, 0)$. Attackers first guess δ' and calculate $\delta = \Delta s_1(X_2^{(16)}[0] \oplus RK_{31}[0])$ by using our property, because $\Delta Y_0^{(14)}$ does not have constant byte. In this time, δ is calculated as follows:

$$\delta = m_0^{-1} \cdot ((M_0^{-1} \Delta C_1)[0] \oplus m'_0 \delta' \oplus \Delta s_0(C_0[0] \oplus RK_{34}[0]),$$
$$\delta = m_1^{-1} \cdot ((M_0^{-1} \Delta C_1)[1] \oplus m'_1 \delta' \oplus \Delta s_1(C_0[1] \oplus RK_{34}[1]),$$
$$\delta = m_2^{-1} \cdot ((M_0^{-1} \Delta C_1)[2] \oplus m'_2 \delta' \oplus \Delta s_0(C_0[2] \oplus RK_{34}[2]),$$
$$\delta = m_3^{-1} \cdot ((M_0^{-1} \Delta C_1)[3] \oplus m'_3 \delta' \oplus \Delta s_1(C_0[3] \oplus RK_{34}[3]),$$

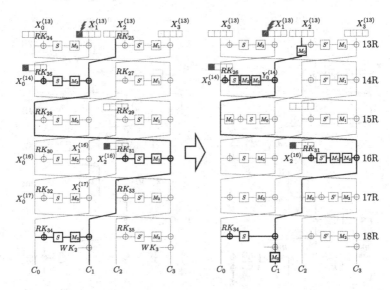

Fig. 7. The 2-nd step of the DFA exploiting faults induced at the 13 round against modified CLEFIA

where m_0^{-1}, m_1^{-1}, m_2^{-1} and m_3^{-1} is constant which is calculated from M_1M_0, and m_0', m_1', m_2' and m_3' is constant which is calculated from M_2M_0. If we use the correct (RK_{34}, δ'), four δ have the same value. For wrong (RK_{34}, δ'), the probability that four δ have the same value is 2^{-24}. Then remaining key space of RK_{34} becomes $2^{40} \times 2^{-24} = 2^{16}$ by using one pair, because key space of RK_{34} is 2^{32} and the number of candidates of δ' is 2^8. Next we use another pair, and guess δ' again because δ' is a value depending on each pair. Similarly, we calculate δ and check that four δ have the same value. Then remaining key space of RK_{34} becomes small enough, because remaining key space of RK_{34} is 2^{16} and the number of candidates of δ' is 2^8, In this time, several wrong keys of RK_{34} remain in the candidates, but we can recover the correct key in the 3-rd step.

We show the attack procedure. Attackers first guess δ' and $RK_{34}[0]$ and calculate two differences δ by using two left faulty texts and the correct ciphertext. The complexity is 3×2^{16} and the complexity is dominant complexity. Next, attackers get $RK_{34}[1]$ by using the s_1-table. Here the input of the s_1-table is $x = C_0[1]$, $\bar{x} = \bar{C}_0[1]$ and $y = m_1\delta \oplus (M_0^{-1}\Delta C_1)[1] \oplus m_1'\delta'$. Similarly, attackers get $RK_{34}[2]$ and $RK_{34}[3]$ by using the s_0 and s_1 tables, respectively. Finally, attackers recover RK_{34}.

Similarly, we recover RK_{35} by using right faulty texts and the correct ciphertext. In this time, we pay attention to the C_3'-path and apply our property. The complexity is about 3×2^{16}, thus the total complexity in the 2-nd step is 6×2^{16}. The attack procedure after this step is the same method as that in Sect. 4.1, but attackers can reduce the key space of RK_{34} to enough small candidates in the 2-nd step. Consequently, the time complexity of 3-rd step becomes only 6×2^8. The total complexity of the DFA against modified CLEFIA is about 2^{16}, hence

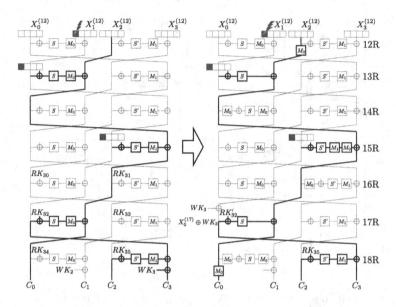

Fig. 8. The DFA exploiting faults induced at the 12-th round

modified CLEFIA has no advantage about the security for the DFA exploiting faults at the 13-th round.

4.4   DFA Exploiting Faults at the 12-th Round

In this section, we show a new DFA exploiting faults induced at the 12-th round. Our attack consists of four steps. In the 1-st step, attackers get a correct cipher-text corresponding to a plaintext. Next attackers induce four random byte faults to $X_1^{(12)}$ which is calculated from the same plaintext and secret key, and get four faulty ciphertexts. The position where several faults are induced can be any byte position, but four faults must be induced at the same byte position. We denote these faulty ciphertexts by left faulty texts. Similarly, attackers induce four faults at the data $X_3^{(12)}$ and get four faulty ciphertexts. We denote these faulty ciphertexts by right faulty texts. In this paper, for simplicity, we assume that the faults are induced to $X_1^{(12)}[0]$ and $X_3^{(12)}[0]$. In the 2-nd step, attackers recover RK_{35} and reduce the key space of RK'_{32} at the same time by using left faulty texts and the correct ciphertext. In this time, we apply our property to the C_0-path. Similarly, attackers recover RK_{34} and reduce the key space of RK'_{33} at the same time by using right faulty texts and the correct ciphertext. In this time we apply our property to the C_2-path. The complexity is 10×2^{40} and this complexity is dominant complexity to recover the secret key. In the 3-rd step, we recover RK_{30} and RK_{31}, and the complexity is 10×2^{16}. Finally we recover the secret key in the final step, where the complexity is negligible.

   We show the 2-nd step, namely, how to recover RK_{35} and reduce the key space of RK'_{32} by using left faulty texts and the correct ciphertext. We pay attention to the C_0-path and apply our property (see Fig. 8). First we guess RK_{35} and calculate $X_0^{(17)}[0] \oplus WK_3$. When correct RK_{35} is used, we can reduce the key space of RK_{32}' by using the same method as the 2-nd step in the attack shown in Sect. 4.1. For wrong RK_{35} and RK_{32}', the probability that all δ have the same value is 2^{-16} which is the same probability as the 2-nd step in the attack shown in Sect. 4.1. When attackers use four faults, the probability that wrong keys remain in round key candidates is $2^{32+24} \times (2^{-16})^4 = 2^{-8}$. Then about 2^8 round key RK_{32}' and a few RK_{35} remain in the candidates, where $(RK_{32}'[1], RK_{32}'[2], RK_{32}'[3])$ are correct values and $RK_{32}'[0]$ is any value from 0 to 255. The complexity is 5×2^{40} because the complexity to guess RK_{35} is 5×2^{32} and that to guess each byte of RK_{32}' is 5×2^8. Similarly, we recover RK_{34} and reduce the key space of RK_{33}' by using right faulty texts and the correct ciphertext. Since the complexity is 5×2^{40}, the total complexity in the 2-nd step is 10×2^{40}. We omit the explanation due to the space limitation, but the attack after this step is the same method as the attack in Sect. 4.1.

To confirm the feasibility, we implemented the simulation of our attack. We confirmed that the time to recover the RK_{35} and reduce the key space of RK_{32}' is about 1 day on 8 threads in an Intel Core-i7 3770 3.4GHz desktop machine. Then the secret key is recovered in a few days with that machine. We argue that this attack is still realistic and protecting last six rounds of CLEFIA is not enough against DFAs.

Finally, we describe the security of modified CLEFIA which uses the DSM using three matrices. Similar to the DFA exploiting faults at the 13-th round against modified CLEFIA, attackers first guess $\delta' = \Delta s_0(X_0^{(13)}[0] \oplus RK_{24}[0])$ and recover RK_{34}. Since the number of candidates of δ is 2^8, the complexity in 2-nd step is about 2^{48}. Since the total complexity for the DFA against original CLEFIA is about 2^{40}, modified CLEFIA have small advantage than original CLEFIA.

5   Conclusion

In this paper, we proposed a new property of the DSM. By applying our new property to CLEFIA, we can improve DFAs on CLEFIA. Our attack can exploit faults induced at an earlier round, which is the 13-th round and is one rounds earlier than that of existing DFAs. The complexity is 2^{24}. Moreover our attack can exploit faults induced at the 12-th round and it is two rounds earlier than that of existing DFAs. The complexity is 10×2^{40} and it is still practical. Then we argue that protecting last six rounds of CLEFIA is not enough against DFAs. On the other hand, implementers can implement ciphers whose all round functions have protection against DFAs. In this case, we can not execute our DFAs.

References

1. The 128-Bit Blockcipher CLEFIA Security and Performance Evaluations Revision 1.0. Sony Corporation (2007)

2. Ali, S.S., Mukhopadhyay, D.: Protecting Last Four Rounds of CLEFIA is Not Enough Against Differential Fault Analysis. Cryptology ePrint Archive, Report 2012/286 (2012)
3. Ali, S.S., Mukhopadhyay, D.: Improved Differential Fault Analysis of CLEFIA. In: FDTC (2013)
4. Biham, E., Shamir, A.: Differential Cryptanalysis of DES-like Cryptosystems. J. Cryptology 4(1), 3–72 (1991)
5. Chen, H., Wu, W., Feng, D.: Differential Fault Analysis on CLEFIA. In: Qing, S., Imai, H., Wang, G. (eds.) ICICS 2007. LNCS, vol. 4861, pp. 284–295. Springer, Heidelberg (2007)
6. Derbez, P., Fouque, P.-A., Leresteux, D.: Meet-in-the-Middle and Impossible Differential Fault Analysis on AES. In: Preneel, B., Takagi, T. (eds.) CHES 2011. LNCS, vol. 6917, pp. 274–291. Springer, Heidelberg (2011)
7. Kim, C.H.: Efficient Methods for Exploiting Faults Induced at AES Middle Rounds. Cryptology ePrint Archive, Report 2011/349 (2011)
8. Li, Y., Wu, W., Zhang, L.: Improved Integral Attacks on Reduced-Round CLEFIA Block Cipher. In: Jung, S., Yung, M. (eds.) WISA 2011. LNCS, vol. 7115, pp. 28–39. Springer, Heidelberg (2012)
9. Mala, H., Dakhilalian, M., Shakiba, M.: Impossible Differential Attacks on 13-Round CLEFIA-128. J. Comput. Sci. Technol. 26(4), 744–750 (2011)
10. Matsui, M.: Linear Cryptanalysis Method for DES Cipher. In: Helleseth, T. (ed.) EUROCRYPT 1993. LNCS, vol. 765, pp. 386–397. Springer, Heidelberg (1994)
11. Matsui, M., Yamagishi, A.: A New Method for Known Plaintext Attack of FEAL Cipher. In: Rueppel, R.A. (ed.) EUROCRYPT 1992. LNCS, vol. 658, pp. 81–91. Springer, Heidelberg (1993)
12. Phan, R.C.-W., Yen, S.-M.: Amplifying Side-Channel Attacks with Techniques from Block Cipher Cryptanalysis. In: Domingo-Ferrer, J., Posegga, J., Schreckling, D. (eds.) CARDIS 2006. LNCS, vol. 3928, pp. 135–150. Springer, Heidelberg (2006)
13. Sasaki, Y., Li, Y., Sakamoto, H., Sakiyama, K.: Coupon Collector's Problem for Fault Analysis Against AES — High Tolerance for Noisy Fault Injections. In: Sadeghi, A.-R. (ed.) FC 2013. LNCS, vol. 7859, pp. 213–220. Springer, Heidelberg (2013)
14. Shirai, T., Preneel, B.: On Feistel Ciphers Using Optimal Diffusion Mappings Across Multiple Rounds. In: Lee, P.J. (ed.) ASIACRYPT 2004. LNCS, vol. 3329, pp. 1–15. Springer, Heidelberg (2004)
15. Shirai, T., Shibutani, K.: Improving Immunity of Feistel Ciphers against Differential Cryptanalysis by Using Multiple MDS Matrices. In: Roy, B., Meier, W. (eds.) FSE 2004. LNCS, vol. 3017, pp. 260–278. Springer, Heidelberg (2004)
16. Shirai, T., Shibutani, K.: On Feistel Structures Using a Diffusion Switching Mechanism. In: Robshaw, M. (ed.) FSE 2006. LNCS, vol. 4047, pp. 41–56. Springer, Heidelberg (2006)
17. Shirai, T., Shibutani, K., Akishita, T., Moriai, S., Iwata, T.: The 128-Bit Blockcipher CLEFIA (Extended Abstract). In: Biryukov, A. (ed.) FSE 2007. LNCS, vol. 4593, pp. 181–195. Springer, Heidelberg (2007)
18. Takahashi, J., Fukunaga, T.: Improved Differential Fault Analysis on CLEFIA. In: FDTC, pp. 25–34. IEEE Computer Society (2008)
19. Takahashi, J., Fukunaga, T.: Differential Fault Analysis on CLEFIA with 128, 192, and 256-Bit Keys. IEICE Transactions 93-A(1), 136–143 (2010)
20. Tezcan, C.: The Improbable Differential Attack: Cryptanalysis of Reduced Round CLEFIA. In: Gong, G., Gupta, K.C. (eds.) INDOCRYPT 2010. LNCS, vol. 6498, pp. 197–209. Springer, Heidelberg (2010)
21. Jie Zhao, X., Wang, T., Zhe Gao, J.: Multiple Bytes Differential Fault Analysis on CLEFIA. Cryptology ePrint Archive, Report 2010/078 (2010)

Improvement of Faugère *et al.*'s Method to Solve ECDLP

Yun-Ju Huang[1], Christophe Petit[2,*],
Naoyuki Shinohara[3], and Tsuyoshi Takagi[4]

[1] Graduate School of Mathematics, Kyushu University
y-huang@math.kyushu-u.ac.jp
[2] UCL Crypto Group
[3] NICT
[4] Institute of Mathematics for Industry, Kyushu University

Abstract. Solving the elliptic curve discrete logarithm problem (ECDLP) by using Gröbner basis has recently appeared as a new threat to the security of elliptic curve cryptography and pairing-based cryptosystems. At Eurocrypt 2012, Faugère, Perret, Petit and Renault proposed a new method using a multivariable polynomial system to solve ECDLP over finite fields of characteristic 2. At Asiacrypt 2012, Petit and Quisquater showed that this method may beat generic algorithms for extension degrees larger than about 2000.

In this paper, we propose a variant of Faugère *et al.*'s attack that practically reduces the computation time and memory required. Our variant is based on the idea of symmetrization. This idea already provided practical improvements in several previous works for composite-degree extension fields, but its application to prime-degree extension fields has been more challenging. To exploit symmetries in an efficient way in that case, we specialize the definition of factor basis used in Faugère *et al.*'s attack to replace the original polynomial system by a new and simpler one. We provide theoretical and experimental evidence that our method is faster and requires less memory than Faugère *et al.*'s method when the extension degree is large enough.

Keywords: Elliptic curve, Discrete logarithm problem, Index calculus, Multivariable polynomial system, Gröbner basis.

1 Introduction

In the last two decades, elliptic curves have become increasingly important in cryptography. Elliptic curve cryptography requires shorter keys than factorization-based cryptography. Additionally, elliptic curve implementations have become increasingly efficient, and many cryptographic schemes have been proposed based on the hardness of the elliptic curve discrete logarithm problem (ECDLP) or another elliptic curve problem. These reasons led the American

---

* F.R.S.-FNRS posdoctoral fellow at Université catholique de Louvain.

K. Sakiyama and M. Terada (Eds.): IWSEC 2013, LNCS 8231, pp. 115–132, 2013.
© Springer-Verlag Berlin Heidelberg 2013

National Security Agency (NSA) to advocate the use of elliptic curves for public key cryptography in 2009 [1].

Given an elliptic curve E defined over a finite field K, some rational point P of E and a second point $Q \in \langle P \rangle \subset E$, ECDLP requires finding an integer k such that $Q = [k]P$. Elliptic curves used in practice are defined either over a prime field $\mathbb{F}_p$ or over a binary field $\mathbb{F}_{2^n}$. Like any other discrete logarithm problem, ECDLP can be solved with generic algorithms such as Baby-step Giant-step algorithm, Pollard's ρ method and their variants [2,3,4,5]. These algorithms can be parallelized very efficiently, but the parallel versions still have an exponential complexity in the size of the parameters. Better algorithms based on the *index calculus framework* have long been known for discrete logarithm problems over multiplicative groups of finite fields or hyperelliptic curves, but generic algorithms have remained the best algorithms for solving ECDLP until recently.

A key step of an index calculus algorithm for solving ECDLP is to solve the *point decomposition problem*. Given a predefined *factor basis* $\mathcal{F} \subset E$ and a random point $R \in E$, this problem asks the existence of points $P_i \in \mathcal{F}$ such that $R = \sum_i P_i$. In 2004, Semaev introduced the *summation polynomials* (also known as Semaev's polynomials) to solve this problem. The Semaev's polynomial s_r is a polynomial in r variables such that $s_r(x_1, \ldots, x_r) = 0$ if and only if there exist r points $P_i := (x_i, y_i) \in E$ such that $\sum_{i=1}^{r} P_i = O$. For a factor basis $\mathcal{F}_V := \{(x, y) | x \in V\}$ where $V \subset K$, the point decomposition problem now amounts to computing all x_i satisfying $s_{r+1}(x_1, \cdots, x_r, x(R)) = 0$ for the x-coordinate $x(R)$ of the given point R. Semaev's polynomials therefore reduce the decomposition problem on the elliptic curve E to algebraic problem over the base field K.

Solving Semaev's polynomials is not a trivial task in general, in particular if K is a prime field. For extension fields $K = \mathbb{F}_{q^n}$, Gaudry and Diem [6,7] independently proposed to define V as the subfield $\mathbb{F}_q$ and to apply a *Weil descent* to further reduce the resolution of Semaev's polynomials to the resolution of a polynomial system of equations over $\mathbb{F}_q$. Diem generalized these ideas by defining V as a vector subspace of $\mathbb{F}_{q^n}$ [8]. Using generic complexity bounds on the resolution of polynomial systems, these authors provided attacks that can beat generic algorithms and can even have subexponential complexity for specific families of curves [6]. At Eurocrypt 2012, Faugère, Perret, Petit and Renault re-analized Diem's attack [8] in the case $\mathbb{F}_{2^n}$, and showed that the systems arising from the Weil descent on Semaev's polynomials are much easier to solve than generic systems [9]. Later at Asiacrypt 2012, Petit and Quisquater provided heuristic evidence that ECDLP is subexponential for that very important family of curves, and would beat generic algorithms when n is larger than about 2000 [10].

Even though these recent results suggest that ECDLP is weaker than previously expected for binary curves, the attacks are still far from being practical. This is mainly due to the large memory and time required to solve the polynomial systems arising from the Weil descent in practice. In particular, the experimental results presented in [10] for primes n were limited to $n = 17$. In order to validate the heuristic assumptions taken in Petit and Quisquater's analysis and to estimate the exact security level of binary elliptic curves in practice, experiments on larger parameters are definitely required.

Hybrid methods (involving a trade-off between exhaustive search and polynomial system solving) have been proposed to practically improve the resolution of the polynomial systems [11]. More importantly, the special structure of these systems can be exploited. When n is composite and the Weil descent is performed on an intermediary subfield, Gaudry already showed in [7] how the symmetry of Semaev's polynomials can be exploited to accelerate the resolution of the polynomial system in practice. In that case, the whole system can be re-written with new variables corresponding to the fundamental symmetric polynomials, therefore reducing the degrees of the equations and improving their resolution. In the particular cases of twisted Edward curves and twisted Jacobi curves, Faugère *et al.* also exploited additional symmetry coming from the existence of a rational 2-torsion point to further reduce the degrees of the equations [12].

In this paper, we focus on Diem's version of index calculus for ECDLP over a binary field of prime extension degree n [8,9,10]. In that case, the Weil descent is performed on a vector space that is not a subfield of $\mathbb{F}_{2^n}$, and the resulting polynomial system cannot be re-written in terms of symmetric variables only. We therefore introduce a different method to take advantage of symmetries even in the prime degree extension case. Our re-writing of the system involves both symmetric and non-symmetric variables. The total number of variables is increased compared to [9,10], but we limit this increase as much as possible thanks to an appropriate choice of the vector space V. On the other hand, the use of symmetric variables in our system allows reducing the degrees of the equations significantly. Our experimental results show that our systems can be solved faster than the original systems of [12, 21] as long as n is large enough.

Notations. In this work, we are interested in solving the elliptic curve discrete logarithm problem on a curve E defined over a finite field $\mathbb{F}_{2^n}$, where n is a prime number. We denote by $E_{\alpha,\beta}$ the elliptic curve over $\mathbb{F}_{2^n}$ defined by the equation $y^2 + xy = x^3 + \alpha x^2 + \beta$. For a given point $P \in E$, we use $x(P)$ and $y(P)$ to indicate the x-coordinate and y-coordinate of P respectively. From now on, we use the specific symbols P, Q and k for the parameters and solution of the ECDLP: $P \in E$, $Q \in \langle P \rangle$, and k is the smallest non-negative integer such that $Q = [k]P$. We identify the field $\mathbb{F}_{2^n}$ as $\mathbb{F}_2[\omega]/h(\omega)$, where h is an irreducible polynomial of degree n. Any element $e \in \mathbb{F}_{2^n}$ can then be represented as $poly(e) := c_0 + c_1\omega + ... + c_{n-1}\omega^{n-1}$ where $c_i \in \mathbb{F}_2$. For any set S, we use the symbol $\#S$ to mean the order of S.

Outline. The remaining of this paper is organized as follows. In Section 2, we recall previous index calculus algorithms for ECDLP, in particular Faugère *et al.*'s attack on binary elliptic curves and previous work exploiting the symmetry of Semaev's polynomials when the extension degree is composite. In Section 3, we describe our variant of Faugère *et al.*'s attack taking advantage of the symmetries even when the extension degree is prime. In Section 4, we provide experimental results supporting our method with respect to Faugère *et al.*'s original attack. Finally in Section 5, we conclude the paper and we introduce further work.

2   Index Calculus for Elliptic Curves

2.1   The Index Calculus Method

For a given point $P \in E_{\alpha,\beta}$, let Q be a point in $\langle P \rangle$. The index calculus method can be adapted to elliptic curves to compute the discrete logarithm of Q with respect to P.

Algorithm 1. Index Calculus for ECDLP [13]

    Input: elliptic curve $E_{\alpha,\beta}$, point $P \in E_{\alpha,\beta}$, point $Q \in \langle P \rangle$
1 $F \longleftarrow$ a subset of $E_{\alpha,\beta}$
2 $M \longleftarrow$ matrix with $\#F + 2$ columns
3 **while** $Rank(M) < \#F + 1$ **do**
4 | $R \longleftarrow [a]P + [b]Q$ where a and b are random integers in $(0, \#\langle P \rangle)$
5 | $sol_m \longleftarrow$ Decompose(R, F)
6 | $M \longleftarrow$ AddRelationToMatrix(sol_m)
7 **end**
8 $M \longleftarrow$ ReducedRowEchelonForm(M)
9 $a', b' \longleftarrow$ last two column entries of last row
10 $k \longleftarrow -a'/b'$
    Output: k, where $Q = [k]P$

As shown in Algorithm 1, we first select a *factor base* $F \subset E_{\alpha,\beta}$ and we perform a *relation search* expressed as the loop between the line 3 and 7 of Algorithm 1. This part is currently the efficiency bottleneck of the algorithm. For each step in the loop, we compute $R := [a]P + [b]Q$ for random integers a and b and we apply the **Decompose** function on R to find all tuples (sol_m) of m elements $P_{j_\ell} \in F$ such that $P_{j_1} + P_{j_2} + \cdots + P_{j_m} + R = O$. Note that we may obtain several decompositions for each point R. In the line 6, the **AddRelationToMatrix** function encodes every decomposition of a point R into a row vector of the matrix M. More precisely, the first $\#F$ columns of M correspond to the elements of F, the last two columns correspond to P and Q, and the coefficients corresponding to these points are encoded in the matrix. In the line 7, the **ReducedRowEchelonForm** function reduces M into a row echelon form. When the rank of M reaches $\#F + 1$, the last row of the reduced M is of the form $(0, \cdots, 0, a', b')$, which implies that $[a']P + [b']Q = O$. From this relation, we obtain $k = -a'/b' \bmod \#\langle P \rangle$.

A straightforward method to implement the **Decompose** function would be to exhaustively compute the sums of all m-tuples of points in F and to compare these sums to R. However, this method would not be efficient enough.

2.2   Semaev's Polynomials

Semaev's polynomials [13] allow replacing the complicated addition law involved in the point decomposition problem by a somewhat simpler polynomial equation over $\mathbb{F}_{2^n}$.

Definition 1.   *The m-th Semaev's polynomial s_m for $E_{\alpha,\beta}$ is defined as follows: $s_2 := x_1 + x_2$, $s_3 := (x_1 x_2 + x_1 x_3 + x_2 x_3)^2 + x_1 x_2 x_3 + \beta$, and $s_m := Res_X(s_{j+1}(x_1, ..., x_j, X), s_{m-j+1}(x_{j+1}, ..., x_m, X))$ for $m \geq 4$, $2 \leq j \leq m - 2$.*

The polynomial s_m is symmetric and it has degree 2^{m-2} with respect to each variable. Definition 1 provides a straightforward method to compute it. In practice, computing large Semaev's polynomials may not be a trivial task, even if the symmetry of the polynomials can be used to accelerate it [11]. Semaev's polynomials have the following property:

Proposition 1.   *We have $s_m(x_1, x_2, ..., x_m) = 0$ if and only if there exist $y_i \in \mathbb{F}_{2^n}$ such that $P_i = (x_i, y_i) \in E_{\alpha,\beta}$ and $P_1 + P_2 + ... + P_m = O$.*

In Semaev's seminal paper [13], he proposed to choose the factor base F in Algorithm 1 as

$$F_V := \{(x, y) \in E_{\alpha,\beta} | x \in V\}$$

where V is some subset of the base field of the curve. According to Proposition 1, finding a decomposition of a given point $R = [a]P + [b]Q$ is then reduced to first finding $x_i \in V$ such that

$$s_{m+1}(x_1, x_2, ..., x_m, x(R)) = 0,$$

and then finding the corresponding points $P_j = (x_j, y_j) \in F_V$.

A straightforward **Decompose** function using Semaev's polynomials is described in Algorithm 2. In this algorithm, Semaev's polynomials are solved by a

---

Algorithm 2. Decompose function with s_{m+1}

   Input: $R = [a]P + [b]Q$, factor base F_V
1 $Set_m \longleftarrow \{e \in F_V^m\}$
2 $sol_m \longleftarrow \{\}$
3 **for** $e = \{P_1, P_2, .., P_m\} \in Set_m$ **do**
4 | **if** $s_{m+1}(x(P_1), x(P_2), ..., x(P_m), x(R)) = 0$ **then**
5 | | **if** $P_1 + P_2 + ... + P_m + R = O$ **then**
6 | | | $sol_m \longleftarrow sol_m \cup \{e\}$
7 | | **end**
8 | **end**
9 **end**
   Output: sol_m contains the decomposition elements of R w.r.t. F_V

---

naive exhaustive search method. Since every x-coordinate corresponds to at most two points on the elliptic curve $E_{\alpha,\beta}$, each solution of $s_{m+1}(x_1, x_2, ..., x_m, x(R)) = 0$ may correspond to up to 2^m possible solutions in $E_{\alpha,\beta}$. These potential solutions are tested in the line 5 of Algorithm 2. As such, Algorithm 2 still involves some exhaustive search and can clearly not solve ECDLP faster than generic algorithms.

2.3   Method of Faugère *et al.*

At Eurocrypt 2012, following similar approaches by Gaudry [7] and Diem [6,8], Faugère *et al.* provided V with the structure of a vector space, to reduce the resolution of Semaev's polynomial to a system of multivariate polynomial equations. They then solved this system using Gröbner basis algorithms [9].

More precisely, Faugère *et al.* suggested to fix V as a random vector subspace of $\mathbb{F}_{2^n}/\mathbb{F}_2$ with dimension n'. If $\{v_1, \ldots, v_{n'}\}$ is a basis of this vector space, the resolution of Semaev's polynomial is then reduced to a polynomial system as follows. For any fixed $P' \in F_V$, we can write $x(P')$ as

$$x(P') = \bar{c}_1 v_1 + \bar{c}_2 v_2 + ... + \bar{c}_{n'} v_{n'}$$

where $\bar{c}_\ell \in \mathbb{F}_2$ are known elements. Similarly, we can write all the variables $x_j \in V$ in $s_{m+1}|_{x_{m+1}=x(R)}$ as

$$\begin{cases} x_j = c_{j,1}v_1 + c_{j,2}v_2 + \ldots + c_{j,n'}v_{n'}, & 1 \leq j \leq m, \\ x_{m+1} = r_0 + r_1\omega + \ldots + r_{n-1}\omega^{n-1}, \end{cases}$$

where $c_{j,\ell}$ are binary variables and $r_\ell \in \mathbb{F}_2$ are known. Using these equations to substitute the variables x_j in s_{m+1}, we obtain an equation

$$s_{m+1} = f_0 + f_1\omega + \ldots + f_{n-1}\omega^{n-1},$$

where $f_0, f_1, ..., f_{n-1}$ are polynomials in the binary variables $c_{j,l}$, $1 \leq j \leq m$, $1 \leq l \leq n'$.

We have $s_{m+1}|_{x_{m+1}=x(R)} = 0$ if and only if each binary coefficient polynomial f_l is equal to 0. Solving Semaev's polynomial s_{m+1} is now equivalent to solving the binary multivariable polynomial system

$$\begin{cases} f_0(c_{1,1}, \ldots, c_{0,l}, \ldots, c_{m,n'}) = 0, \\ f_1(c_{1,1}, \ldots, c_{1,l}, \ldots, c_{m,n'}) = 0, \\ \quad \vdots \\ f_m(c_{1,1}, \ldots, c_{m,l}, \ldots, c_{m,n'}) = 0, \end{cases} \tag{1}$$

in the variables $c_{j,\ell}$, $1 \leq j \leq m, 1 \leq \ell \leq n'$.

The **Decompose** function using this system is described in Algorithm 3. It is denoted as Imp_{FPPR} in this work. We first substitute x_{m+1} with $x(R)$ in

---

Algorithm 3. Decompose function with binary multivariable polynomial system (Imp_{FPPR}) [9]

Input: $R = [a]P + [b]Q$, factor base F_V

1 $F \longleftarrow \texttt{TransFromSemaevToBinary}(s_{m+1}\,|_{x_{m+1}=x(R)})$

2 $GB(F) \longleftarrow \texttt{GroebnerBasis}(F, \prec_{lex})$

3 $sol(F) \longleftarrow \texttt{GetSolutionFromGroebnerBasis}(GB(F))$

4 $sol_m \longleftarrow \{\}$

5 **for** $e = \{P_1, P_2, .., P_m\} \in sol(F)$ **do**

6   **if** $P_1 + P_2 + ... + P_m + R = O$ **then**

7        $sol_m \longleftarrow sol_m \cup \{e\}$

8   **end**

9 **end**

Output: sol_m contains the decomposition elements of R w.r.t. F_V

---

s_{m+1}. The **TransFromSemaevToBinaryWithSym** function transforms the equation $s_{m+1}\,|_{x_{m+1}=x(R)}= 0$ into System (1) as described above. To solve this system, we compute its Gröbner basis with respect to a lexicographic ordering using an algorithm such as F_4 or F_5 algorithm [14,15]. A Gröbner basis for a lexicographic ordering always contains some univariate polynomial (the polynomial 1 when there is no solution), and the solutions of F can be obtained from the roots of this polynomial. However, since it is much more efficient to compute a Gröbner basis for a graded-reversed lexicographic order than for a lexicographic ordering, a Gröbner basis of F is first computed for a graded-reverse lexicographic ordering and then transformed into a Gröbner basis for a lexicographic ordering using FGLM algorithm [16].

After getting the solutions of F, we find the corresponding solutions over $E_{\alpha,\beta}$. As before, this requires to check whether $P_1 + P_2 + ... + P_m + R = O$ for all the potential solutions in the line 6 of Algorithm 3.

Although Faugère *et al.*'s approach provides a systematic way to solve Semaev's polynomials, their algorithm is still not practical. Petit and Quisquater estimated that the method could beat generic algorithms for extension degrees n larger than about 2000 [10]. This number is much larger than the parameter $n = 160$ that is currently used in applications. In fact, the degrees of the equations in F grow quadratically with m, and the number of monomial terms in the equations is exponential in this degree. In practice, the sole computation of the Semaev's polynomial s_{m+1} seems to be a challenging task for m larger than 7. Because of the large computation costs (both in time and memory), no experimental result has been provided yet when n is larger than 20.

In this work, we provide a variant of Faugère *et al.*'s method that practically improves its complexity. Our method exploits the symmetry of Semaev's polynomials to reduce both the degree of the equations and the number of monomial terms appearing during the computation of a Gröbner basis of the system F.

2.4   Use of Symmetries in Previous Works

The symmetry of Semaev's polynomials has been exploited in previous works, but always for finite fields $\mathbb{F}_{p^n}$ with *composite* extension degrees n. The approach was already described by Gaudry [7] as a mean to accelerate the Gröbner basis computations. The symmetry of Semaev's polynomials has also been used by Joux and Vitse's to establish new ECDLP records for composite extension degree fields [11,17]. Extra symmetries resulting from the existence of a rational 2-torsion point have also been exploited by Faugère *et al.* for twisted Edward curves and twisted Jacobi curves [12]. In all these approaches, exploiting the symmetries of the system allows reducing the degrees of the equations and the number of monomials involved in the Gröbner basis computation, hence it reduces both the time and the memory costs.

To exploit the symmetry in ECDLP index calculus algorithms, we first rewrite Semaev's polynomial s_{m+1} with the elementary symmetric polynomials.

Definition 2.   *Let $x_1, x_2, ..., x_m$ be m variables, then the elementary symmetric polynomials are defined as*

$$\begin{cases} \sigma_1 := \sum_{1 \le j_1 \le m} x_{j_1} \\ \sigma_2 := \sum_{1 \le j_1 < j_2 \le m} x_{j_1} x_{j_2} \\ \sigma_3 := \sum_{1 \le j_1 < j_2 < j_3 \le m} x_{j_1} x_{j_2} x_{j_3} \\ \quad \vdots \\ \sigma_m := \prod_{1 \le j \le m} x_j \end{cases} \tag{2}$$

Any symmetric polynomial can be written as an algebraic combination of these elementary symmetric polynomials. We denote the symmetrized version of Semaev's polynomial s_m by s'_m. For example for the curve $E_{\alpha,\beta}$ in characteristic 2, we have

$$s_3 = (x_1 x_2 + x_1 x_3 + x_2 x_3)^2 + x_1 x_2 x_3 + \beta,$$

where x_3 is supposed to be fixed to some $x(R)$. The elementary symmetric polynomials are

$$\sigma_1 = x_1 + x_2,$$
$$\sigma_2 = x_1 x_2.$$

The symmetrized version of s_3 is therefore

$$s'_3 = (\sigma_2 + \sigma_1 x_3)^2 + \sigma_2 x_3 + \beta.$$

Since x_3 is fixed and the squaring is a linear operation over $\mathbb{F}_2$, we see that symmetrization leads to a much simpler polynomial.

Let us now assume that n is a composite number with a non-trivial factor n'. In this case, we can fix the vector space V as the subfield $\mathbb{F}_{p^{n'}}$ of $\mathbb{F}_{p^n}$. We note that all arithmetic operations are closed on the elements of V for this special choice. In particular, we have

$$if \ x_i \in V \ then \ \sigma_i \in V \ . \tag{3}$$

Let now $\{1, \omega_2, \ldots, \omega_{n/n'}\}$ be a basis of $\mathbb{F}_{p^n}/\mathbb{F}_{p^{n'}}$. We can write

$$\sigma_j = d_{j,0} \; for \; 1 \leq j \leq m,$$
$$x_{m+1} = r_1 + r_2 \omega_2 + \ldots + r_{n/n'} \omega_{n/n'},$$

where $r_\ell \in \mathbb{F}_{p^n}$ are known and the variables $d_{j,0}$ are defined over $\mathbb{F}_{p^{n'}}$. These relations can be substituted in the equation $s'_{m+1} \mid_{x_{m+1}=x(R)} = 0$ to obtain a system of n/n' equations in the m variables $d_{j,0}$ only. Since the total degree and the degree of s'_m with respect to each symmetric variable σ_i are lower than those of s_m with respect to all non-symmetric variables x_i, the degrees of the equations in the resulting system are also lower and the system is easier to solve. As long as $n/n' \approx m$, the system has a reasonable chance to have a solution.

Given a solution $(\sigma_1, \ldots, \sigma_m)$ for this system, we can recover all possible corresponding values for the variables $x_1, \ldots, x_m$ (if there is any) by solving the system given in Definition 2, or equivalently by solving the symmetric polynomial equation

$$x^m + \sum_{i=1}^m \sigma_i x^{m-i} = x^m + \sigma_1 x^{m-1} + \sigma_2 x^{m-2} + \ldots + \sigma_m.$$

Note that the existence of a non-trivial factor of n and the special choice for V are crucial here. Indeed, they allow building a new system that only involves symmetric variables and that is significantly simpler to solve than the previous one.

3   Using Symmetries with Prime Extension Degrees

When n is prime, the only subfield of $\mathbb{F}_{2^n}$ is $\mathbb{F}_2$, but choosing $V = \mathbb{F}_2$ would imply to choose $m = n$, hence to work with Semaev's polynomial s_{n+1} which would not be practical when n is large. In Diem's and Faugère *et al.*'s attacks [9,8], the set V is therefore a generic vector subspace of $\mathbb{F}_{2^n}/\mathbb{F}_2$ with dimension n'. In that case, Implication (3) does not hold, but we now show how to nevertheless take advantage of symmetries in Semaev's polynomials.

3.1   A New System with Both Symmetric and Non-symmetric Variables

Let n be an arbitrary integer (possibly prime) and let V be a vector subspace of $\mathbb{F}_{2^n}/\mathbb{F}_2$ with dimension n'. Let $\{v_1, \ldots, v_{n'}\}$ be a basis of V. We can write

$$\begin{cases} x_j = c_{j,1} v_1 + c_{j,2} v_2 + \ldots + c_{j,n'} v_{n'}, \; for \; 1 \leq j \leq m \\ x_{m+1} = r_0 + r_1 \omega + \ldots + r_{n-1} \omega^{n-1}, \end{cases}$$

where $c_{j,\ell}$ with $1 \leq j \leq m$ and $1 \leq \ell \leq n'$ are variables but r_ℓ, $1 \leq \ell \leq n$ are known elements in $\mathbb{F}_2$.

Like in the composite extension degree case, we can use the elementary symmetric polynomials to write Semaev's polynomial s_{m+1} as a polynomial s'_{m+1} in the variables σ_j only. However since V is not a field anymore, constraining x_j in V does not constrain σ_j in V anymore. Since $\sigma_j \in \mathbb{F}_{2^n}$, we can however write

$$
\begin{cases}
\sigma_1 = d_{1,0} + d_{1,1}\omega + \ldots + d_{1,n-1}\omega^{n-1}, \\
\sigma_2 = d_{2,0} + d_{2,1}\omega + \ldots + d_{2,n-1}\omega^{n-1}, \\
\quad\vdots \\
\sigma_m = d_{m,0} + d_{m,1}\omega + \ldots + d_{m,n-1}\omega^{n-1}.
\end{cases}
$$

where $d_{j,\ell}$ with $1 \le j \le m$ and $1 \le \ell \le n$ are binary variables. Using these equations, we can substitute σ_j in s'_{m+1} to obtain

$$
s'_{m+1} = f'_0 + f'_1\omega + \ldots + f'_{n-1}\omega^{n-1}
$$

where $f'_0, f'_1, \ldots, f'_{n-1}$ are polynomials in the binary variables $d_{j,\ell}$. Applying a Weil descent on the symmetrized Semaev's polynomial equation $s'_m = 0$, we therefore obtain a polynomial system

$$
f'_0 = f'_1 = \ldots = f'_{n-1} = 0
$$

in the mn binary variables $d_{j,\ell}$.

The variables $d_{j,\ell}$ must also satisfy certain constraints provided by System (2). More precisely, substituting both the x_j and the σ_j variables for binary variables in the equation

$$
\sigma_j = \sum_{\substack{I \subset \{1,\ldots,m\} \\ \#I=j}} \prod_{k \in I} x_k \ ,
$$

we obtain

$$
d_{j,0} + d_{j,1}\omega + \ldots + d_{j,n-1}\omega^{n-1} = \sigma_j = \sum_{\substack{I \subset \{1,\ldots,m\} \\ \#I=j}} \prod_{k \in I} \sum_{\ell=1}^{n'} c_{k,\ell} v_\ell
$$

$$
= g_{j,0} + g_{j,1}\omega + \ldots + g_{j,n-1}\omega^{n-1}
$$

where $g_{j,\ell}$ are polynomials in the mn' binary variables $c_{i,\ell}$ only. In other words, applying a Weil descent on each equation of System (2), we obtain mn new equations

$$
d_{j,\ell} = g_{j,\ell}
$$

in the $mn + mn'$ binary variables $c_{j,\ell}$ and $d_{j,\ell}$. The resulting system

$$
\begin{cases}
f'_j = 0, & 1 \le j \le n, \\
d_{j,\ell} = g_{j,\ell}, & 1 \le j \le m, 1 \le \ell \le n,
\end{cases}
$$

has $mn + n$ equations in $mn + mn'$ binary variables. As before, the system is expected to have solutions if $mn' \approx n$, and it can then be solved using a Gröbner basis algorithm.

In comparison with the simpler method of Faugère *et al.* (denoted as FPPR) [9], the number of variables is multiplied by a factor roughly $(m + 1)$. However, the degrees of our equations are also decreased thanks to the symmetrization, and this may decrease the degree of regularity of the system. In order to compare the time and memory complexities of both approaches, let D_{FPPR} and D_{Ours} be the degrees of regularity of the corresponding systems. The time and memory costs are respectively roughly N^{2D} and N^{3D}, where N is the number of variables and D is the degree of regularity . Assuming that neither D_{FPPR} nor D_{Ours} depends on n (as suggested by Petit and Quisquater's experiments [10]), that $D_{Ours} < D_{FPPR}$ (thanks to the use of symmetric variables) and that m is small enough, then the extra $(m + 1)$ factors in the number of variables will be a small price to pay for large enough parameters. In practice, experiments are limited to very small n and m values. For these small parameters, we could not observe any significant advantage of this variant with respect to Faugère *et al.*'s original method. However, the complexity can be improved even further in practice with a clever choice of vector space.

3.2 A Special Vector Space

In the prime degree extension case, V cannot be a subfield, hence the symmetric variables σ_j are not restricted to V. This led us to introduce mn variables $d_{j,\ell}$ instead of mn' variables only in the composite extension degree case. However, we point out that some vector spaces may be "closer to a subfield" than other ones. In particular if V is generated by the basis $\{1, \omega, \omega^2, \ldots, \omega^{n'-1}\}$, then we have

$$if \ x_j \in V \ then \ \sigma_2 \in V'$$

where $V' \supset V$ is generated by the basis $\{1, \omega, \omega^2, \ldots, \omega^{2n'-2}\}$.

More generally, we can write

$$\begin{cases} \sigma_1 = d_{1,0} + d_{1,1}\omega + \ldots + d_{1,n'-1}\omega^{n'-1}, \\ \sigma_2 = d_{2,0} + d_{2,1}\omega + \ldots + d_{2,2n'-2}\omega^{2n'-2}, \\ \quad \vdots \\ \sigma_m = d_{m,0} + d_{m,1}\omega + \ldots + d_{m,n-m}\omega^{n-m}. \end{cases}$$

Applying a Weil descent on $s'_{m+1}\mid_{x_{m+1}=x(R)}$ and each equation of System (2) as before, we obtain a new polynomial system

$$\begin{cases} f'_j = 0, & 1 \le j \le n, \\ d_{j,\ell} = g_{j,\ell}, & 1 \le j \le m, 0 \le \ell \le j(n'-1), \end{cases}$$

in $n + (n'-1)\frac{m(m+1)}{2} + m$ equations and $n'm + (n'-1)\frac{m(m+1)}{2} + m$ variables.

When m is large and $mn' \approx n$, the number of variables is decreased by a factor 2 if we use our special choice of vector space instead of a random one. For $m = 4$ and $n \approx 4n'$, the number of variables is reduced from about $5n$ to about $7n/2$. For $m = 3$ and $n \approx 3n'$, the number of variables is reduced from about $4n$ to about $3n$ thanks to our special choice for V. In practice, this improvement turns out to be significant.

Table 1. Comparison for different multivariate polynomial system

	s_{m+1}	s'_{m+1}	s'_{m+1} with specific V
variables number	mn'	$mn' + mn$	$mn' + (n' - 1)\frac{m(m+1)}{2} + m$
polynomials number	n	$n + mn$	$n + (n' - 1)\frac{m(m+1)}{2} + m$
degree of regularity	7 or 6	4 or 3	4 or 3

3.3 New Decomposition Algorithm

Our new algorithm for the decomposition problem is described in Algorithm 4. It is denoted as Imp_{Ours} in this work. The only difference between Imp_{FPPR}

Algorithm 4. Decompose function with binary multivariable polynomial system and symmetric elementary functions (Imp_{Ours})

Input: $R = [a]P + [b]Q$, factor base F_V
1 $F \longleftarrow$ TransFromSemaevToBinaryWithSym($s_{m+1} \mid_{x_{m+1}=x(R)}$)
2 $F_- \longleftarrow$ GroebnerBasis(F)
3 $sol(F) \longleftarrow$ GetSolutionFromGroebnerBasis(F_-)
4 $sol_m \longleftarrow \{\}$
5 **for** $e = \{P_1, P_2, .., P_m\} \in sol(F)$ **do**
6 $\quad$ **if** $P_1 + P_2 + ... + P_m + R = O$ **then**
7 $\quad\quad$ | $sol_m \longleftarrow sol_m \cup \{e\}$
8 $\quad$ **end**
9 **end**
Output: sol_m contains the decomposition elements of R w.r.t. F_V

---

and Imp_{Ours} comes from a different transformation function in the line 1 of Algorithm 4. Although the system solved in Imp_{Ours} contains more variables and equations than the system solved in Imp_{FPPR}, the degrees of the equations are smaller and they involve less monomial terms. We now describe our experimental results.

4 Experimental Results

To validate our analysis and experimentally compare our method with Faugère *et al.*'s previous work, we implemented both algorithms in Magma. All our experiments were conducted on a CPU with four AMD Opteron Processor 6276 with 16 cores, running at 2.3GHz with a L3 cache of 16MB. The Operating System was CentOS 6.3 with Linux kernel version 2.6.32-279.14.1.el6.x86_64 and 512GB memory. The programming platform was Magma V2.18-9 in its 64-bit version. Gröbner basis were computed with the *GroebnerBasis* function of Magma. Our implementations of Imp_{FPPR} and Imp_{Ours} share the same program, except that the former uses Algorithm 3 and the latter uses Algorithm 4 to set up the binary multivariate system. We first focus on the relation search, then we describe experimental results for a whole ECDLP computation.

4.1 Relation Search

The relation search is the core of both Faugère *et al.*'s algorithm and our variant. In our experiments, we considered a fixed randomly chosen curve $E_{\alpha,\beta}$, a fixed ECDLP with respect to P, and a fixed $m = 3$ for all values of the parameters n and n'. For random integers a and b, we used both Faugère *et al.*'s method and our approach to find factor basis elements $P_j \in F_V$ such that $P_1 + \cdots + P_m = [a]P + [b]Q$.

We focused on $m = 3$ (fourth Semaev's polynomial) in our experiments. Indeed, there is no hope to solve ECDLP faster than with generic algorithms using $m = 2$ because of the linear algebra stage at the end of the index calculus algorithm[1]. On the other hand, the method appears unpractical for $m = 4$ even for very small values of n because of the exponential increase with m of the degrees in Semaev's polynomials.

The experimental results are given in Table 2 and 3. For most values of the parameters n and n', the experiment was repeated 200 times and average values are presented in the table. For large values $n' = 6$, the experiment was only repeated 3 times due to the long execution time.

We noticed that the time required to solve one system varied significantly depending on whether it had solutions or not. Table 2 and 3 therefore present results for each case in separate columns. The table contains the following information: D_{reg} is the maximum degree appearing when solving the binary system with Magma's Gröbner basis routine; var is the number of $\mathbb{F}_2$ variables of the system; poly and mono are the number of polynomials and monomials in the system; rel is the average number of solutions obtained (modulo equivalent solutions through symmetries); t_{trans} and t_{groe} are respectively the time (in seconds) needed to transform the polynomial s_{m+1} into a binary system and to compute a Gröbner basis of this system; mem is the memory required by the experiment (in MB).

The experiments show that the degrees of regularity of the systems occurring during the relation search are decreased from values between 6 and 7 in Faugère *et al.*'s method to values between 3 and 4 in our method. This is particularly important since the complexity of Gröebner basis algorithms is exponential in this degree. However, this huge advantage of our method comes at the cost of a significant increase in the number of variables, which itself tends to increase the complexity of Gröbner basis algorithms. Our experimental results confirm the analysis of Section 3: while our method may require more memory and time for small parameters (n, n'), it becomes more efficient than Faugère *et al.*'s method when the parameters increase. We remark that although the time required to

---

[1] In fact, even $m = 3$ would require a double large prime variant of the index calculus algorithm described above in order to beat generic discrete logarithm algorithms [7].

Table 2. Comparison of the relation search ($m = 3$, $n' = 3, 4$) with two strategies, Imp_{FPPR} and Imp_{Ours}. D_{reg}, var, poly and mono are the degree of regularity, the number of variables, the number of polynomials and the number of monomials in the system. t_{trans} and t_{groe} are the transformation time and solving Gröbner basis time (seconds). men is the memory consumptions for solving the system (MB).

	n	n'	sol: yes							sol: no						
			D_{reg}	var	poly	mono	t_{trans}	t_{groe}	mem	D_{reg}	var	poly	mono	t_{trans}	t_{groe}	mem
Imp_{FPPR}	17	3	6	9	17	2070.59	3.95	1.08	21.51	6	9	17	2149.37	4.50	0.09	23.40
Imp_{Ours}	17	3	3	24	32	826.12	0.67	1.14	14.86	3	24	32	867.87	0.72	0.24	16.26
Imp_{FPPR}	19	3	6	9	19	2305.76	4.44	1.08	27.55	6	9	19	2401.07	4.97	0.11	29.59
Imp_{Ours}	19	3	3	24	34	912.57	0.75	1.13	19.75	3	24	34	962.67	0.79	0.31	20.90
Imp_{FPPR}	23	3	6	9	23	2792.97	5.47	1.06	29.10	6	9	23	2908.92	6.18	0.12	32.25
Imp_{Ours}	23	3	3	24	38	1079.60	0.91	1.04	15.59	3	24	38	1147.65	0.97	0.14	16.68
Imp_{FPPR}	29	3	6	9	29	3509.17	6.94	1.02	38.85	6	9	29	3669.15	7.75	0.07	43.14
Imp_{Ours}	29	3	3	24	44	1329.85	1.15	0.95	17.16	3	24	44	1427.97	1.22	0.17	18.43
Imp_{FPPR}	31	3	6	9	31	3739.76	7.38	1.03	41.12	5	9	31	3922.40	8.38	0.06	46.33
Imp_{Ours}	31	3	3	24	46	1428.49	1.24	0.90	17.59	3	24	46	1515.79	1.30	0.04	18.87
Imp_{FPPR}	37	3	6	9	37	4450.86	8.90	1.00	48.88	6	9	37	4677.23	9.99	0.06	54.81
Imp_{Ours}	37	3	3	24	52	1673.42	1.48	0.88	19.23	3	24	52	1800.79	1.58	0.05	20.85
Imp_{FPPR}	41	3	6	9	41	4921.38	9.81	0.98	54.35	6	9	41	5182.97	11.17	0.06	61.70
Imp_{Ours}	41	3	3	24	56	1847.03	1.64	0.87	20.58	3	24	56	1983.08	1.75	0.05	22.60
Imp_{FPPR}	43	3	6	9	43	5175.86	10.47	0.99	57.69	6	9	43	5436.94	11.73	0.05	64.74
Imp_{Ours}	43	3	3	24	58	1931.96	1.76	0.87	21.28	3	24	58	2076.11	1.86	0.05	23.24
Imp_{FPPR}	47	3	6	9	47	5631.62	11.29	1.00	63.77	5	9	47	5947.98	12.85	0.06	72.47
Imp_{Ours}	47	3	3	24	62	2116.38	1.92	0.83	23.17	3	24	62	2263.80	2.02	0.06	25.32
Imp_{FPPR}	53	3	6	9	53	6358.94	12.86	1.03	72.06	5	9	53	6706.36	14.57	0.07	81.22
Imp_{Ours}	53	3	3	24	68	2348.50	2.12	0.79	24.89	2	24	68	2541.59	2.28	0.04	27.52
Imp_{FPPR}	17	4	7	12	17	8997.76	15.47	6.81	58.16	7	12	17	9028.92	16.53	1.20	55.37
Imp_{Ours}	17	4	3	33	38	1622.88	1.31	3.91	31.52	3	33	38	1641.84	1.33	2.23	24.88
Imp_{FPPR}	19	4	7	12	19	9915.47	17.04	6.88	67.24	7	12	19	10072.64	17.85	1.54	64.78
Imp_{Ours}	19	4	3	33	40	1823.58	1.51	3.26	32.97	3	33	40	1823.69	1.46	1.57	27.11
Imp_{FPPR}	23	4	6	12	23	12059.19	21.06	6.83	95.66	6	12	23	12201.94	22.31	4.67	91.23
Imp_{Ours}	23	4	3	33	44	2173.29	1.83	3.19	29.63	3	33	44	2173.69	1.81	1.72	22.75
Imp_{FPPR}	29	4	6	12	29	15048.54	26.63	6.56	125.32	6	12	29	15361.50	27.80	1.37	129.78
Imp_{Ours}	29	4	3	33	50	2652.74	2.30	3.11	32.95	3	33	50	2716.43	2.29	1.06	27.88
Imp_{FPPR}	31	4	6	12	31	16130.71	28.94	3.37	136.23	6	12	31	16443.60	30.19	1.56	142.69
Imp_{Ours}	31	4	3	33	52	2839.32	2.49	3.20	35.30	3	33	52	2907.78	2.48	1.24	29.22
Imp_{FPPR}	37	4	6	12	37	19466.94	35.03	2.43	172.56	6	12	37	19611.72	35.68	0.88	176.13
Imp_{Ours}	37	4	3	33	58	3314.88	2.93	2.45	31.32	3	33	58	3437.06	2.96	0.49	32.45
Imp_{FPPR}	41	4	6	12	41	21095.65	37.58	2.79	189.16	6	12	41	21756.74	39.80	0.84	201.77
Imp_{Ours}	41	4	3	33	62	3668.86	3.24	2.23	33.84	3	33	62	3783.47	3.33	0.56	35.49
Imp_{FPPR}	43	4	6	12	43	22472.30	40.59	2.24	207.05	6	12	43	22868.33	41.39	0.85	210.59
Imp_{Ours}	43	4	3	33	64	3857.07	3.41	2.23	35.02	3	33	64	3965.76	3.48	0.60	36.51
Imp_{FPPR}	47	4	6	12	47	24264.24	43.37	2.10	225.73	6	12	47	24955.58	46.01	0.66	239.89
Imp_{Ours}	47	4	3	33	68	4197.12	3.73	2.12	37.93	3	33	68	4336.85	3.84	0.67	39.78
Imp_{FPPR}	53	4	6	12	53	27655.34	50.63	1.86	272.55	6	12	53	28043.51	52.26	0.37	279.83
Imp_{Ours}	53	4	3	33	74	4701.09	4.19	1.75	40.46	3	33	74	4824.09	4.36	0.46	42.63

Table 3. Comparison of the relation search ($m = 3$, $n' = 5, 6$) with two strategies, Imp_{FPPR} and Imp_{Ours}

	n	n'	sol: yes							sol: no						
			D_{reg}	var	poly	mono	t_{trans}	t_{groe}	mem	D_{reg}	var	poly	mono	t_{trans}	t_{groe}	mem
Imp_{FPPR}	17	5	7	15	17	29408.19	46.53	218.87	723.08	7	15	17	29562.07	48.06	59.82	725.07
Imp_{Ours}	17	5	4	42	44	2680.14	2.21	485.10	596.46	4	42	44	2687.94	2.16	136.93	492.88
Imp_{FPPR}	19	5	7	15	19	32812.55	50.50	91.61	401.17	7	15	19	32300.00	54.03	41.80	348.01
Imp_{Ours}	19	5	4	42	46	3264.00	1.97	516.67	619.63	4	42	46	2922.50	2.67	182.92	492.82
Imp_{FPPR}	23	5	7	15	23	40168.90	64.67	70.46	475.55	7	15	23	39659.80	65.07	55.75	381.39
Imp_{Ours}	23	5	4	42	50	3572.00	3.01	157.86	323.60	4	42	50	3619.30	3.07	17.83	253.16
Imp_{FPPR}	29	5	7	15	29	50156.00	81.75	109.40	587.39	7	15	29	50403.80	80.99	50.75	530.53
Imp_{Ours}	29	5	4	42	56	4414.90	3.67	140.47	372.59	4	42	56	4356.70	3.82	20.03	278.07
Imp_{FPPR}	31	5	7	15	31	53222.10	84.08	70.64	547.86	7	15	31	53415.30	85.50	53.56	410.47
Imp_{Ours}	31	5	4	42	58	4781.80	3.99	130.07	362.76	4	42	58	4800.60	4.13	20.98	279.23
Imp_{FPPR}	37	5	7	15	37	63941.80	101.06	158.23	828.44	7	15	37	64215.10	103.29	88.29	690.51
Imp_{Ours}	37	5	3	42	64	5586.20	4.85	11.68	118.00	3	42	64	5496.80	4.87	6.85	57.52
Imp_{FPPR}	41	5	6	15	41	70895.30	113.85	230.40	889.70	7	15	41	71215.80	114.09	69.12	930.24
Imp_{Ours}	41	5	3	42	68	6042.50	5.33	13.26	126.19	3	42	68	5986.60	5.34	8.53	58.99
Imp_{FPPR}	43	5	6	15	43	75145.70	118.87	75.46	600.95	6	15	43	74671.20	118.31	39.69	615.72
Imp_{Ours}	43	5	3	42	70	6223.40	5.41	11.35	89.33	3	42	70	6470.90	5.74	8.21	56.86
Imp_{FPPR}	47	5	6	15	47	81488.60	128.63	65.03	674.87	6	15	47	81215.20	131.95	45.34	693.31
Imp_{Ours}	47	5	3	42	74	7043.30	6.07	9.57	109.38	3	42	74	7183.40	6.26	4.71	60.15
Imp_{FPPR}	53	5	6	15	53	91642.50	147.66	80.76	810.08	6	15	53	92314.60	150.41	23.31	814.76
Imp_{Ours}	53	5	3	42	80	8034.10	6.83	6.68	59.58	3	42	80	7849.50	6.96	1.36	59.91
Imp_{FPPR}	23	6	7	18	23	107008.67	163.45	3888.70	6656.13	7	18	23	105744.33	156.11	3309.43	5025.06
Imp_{Ours}	23	6	4	51	56	5270.00	4.36	5150.12	4791.31	4	51	56	5510.33	4.42	3082.15	4428.07
Imp_{FPPR}	29	6	7	18	29	136465.67	198.99	4511.74	6685.01	7	18	29	137194.33	204.07	1681.27	6528.03
Imp_{Ours}	29	6	4	51	62	6093.33	5.67	2848.46	3368.01	4	51	62	6263.33	5.76	932.65	2681.20
Imp_{FPPR}	31	6	7	18	31	145504.00	209.98	4664.25	7336.11	7	18	31	145700.33	206.29	1205.29	7276.85
Imp_{Ours}	31	6	4	51	64	6538.33	5.82	2811.99	3257.82	4	51	64	6916.67	6.09	1049.14	2616.21
Imp_{FPPR}	37	6	7	18	37	171914.33	248.24	4733.79	9777.27	7	18	37	175419.00	256.90	1126.29	9812.93
Imp_{Ours}	37	6	4	51	70	8223.00	6.77	1101.04	1327.00	4	51	70	8459.33	7.10	146.14	927.36
Imp_{FPPR}	41	6	7	18	41	189028.67	279.05	1045.53	4416.99	7	18	41	192778.33	266.44	653.92	3062.68
Imp_{Ours}	41	6	4	51	74	9297.67	7.87	953.60	1361.59	4	51	74	9246.00	8.31	87.61	896.38
Imp_{FPPR}	43	6	7	18	43	203094.33	298.13	1444.41	4288.28	7	18	43	199325.67	280.46	787.02	3796.57
Imp_{Ours}	43	6	4	51	76	9899.33	8.02	920.13	1340.39	4	51	76	8958.33	8.33	87.18	918.05
Imp_{FPPR}	47	6	7	18	47	222208.67	326.22	1278.79	4524.33	7	18	47	221999.67	326.08	463.62	3287.07
Imp_{Ours}	47	6	4	51	80	10789.00	9.06	858.66	1296.09	4	51	80	10438.33	9.24	80.54	919.39
Imp_{FPPR}	53	6	7	18	53	245891.33	366.92	2967.03	7311.44	7	18	53	248212.33	359.03	1857.65	6677.92
Imp_{Ours}	53	6	3	51	86	11748.00	10.48	34.82	151.04	3	51	86	11744.00	10.70	31.21	151.02

solve the system may be larger with our method than with Faugère *et al.*'s method for small parameters, the time required to *build* this system is always smaller. This is due to the much simpler structure of s'_{m+1} compared to s_{m+1} (lower degrees and less monomial terms).

4.2   Whole ECDLP Computation

In a next step, we also implemented the whole ECDLP algorithm with the two strategies Imp_{FPPR} and Imp_{Ours}. For n in $\{7, 11, 13, 17, 19\}$, we ran the whole attack using $m = 3$ and several values for n'. The orders of the curves we picked in our experiments are shown in Table 4 together with the experimental results for the best value of n', which turned out to be 3 in all cases. Timings provided

Table 4. Comparison of two ECDLP strategies, Imp_{FPPR} and Imp_{Ours}. The last two columns are computing time in seconds.

n	$\#E_{\alpha,\beta}$	Imp_{FPPR}	Imp_{Ours}
7	4*37	1.574	0.864
11	4*523	8.625	6.702
13	4*2089	49.698	31.058
17	4*32941	2454.470	1364.742
19	4*131431	22474.450	9962.861

Table 5. Trade-off for choosing m and n'. N: total number of variables. D: degree of regularity.

	probability to get an answer $\frac{2^{mn'}}{m!2^n}$	complexity $N^{\omega D}$
m increases	probability increases	D increases, N increases.
n' increases	probability increases	N increases.

in the table are in seconds and averaged over 20 experiments. Table 4 clearly shows that our method (Imp_{Ours}) is more efficient than Faugère *et al.*'s method (Imp_{FPPR}).

It may look strange that $n' = 3$ leads to optimal timings at first sight. Indeed, the ECDLP attacks described above use $mn' \approx n$ and a constant value for n' leads to a method close to exhaustive search. However, this is consistent with the observation already made in [9,10] that exhaustive search is more efficient than index calculus for small parameters. Table 5 also shows that while increasing n' increases the probability to have solutions, it also increases the complexity of the Gröebner basis algorithm. This increase turns out to be significant for small parameters.

5   Conclusion and Future Work

In this paper, we proposed a variant of Faugère *et al.*'s attack on the binary elliptic curve discrete logarithm problem (ECDLP). Our variant takes advantage of the symmetry of Semaev's polynomials to compute relations more efficiently. While symmetries had also been exploited in similar ECDLP algorithms for curves defined over finite fields with *composite* extension degrees, our method is the first one in the case of extension fields with *prime* extension degrees, which is the most interesting case for applications.

At Asiacrypt 2012, Petit and Quisquater estimated that Faugère *et al.*'s method would beat generic discrete logarithm algorithms for any extension degree larger than roughly 2000. We provided heuristic arguments and experimental data showing that our method reduces both the time and the memory required to compute a relation in Faugère *et al.*'s method, unless the parameters are very small. Our results therefore imply that Petit and Quisquater's bound can be lowered a little.

Our work raises several interesting questions. On a theoretical side, it would be interesting to prove that the degrees of regularity of the systems appearing in the relation search will not rise when n increases (in all our experiments for various parameter sizes, they were equal to either 3 or 4). It would also be interesting to provide a more precise analysis of our method and to precisely estimate for which values of the parameters it will become better than Faugère *et al.*'s method.

On a practical side, it would be interesting to improve the resolution of the systems even further. One idea in that direction is pre-computation. The relation search involves solving a large number of closely related systems, where only the value $x(R)$ changes from one system to the other. The transformation of Semaev's polynomial into a binary multivariate system could therefore be done in advance, and its cost be neglected. In fact, even the resolution of the system could potentially be improved using special Gröebner basis algorithms such as F_4 trace [18,14]. A second direction on the practical side is parallelization. A powerful feature of Pollard's ρ method and its variants is their highly-parallelized structure. Since our method saves memory compared to Faugère *et al.*'s method, it is also more suited to parallelization.

Using Gröbner basis algorithms to solve ECDLP is a very recent idea. We expect that the index calculus algorithms that have recently appeared in the literature will be subject to further theoretical improvements and practical optimizations in a close future.

References

1. National Security Agency: The case for elliptic curve cryptography (January 2009), http://www.nsa.gov/business/programs/elliptic_curve.shtml
2. Shanks, D.: Class number, A theory of factorization, and genera. In: 1969 Number Theory Institute (Proc. Sympos. Pure Math., vol. XX, State Univ. New York, Stony Brook, N.Y., 1969), Providence, R.I., pp. 415–440 (1971)
3. Pollard, J.M.: A Monte Carlo method for factorization. BIT Numerical Mathematics 15(3), 331–334 (1975)
4. Brent, R.P.: An improved Monte Carlo factorization algorithm. BIT Numerical Mathematics 20, 176–184 (1980)
5. Pollard, J.M.: Kangaroos, monopoly and discrete logarithms. Journal of Cryptology 13, 437–447 (2000)
6. Diem, C.: An index calculus algorithm for plane curves of small degree. In: Hess, F., Pauli, S., Pohst, M. (eds.) ANTS 2006. LNCS, vol. 4076, pp. 543–557. Springer, Heidelberg (2006)
7. Gaudry, P.: Index calculus for abelian varieties of small dimension and the elliptic curve discrete logarithm problem. Journal of Symbolic Computation 44(12), 1690–1702 (2009)
8. Diem, C.: On the discrete logarithm problem in elliptic curves. Compositio Mathematica 147, 75–104 (2011)
9. Faugère, J.-C., Perret, L., Petit, C., Renault, G.: Improving the complexity of index calculus algorithms in elliptic curves over binary fields. In: Pointcheval, D., Johansson, T. (eds.) EUROCRYPT 2012. LNCS, vol. 7237, pp. 27–44. Springer, Heidelberg (2012)

10. Petit, C., Quisquater, J.-J.: On polynomial systems arising from a Weil descent. In: Wang, X., Sako, K. (eds.) ASIACRYPT 2012. LNCS, vol. 7658, pp. 451–466. Springer, Heidelberg (2012)
11. Joux, A., Vitse, V.: Elliptic curve discrete logarithm problem over small degree extension fields. Journal of Cryptology, 1–25 (2011)
12. Faugère, J.C., Gaudry, P., Huot, L., Renault, G.: Using symmetries in the index calculus for elliptic curves discrete logarithm. IACR Cryptology ePrint Archive 2012, 199 (2012)
13. Semaev, I.: Summation polynomials and the discrete logarithm problem on elliptic curves. IACR Cryptology ePrint Archive 2004, 31 (2004)
14. Faugère, J.C.: A new efficient algorithm for computing Gröbner bases (F_4). Journal of Pure and Applied Algebra 139(1-3), 61–88 (1999)
15. Faugère, J.C.: A new efficient algorithm for computing Gröbner bases without reduction to zero (F_5). In: Proceedings of the 2002 International Symposium on Symbolic and Algebraic Computation, ISSAC 2002, pp. 75–83. ACM, New York (2002)
16. Faugère, J., Gianni, P., Lazard, D., Mora, T.: Efficient computation of zero-dimensional Gröbner bases by change of ordering. Journal of Symbolic Computation 16(4), 329–344 (1993)
17. Joux, A., Vitse, V.: Cover and decomposition index calculus on elliptic curves made practical - application to a previously unreachable curve over $\mathbb{F}_{p^6}$. In: Pointcheval, D., Johansson, T. (eds.) EUROCRYPT 2012. LNCS, vol. 7237, pp. 9–26. Springer, Heidelberg (2012)
18. Joux, A., Vitse, V.: A variant of the F4 algorithm. In: Kiayias, A. (ed.) CT-RSA 2011. LNCS, vol. 6558, pp. 356–375. Springer, Heidelberg (2011)

Statistics on Encrypted Cloud Data

Fu-Kuo Tseng, Yung-Hsiang Liu, Rong-Jaye Chen,
and Bao-Shuh Paul Lin

National Chiao-Tung University,
No.1001, Daxue Road, Hsinchu City 300, Taiwan
{fktseng,liuyh,rjchen}@cs.nctu.edu.tw,
bplin@mail.nctu.edu.tw

Abstract. As an increasing number of data is to be processed, out-sourcing data to the cloud environment becomes an appealing proposal to heighten the computation/storage efficiency, while avoiding costly and complicated system construction. However, it is necessary to encrypt the outsourced data to prevent the breaches of both data confidentiality and privacy. Most of the statistical procedures deal with the data in the cleartext form, making it hard to directly apply them to the data in the encrypted form. In this paper, we present a statistical framework to securely and efficiently obtain the statistics on encrypted cloud data through real-time processing. We build our framework on top of the searchable public-key encryption and provide detailed transformation of the statistical procedures for the plain data to those for the encrypted data. We provide detailed descriptions and examples of these transformed statistical procedures. Finally, we provide security analysis and performance evaluation of these transformed procedures and demonstrate the effectiveness and efficiency of the proposed framework.

Keywords: statistics, encrypted cloud data, efficient transformation of statistical procedures, statistical framework, online storage services.

1 Introduction

Statistics pertains to the collection, analysis, interpretation and presentation of data population. The purpose of statistics is to summarize the data population by sampling and to draw inferences about the population from the observed samples. The application areas of statistics cover a wide variety of disciplines such as economics, physics, computer science, and business. As an increasingly great amount of data need to be processed, outsourcing data to the cloud environment becomes an appealing proposal to increase the computation and storage efficiency as well as avoiding costly system construction and maintenance [1, 2].

While cloud computing brings in a promising future for big data statistics, it also brings along security and privacy risks, which should be considered seriously before adopting this paradigm shift [3, 4]. There are encryption techniques available [5, 6] to protect cloud data and services by transforming the plain data into an unintelligible form. However, most of the statistical methods deal with the

K. Sakiyama and M. Terada (Eds.): IWSEC 2013, LNCS 8231, pp. 133–150, 2013.
© Springer-Verlag Berlin Heidelberg 2013

data in the plain-text form, making it hard to directly apply these methods to the data in the ciphertext form. The primitive approach requires users to download all the encrypted cloud data files, decrypts them, and manipulates them to obtain the demanded statistical results. This approach is time-consuming and does not scale well. Therefore, there is a high demand for a secure *yet efficient* scheme to draw intended statistical results from the encrypted data collection.

On the one hand, homomorphic encryption schemes allow service providers to perform certain algebraic operations on the encrypted data without learning the underlying data content. Partially homomorphic encryption allows only one operation (either addition or multiplication) on encrypted data, while fully homomorphic encryption supports both addition and multiplication operations. These two operations in fully homomorphic encryption schemes can be employed to design the converting statistical procedures for the encrypted data to obtain the intended statistics. However, fully homomorphic encryption is still away from practical uses in terms of ciphertext size and computation efficiency [7, 8].

On the other hand, searchable public-key encryption schemes empower data senders to produce the searchable keywords of an encrypted data for a receiver by encryption. Later, the receiver can generate appropriate encrypted trapdoors of demanded search predicates to enable the service provider to efficiently collect and return the encrypted data by these searchable keywords. The service providers can only tell whether the searchable keywords of a data satisfy the encrypted trapdoors. They know neither the content of the searchable keywords nor the encrypted trapdoors. A number of efficient searchable public-key encryption schemes have been proposed to enable a rich set of search predicates such as equality, range, subset predicates and the conjunction of these predicates [9, 10].

Contribution. In this paper, we present a statistical framework to obtain the statistics on encrypted cloud data securely and efficiently. We build our framework on top of the searchable public-key encryption named hidden vector encryption (HVE) and provide detailed transformation of common statistical procedures for the plain data to the equivalent ones for the encrypted data. Finally, we not only provide security and performance analysis of the proposed framework but also make comparison with the primitive approach to demonstrate the effectiveness and efficiency of the proposed framework.

The rest of the paper is structured as follows. Related works are described in Section 2, while problem formulation is presented in Section 3. Next, our novel design is detailed in Section 4. Then the security analysis and performance evaluation are presented in Section 5. Finally, Section 6 concludes this paper by reiterating our contributions and addressing possible future work.

2   Related Works

In this section, we briefly review the descriptive and inferential statistics. Following that, we discuss homomorphic encryption schemes. Then we introduce the privacy-preserving retrieval of unencrypted and encrypted dataset. Finally, we elaborate on the building block of the proposed statistical framework.

2.1 Descriptive and Inferential Statistics

Descriptive statistics focuses on describing the summaries of a data collection. These summaries can be either the graphical presentations or the calculation of the summary statistics. The graphical presentations involve the use of the visual arts like tables or graphs to exhibit the data collection, while the calculation of the summary statistics usually associates with the counting techniques to obtain the measures of the data collection. The visual arts include the frequency table, histogram, scatter diagram, bar chart and pie chart. The statistical measures include the measures of central tendency, dispersion, skewness and kurtosis.

Inferential statistics employs sampling to estimate the characteristics of the whole population. Good sampling can use relatively small sampled data to measure the entire population. In addition, inferential statistics utilizes the sampled data to draw inferences about the population. These inferences include the parameters of the population by estimation techniques, the true/false assertion of the population by hypothesis tests, the relationships within data by regression analysis, and the dependency of two sets of data by correlation analysis. These inferences can be further extended to estimate the unobserved events or forecast the other population using extrapolation/interpolation techniques [11, 12, 13].

2.2 Homomorphic Encryption

Homomorphic encryption schemes allow service providers to perform certain algebraic operations on the encrypted data without learning the corresponding data content. The content of the encrypted operational result matches the result of the operation performed on the corresponding plain data. Partially homomorphic encryption allows only one operation (either addition or multiplication) on encrypted data, while fully homomorphic encryption supports both addition and multiplication operations. Homomorphic encryption schemes have been utilized to ensure the confidentiality of the processed data. The addition and multiplication in fully homomorphic encryption schemes can be employed to design the converting statistical procedures for the encrypted data to obtain the related statistics. However, fully homomorphic encryption is still away from practical uses in terms of ciphertext size and computation efficiency [7, 8].

2.3 Privacy-Preserving Retrieval of Unencrypted Dataset

In cryptography, Private Information Retrieval (PIR) protocols enable users to retrieve data from a server storing *unencrypted* dataset without revealing which data is retrieved [14, 15]. A trivial but less efficient PIR is to download the entire dataset and issue queries locally. Oblivious Transfer (OT), a stronger notion of PIR, further requires that the user should not obtain unnecessary data from the server [16]. If given an implementation of OT, secure multiparty computation (MPC) is possible to compute a function whose inputs are from different parties without revealing one's input to the others [17]. To obtain statistics on the encrypted data, one should ask servers to search on the encrypted data based on the specified predicate. PIR and OT protocols are not applicable for this purpose because the intended (encrypted) data should be pointed out to the server, which is not possible because the data is encrypted and no search procedures are provided. On the other hand, secure MPC protocols require interactions among involved servers. Because we would like to obtain statistics in an efficient and non-interaction manner, MPC protocols are not suitable to our application.

On the other hand, in database systems, there are methods such as randomization and k-anonymization for privacy-preserving data mining [18]. Typically, these methods make use of the transformation on the dataset to preserve data and search privacy, which reduces the granularity of this dataset. The randomization methods transform dataset by adding noise to mask the attribute values. However, only the aggregate distributions can be preserved and individual attribute values of the dataset cannot be recovered. In addition, the k-anonymization usually reduces the granularity/accuracy of data by generalization and suppression mechanisms. These mechanisms require preprocessing in order to preserve data anonymity. Because our design goals aim at real-time retrieval of both descriptive (usually connected with counting techniques) and inferential (usually associated with sampling techniques) statistics, the data preprocessing and lower data granularity/utility do not fit well for our application.

2.4   Privacy-Preserving Retrieval of Encrypted Dataset

Searchable encryption enables users to encrypt keywords in such a way that (1) with appropriate trapdoor for a keyword, one can retrieve all the files with this keyword, and (2) without appropriate trapdoor, the retrieval will fail. In addition, the trapdoors can only be generated with the knowledge of a secret and searching reveals nothing about the content of the keyword (which is known as the search pattern) except for the set of encrypted files containing specific encrypted searchable keyword (which is referred to as the access pattern).

There are two types of non-interactive searchable encryption schemes: searchable private-key encryption and searchable public-key encryption. In searchable private-key encryption schemes, the user *possesses* the data and can organize the data in any convenient way, including customized data structures before encryption. Later, the user encrypts the data (and corresponding data structures) using his/her private key and stores them to the server. Only someone with this private key can efficiently access the encrypted data in the server [19, 20]. In searchable public-key encryption, each user can securely generate searchable keywords of a file and delegate encrypted trapdoor of a search query to retrieve the intended data. The sender, who uses the public-key of the receiver to produce searchable keywords of a file, can be different from the receiver who has the corresponding private key. The receiver can generate the trapdoor of the demanded search predicate by his/her own private key to retrieve the interested data. The service providers can employ the received trapdoor and the stored searchable keywords to determine whether the searchable keywords satisfy the predicates in the trapdoor. The contents of the searchable keywords and the encrypted trapdoor remain secret to the service providers [9, 10].

A number of searchable public-key encryption schemes have been proposed to enable a rich set of search predicates including equality, range, subset and the conjunction of these predicates. Boneh *et al.* first proposed the *public-key encryption with keyword search* (PEKS) scheme [9] to support equality search predicates. Later, Park *et al.* proposed *public-key encryption with conjunctive keyword search* scheme to enable the conjunctions of equality predicates. To provide richer predicates, Boneh and Waters [10] devised a hidden vector encryption (HVE) scheme to manage subset, range search predicates, and the conjunction of these predicates. For efficient interactive searchable public-key encryption, iPEKS was proposed to increase storage and search efficiency [21].

In this paper, we adopt the HVE scheme constructed more efficiently by Iovino and Persiano [22] as our building block. In their scheme, each searchable keyword vector S_x is associated with a hidden vector $x = (x_1, x_2, \ldots, x_n) \in \{0,1\}^n$. Each trapdoor T_y is associated with a predicate vector $y = (y_1, y_2, \ldots, y_n) \in \{0, 1, *\}^n$, where the symbol "$*$" denotes the wildcard. Define the predicate function as $P_y(x)$ is 1 if $x_i = y_i$ for all $y_i \neq *$ and 0, otherwise. A encrypted trapdoor T_y matches an searchable vector S_x if $P_y(x) = 1$, namely, if all the non-wildcard symbols of the predicate vector y are the same as those of the keyword vector x. HVE schemes consist of four algorithms: (1) $\texttt{Setup}(1^k, n)$ takes as input a security parameter 1^k and the length of the hidden vector $n=poly(k)$ and outputs the private key SK. (2) $\texttt{SE}(PK, x, 'True')$ uses PK to generate the searchable keywords S_x associated with x with the signaling message '$True$'. (3) $\texttt{Trapdoor}(SK, y)$ produces encrypted trapdoor T_y using SK. (4) $\texttt{Test}(T_y, S_x)$ returns '$True$' if $P_y(x)=1$; otherwise, return a random message.

3   Problem Formulation

This section begins by defining the targeted system model including the system entities and related operations. Following that, the notations used throughout the paper are explained. Finally, we address the security/privacy threats to the search and define our design goals of th eproposed statistical framework.

3.1   System Model

We consider a general enterprise cloud storage architecture containing two system entities. (See Fig. 1)

1. *Cloud Storage Client (CSC)* stores a large number of data in the cloud. These data are either generated on his/her own or sent from other CSCs. CSCs would like to protect the security/privacy of their data and their search queries, while utilizing these protected data efficiently.
2. *Cloud Storage Provider (CSP)* provides search-based store/retrieval services for CSCs. CSPs are assumed to be honest-but-curious. They follow the specified protocols, but may attempt to learn extra information from the information flow for their own purposes. To tackle malicious CSPs, verifiable store and retrieval services can be integrated. [23, 24]

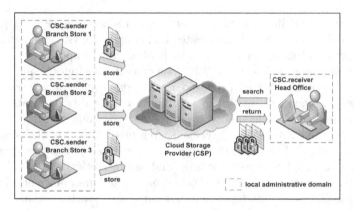

Fig. 1. Cloud Storage Access Model

The CSC is further divided into two roles, the sender and the receiver, based on their operations. The sender (say the branch store) creates and sends encrypted searchable transactions to the CSP. The CSP receives/stores the encrypted searchable transactions. The receiver (say the head office) generates encrypted trapdoors of the search predicates for the intended data and corresponding statistical procedures and sends them to the CSP. The CSP performs search on receiving the encrypted trapdoors and return the requested data or the corresponding statistics. The head office would like to know some facts about the transactions at any time such as the sales of individual items, the performance of the branch stores, the relationship between the two items bought together, and the correlation between the market expenses and the sale revenues.

We assume this enterprise maintains its own universal keyword set $\mathcal{W}$ shared among head office and branch stores. The enterprise keeps a collection $\mathcal{F}$ of the encrypted transaction records from the branch stores in the cloud. The keywords set $\mathcal{W}$ of a transaction record is composed of m keyword fields, each of which has n_i possible values. Thus, we have $n = \sum_{i=1}^{m} n_i$ possible keyword values, denoted as $\{w_{i,j}\}_{1 \leq i \leq m, 1 \leq j \leq n_i}$, to choose for a transaction record. We will use the keyword and keyword values interchangeably throughout the paper. In Table 1, $\mathcal{W}$ contains three keyword fields: 'w_1:amount of money', 'w_2:method of payment' and 'w_3:item bought', thus m is 3. For the 'amount of money' field, there are five possible values: '$w_{1,1}$:amount of money:10', '$w_{1,2}$:amount of money:15' and '$w_{1,3}$:amount of money:20', '$w_{1,4}$:amount of money:25' and '$w_{1,5}$:amount of money:30', thus n_1 is 5. Similarly, n_2 is 4, while n_3 is 4. Therefore, the size of $\mathcal{W}$ is $n = \sum_{i=1}^{3} n_i = 13$. The transaction record also includes supplement data like the signature of the customer and the soft copy of the receipt. Thus, the branch store and the head office can make use of the searchable part of the transaction record to obtain statistics and fetch the corresponding supplement data. The supplement data is protected by symmetric encryptions.

Each transaction record f in $\mathcal{F}$ is given a unique transaction identifier id, denoted as f_{id}, to support the management through these identifiers. For the collection of transaction records, $\mathcal{W}_{id}$ is a set containing the keywords

Table 1. Transaction Record Format

keyword type	quantitative uni-value	qualitative uni-value	qualitative multi-value
	ex:amount of money(\$)	ex:method of payment	ex:bought items
keyword set	$w_{1,1}$ $w_{1,2}$ $w_{1,3}$ $w_{1,4}$ $w_{1,5}$	$w_{2,1}$ $w_{2,2}$ $w_{2,3}$ $w_{2,4}$	$w_{3,1}$ $w_{3,2}$ $w_{3,3}$ $w_{3,4}$
keyword value	10 15 20 25 30	cash credit debit check	beer apple diaper egg
supplement	[image] the signature of the customer		
image data	[image] the soft copy of the corresponding receipt		

specified for f_{id}, while $\mathcal{F}_{w_{i,j}}$ denotes the set of the transaction identifiers where the keyword value $w_{i,j}$ is specified for the corresponding transaction records. In addition, a transaction records can be specified one single value (*uni-value*) or multiple values (*multi-value*) for the keyword field. In Table 1, the keywords in 'w_1:amount of money' and and 'w_2:method of payment' are mutually exclusive, only one value/method can be specified for a transaction. Multiple keyword values in 'w_3:item bought' are selected to express multiple items are involved. For the search queries, the universal set of all possible queries is denoted as $\mathcal{Q}$. One search query contains one or more predicates specified conjunctively. Search predicates include the conjunction of disjunction of predicates of the keyword fields. $\mathcal{Q}_{id}$ denotes the set of the transaction identifiers whose keywords satisfy the $rtrvl\_rule_{id}$, while $|\mathcal{Q}_{id}|$ denotes the size of the set $\mathcal{Q}_{id}$.

3.2   Threat Model

There are mainly four security aspects which our statistical framework cares about in the face of honest-but-curious adversaries: (1) the content of the stored data (*the semantic security*); (2) the content of the searchable keywords (*the attribute hiding*); (3) the content of the encrypted trapdoors (*the search pattern*); (4) the search result on encrypted cloud data including the corresponding statistics (*the access pattern*). In addition, the adversary may the search results to imply what keywords may have been specified for the file or what predicates have been appointed in the trapdoors. The proposed framework should take of these aspects to provide secure and efficient retrieval of encrypted cloud data.

3.3   Goal Model

The design goals are (1) **Search Privacy**: The searchable keywords of a transaction record reveal nothing about the content of the underlying keywords. Similarly, the encrypted trapdoors leak no information about the underlying predicates; (2) **Non-interactive Search**: The retrieval should be processed in a request-and-response manner. No further interaction between the CSP and the CSC is required; (3) **Data utility**: The search should be processed without decrypting the transaction records first. The retrieval includes descriptive and inferential statistics; (4) **Real-time**: The data can be counted in the corresponding statistics once it is stored in the CSPs. No off-line preprocessing is required;

(5) **Efficiency**: The transformed procedures should be carried out efficiently by the CSP to help users retrieve their records and related statistics.

4 Proposed Statistical Framework

Overview. We provide basic statistical procedures to show how our proposed framework enables statistical procedures on encrypted searchable keywords of encrypted transactional data. These procedures include the storing transformation, the retrieval transformation such as the counting and sampling transformation, the transformation of descriptive statistical procedures and inferential statistical procedures. The storing transformation indicates how transaction records are processed to enable the following procedures. The counting and sampling transformation involve basic probability calculation. Following that, the descriptive statistical procedures contain the measures of central tendency, dispersion, skewness and kurtosis. Finally, the inferential statistical procedures deal with the estimate, correlation analysis, regression analysis and hypothesis testing. We detail the transformation from the procedures for the plain records to the ones for the encrypted records. In addition, we provide concrete examples for each of the procedures to explain their effectiveness. The explanation of the transformed procedures applies the dataset as shown in Table 2.

Table 2. Example of Transaction Records (*The Searchable Part*)

keyword set	$w_{1,1}$	$w_{1,2}$	$w_{1,3}$	$w_{1,4}$	$w_{1,5}$	$w_{2,1}$	$w_{2,2}$	$w_{2,3}$	$w_{2,4}$	$w_{3,1}$	$w_{3,2}$	$w_{3,3}$	$w_{3,4}$
keyword value	10	15	20	25	30	cash	credit	debit	check	beer	apple	diaper	egg
str_rule_1	0	0	0	0	1	0	1	0	0	1	1	1	1
str_rule_2	0	0	1	0	0	1	0	0	0	1	1	0	1
str_rule_3	0	1	0	0	0	0	0	0	1	0	1	1	0
str_rule_4	0	0	0	1	0	0	0	1	0	1	0	1	0
str_rule_5	1	0	0	0	0	0	1	0	0	0	1	0	1
str_rule_6	0	0	0	0	1	0	1	0	0	0	1	0	1
str_rule_7	0	0	0	1	0	0	1	0	0	0	1	1	1
str_rule_8	0	0	0	0	1	0	0	0	1	1	1	1	0
str_rule_9	0	1	0	0	0	0	1	0	0	1	1	0	0
str_rule_{10}	0	1	0	0	0	0	0	1	0	0	0	1	1

4.1 Storing Transformation

To store a transaction f_{id} with str_rule_{id}, the branch store executes Storing (See Table 3) to produce transaction vector x. The branch store applies SE(SK, x, '$True$') to generate encrypted searchable keywords S_x. str_rule_{id} is an array of length n. If $w_{i,j}$ is specified in W_{id}, set $str\_rule[i][j]$ as 1 and leave the rest as 0. x_{id} is the transaction vector of length n from str_rule_{id}. (See Table 4)

Table 3. Algorithm - Storing

Algorithm Storing(SK, str_rule)
01: $x \leftarrow$ Rule2BinVec($'storing'$, str_rule,$'DC'$)
02: **return** $S_x \leftarrow$ SE(SK, x, $'True'$)

4.2   Retrieval: Counting and Sampling Transformation

There are two parts for search trapdoors: *rule* and *rule_type*. *rule* specifies which keywords are involved, while *rule_type* denotes the operator of this *rule*. The operator for one keyword field can be 'Q1', 'Q0', 'RG', 'DC', which denotes 'equality(at least)', 'equality(exactly)', 'range' and 'don't care' respectively. $x_{i,j}$ is initially set as '*'. If $w_{i,j}$ is specified in Q_{id}, $rtrvl\_rule[i][j]$ and $x_{i,j}$ are set as 1, and the rest elements (all $rtrl\_rule[i][j']$ and $x_{i,j'}$) are set as 0 for all *rule_type*. In addition, if the operator specified in $rule\_type[i]$ is 'Q0' or 'RG', fill the remaining $x_{i,j'}$ as 0; otherwise, leave all the $x_{i,j'}$ unchanged as '*' (See Table 4).

Table 4. Algorithm - Rule2BinVec

Algorithm Rule2BinVec($action$, $rule$, $rule\_type$)
01: $l \leftarrow 0$, $k \leftarrow 0$, $p \leftarrow 0$, $\{x_i\} \leftarrow 0$, $\{y_i\} \leftarrow *$
02: **if** ($action$ == $'storing'$)
03:    **while** ($k < m$)
04:      $l \leftarrow 0$;
05:      **while** ($l < n_k$) **if**($rule[k][l] \neq NULL$) $x_{k,l} \leftarrow 1$
06:    **end of while**
07: **return** x
08: **else if** ($action$ == $'retrieval'$)
09:    **while** ($k < m$)
10:      $l \leftarrow 0$, $p \leftarrow 0$
11:      **if**($rule\_type[k]$ == 'Q1')
12:        **while** ($l < n_k$)
13:          **if**($rule[k][l] \neq NULL$){ $y_{k,l} \leftarrow 1$, l++ }
14:      **else if**($rule\_type[k]$ == 'Q0')
15:        **while** ($l < n_k$)
16:          **if**($rule[k][l] \neq NULL$){ $y_{k,l} \leftarrow 1$, l++ }
17:          **else** {$y_{k,l} \leftarrow 0$, l++ }
18:        **end of while**
19:      **else if**($rule\_type[k]$ == 'RG')
20:        $set \leftarrow 0$, $\{y_i\} \leftarrow 0$
21:        **while** ($p < n$)
22:          **if**(set == $0 \wedge rule[k][p]$ == 1)
23:            $y_{k,p} \leftarrow 1$, $set \leftarrow 1$, p++
24:          **else if**(set == $1 \wedge rule[k][p]$ == 1)
25:            $y_{k,p} \leftarrow 1$, $set \leftarrow 0$, p++
26:          **if**(set == 1) $y_{k,p} \leftarrow 1$, p++
27:        **end of while**
28:    **end of while**
29: **return** y

Table 5. Algorithm: Counting

Algorithm Counting(SK, cnt_rule, $rule\_type$)
01: $y \leftarrow$ Rule2BinVec('$retrieval$', cnt_rule, $rule\_type$)
02: Let t be the number of 1s in y, $i \leftarrow 0$
03: **if** ($rule\_type$== 'Q1' || $rule\_type$== 'RG')
04:    **while** ($i < t$)
05:      y^i is the y preserving the ith 1 and set the other 1s as *.
06:      $T_{y^i} \leftarrow$ Trapdoor(SK, y^i)
07:      For each S_x of f_{id}, **if** (Test(T_{y^i}, S_x) == '$True$'), $\mathcal{QF} \Leftarrow id$, i ++
08:    **end of while**
09: **else**
10:    $T_y \leftarrow$ Trapdoor(SK, y)
11:    For each S_x of f_{id}, **if** (Test(T_y, S_x) == '$True$'), $\mathcal{QF} \Leftarrow id$, i ++
12: **return** $|\mathcal{QF}|$

Table 6. Frequency Table of Encrypted Transaction Records

rule	rule_type	frequency ($c_{i,j}$)	cumulative frequency ($C_{i,j}$)
$|F_{w_{1,1}}|$	{Q0,DC,DC}	Counting(SK, 10000********)= 1	1
$|F_{w_{1,2}}|$	{Q0,DC,DC}	Counting(SK, 01000********)= 3	4
$|F_{w_{1,3}}|$	{Q0,DC,DC}	Counting(SK, 00100********)= 1	5
$|F_{w_{1,4}}|$	{Q0,DC,DC}	Counting(SK, 00010********)= 2	7
$|F_{w_{1,5}}|$	{Q0,DC,DC}	Counting(SK, 00001********)= 3	10

Counting Transformation. For quantitative uni-value keyword field (See Table 1), there are five possible values. We can build the corresponding frequency table by issuing T_y for each of these five keyword values. (See Table 5,6). In addition, the retrieval of all records, the retrieval of the records with exactly $w_{3,1}$ and $w_{3,3}$, the retrieval of the records with either $w_{3,1}$ or $w_{3,3}$, the retrieval of the records within the range from $w_{3,1}$ to $w_{3,3}$, the retrieval of the records without $w_{3,1}$, and the conditional probability of $w_{3,3}$ given $w_{3,1}$ (See Table 7).

Sampling Transformation. There are two common ways of sampling: random sampling and systematic sampling. Random sampling uses a random number generator to collect samples from the population until the required number is achieved (See Table 8). The first part decides the targeted population like all the transaction records paid by the credit card or all the records whose total amount is smaller than 25. Given random number generator outputs $\{20, 31, 3, 18, 9\}$, thus $s\mathcal{QF}$={$str\_rule_1, str\_rule_2, str\_rule_4, str\_rule_9, str\_rule_{10}$}. Systematic sampling first decides the population and chooses one random record as the first sample. Skip the next $j-1$ records and pick the jth records as the second sample. Keep on including the jth records from the previous sample as one sample until the sample size is achieve. Given the random number generator

Table 7. Frequency Table of Encrypted Transaction Records - Multiple Keywords

rule	rule_type	frequency
$\lvert F \rvert$	{DC,DC,DC}	$\texttt{Counting}(SK,\ y = \text{*************}) = 10$
$\lvert F_{w_{3,1} \wedge w_{3,3}} \rvert$	{DC,DC,Q0}	$\texttt{Counting}(SK,\ y = \text{*********1010}) = 3$
$\lvert F_{w_{3,1} \vee w_{3,3}} \rvert$	{DC,DC,Q1}	$\texttt{Counting}(SK,\ y = \text{*********1*1*}) = 8$
$\lvert F_{w_{3,1} : w_{3,3}} \rvert$	{DC,DC,RG}	$\texttt{Counting}(SK,\ y = \text{*********111*}) = 8$
$\lvert F_{\neg w_{3,1}} \rvert$	{DC,DC,Q1}	$\texttt{Counting}(SK,\ y = \text{*************}) - $ $\texttt{Counting}(SK,\ y = \text{*********1***}) = 5$
$\lvert F_{w_{3,1} \mid w_{3,3}} \rvert$	{DC,DC,Q1}	$\texttt{Counting}(SK,\ y = \text{*********1*1*})/$ $\texttt{Counting}(SK,\ y = \text{***********1*}) = 3/6$

outputs 14 for the first time, so we have str_rule_5 as the first sample. Because we would like to have 5 samples from the population, the skip j is $10/5 = 2$, thus $sQF=\{str\_rule_5, str\_rule_7, str\_rule_9, str\_rule_1, str\_rule_3\}$.

Table 8. Algorithm - Sampling

Algorithm Sampling($smpl\_rule, smpl\_method, smpl\_size$)

▶ Obtain the sample frame under the $smpl\_rule$

01: $y \leftarrow$ Rule2BinVec(*'retrieval'*, $smpl\_rule, rule\_type$)
02: $T_y \leftarrow$ Trapdoor(SK, y)
03: For each S_x, if (Test(T_y, S_x) == *'True'*), $QF \Leftarrow x$

▶ Obtain the sample of size $smpl\_size$ by $smpl\_method$

04: $sQF \leftarrow \emptyset, i \leftarrow 0, j \leftarrow 0, k \leftarrow 1$
05: $y \leftarrow$ Rule2BinVec(*'retrieval'*, $smpl\_rule, rule\_type$)
06: **if**($smpl\_method$ == *'random'*)
07:   **while**($\lvert sQF \rvert <smpl\_size$)
08:     $sQF \Leftarrow QF[\text{Rand}()\%\lvert QF \rvert]$
09:   **else if**($smpl\_method$ == *'systematic'*)
10:     $i=\text{Rand}()\%\lvert QF \rvert, j = \lvert QF \rvert / smpl\_size$
11:     **while**($\lvert sQF \rvert < smpl\_size$)
12:       **if** (sQF == $\emptyset$) $sQF \Leftarrow QF[\text{Rand}()\%\lvert QF \rvert]$
13:       **else** $sQF \Leftarrow QF[(i + j * k{+}{+})\%\lvert QF \rvert]$
14:     **end of while**
15: **return** sQF

4.3   Retrieval: Descriptive Statistics

Descriptive statistics focuses on describing the basic summaries of a data collection. These summaries can be either the graphical presentations or the calculation of the summary statistics. We focus on the calculation of the summary statistics usually associated with the counting techniques to obtain the measures of the data collection. The graphical presentation can be easily plotted once the summary statistics is obtained. The statistics measures include the measure of central tendency, dispersion, skewness and kurtosis.

Measure of Central Tendency. The mean, taken as the sum of the numbers divided by the size of the data collection, represents the central tendency of a collection of numbers. From Table 6, population mean μ is calculated as $\mu = \sum_{j=1}^{n_1}(|F_{w_{1,j}}|*w_{1,j})/|F| = (1 \times 10+3 \times 15+1 \times 20+2 \times 25+3 \times 30)/10 = 21.5$. We can also compute the geometric mean or harmonic mean based on the characteristic of the keyword field. To find the median Me of a data population, find the half of the number of transaction records $|F|/2 = 5$ and find the interval $[w_{1,j}, w_{1,j'}]$ where the median occurs. If there is odd number of records, the median is the one sits right in the middle. If there is an even number of records, then there is no single middle value; the median is then defined as the average of the two middle values. In this example, the median is $(w_{1,3} + w_{1,4})/2 = 22.5$. Finally, to find the mod of a data collection, look up the frequency table and the row(s) with the maximum number of frequency is the mod. If the values are uniformly distributed, the mod does not exist. There can be multiply values for mode like the mode in the example is $w_{1,2}=15$ and $w_{1,5}=30$.

Measure of Dispersion. The range R denotes the difference between the maximum and minimum specified values for one keyword field. In Table 6, R is $30-10=20$ for the keyword field w_1. The interquartile-range (IQR) works in the similar way. The IQR takes the difference between the values situated at the 3/4 and 1/4 position from the sorted values. In the example, the IQR can also be computed as $30-15=15$. The variance represents how far the numbers are spread out with respect to the mean. The population variance of w_1 is $\sigma^2 = \sum_{i=1}^{k} c_i(m_i - \mu)^2/N = \sum_{j=1}^{n_i} |F_{1,j}|(w_{1,j} - \mu)^2/|F| = 55.8\overline{3}$ The standard derivation is the square root of the variance computed as $\sqrt{\sigma^2} = 7.472$.

Measure of Skewness and Kurtosis. The measure of skewness defines the degree to which a frequency distribution is symmetric or not. It is measured by $\beta_1=\sum_{i=1}^{n}(x_i - \mu)^3/(n\sigma^3)=\sum_{j=1}^{n_i} |F_{1,j}|(w_{1,j} - \mu)^3/(|F|\sigma^3)$. If $\beta_1=0$, the distribution is symmetric. Otherwise, the distribution is positively skewed for $\beta_1>0$ or negatively skewed for $\beta_1<0$. The measure of kurtosis defines the degree to which a frequency distribution is flat (low kurtosis) or peaked (high kurtosis) is measured by $\beta_2=\sum_{i=1}^{n}(x_i - \mu^4)/(n\sigma^4)=\sum_{j=1}^{n_i} |F_{1,j}|(w_{1,j} - \mu)^4/(|F|\sigma^4)$. If $\beta_2 > 3$, the distribution is leptokurtosis and if $\beta_2=3$, the distribution is mesokurtosis. Finally, if $0 \leq \beta_2 < 3$, the distribution is platykurtosis.

4.4   Retrieval: Inferential Statistical Procedures

Inferential statistics employs sampling to estimate/assert the characteristics of the whole data population. Inferential statistics includes the parameters of the population by estimation techniques, the true/false assertion of the population by hypothesis tests, the relationships within data by regression analysis, and the dependency of two sets of data by correlation analysis. These inferences can be further extended to estimate the unobserved events or forecast the other population using extrapolation and interpolation techniques.

Estimate. The sample statistics can be obtained in the similar way except for the targeted data population. Use the `sampling` algorithm to decide the targeted population to be estimated. Build the frequency table of the sampled data and perform the `counting` algorithm to obtain the sample mean and sample variance. Use $\bar{x}$ to estimate μ, while use s to estimate σ. Find the sample set sQF as the one in Table 9. Then calculate $\bar{x}=\frac{\sum_{j=1}^{n_1}(|F'_{w_{1,j}}|w_{1,j})}{|sQF|}=\frac{2\times15+20+25+30}{10}=21$ and $s^2=\frac{\sum_{i=1}^{k}F'_{w_{1,j}}(w_{1,j}-\bar{x})^2}{|sQF|-1}=415$, where $|F'_{w_{1,j}}|$ denotes the frequency of the sampled records having $w_{1,j}$. The sample standard derivation is $\sqrt{s^2}=20.372$.

Table 9. Example of Sampled Transaction Records by Random Sampling

keyword set	$w_{1,1}$	$w_{1,2}$	$w_{1,3}$	$w_{1,4}$	$w_{1,5}$	$w_{2,1}$	$w_{2,2}$	$w_{2,3}$	$w_{2,4}$	$w_{3,1}$	$w_{3,2}$	$w_{3,3}$	$w_{3,4}$
keyword value	10	15	20	25	30	cash	credit	debit	check	beer	apple	diaper	egg
str_rule_1	0	0	0	0	1	0	1	0	0	1	1	1	1
str_rule_2	0	0	1	0	0	1	0	0	0	1	1	0	1
str_rule_4	0	0	0	1	0	0	0	1	0	1	0	1	0
str_rule_9	0	1	0	0	0	0	1	0	0	1	1	0	0
str_rule_{10}	0	1	0	0	0	0	0	1	0	0	0	1	1

Statistical Hypothesis Testing. Different from the estimation of the mean and variance. At times, it is required to decide whether a measure meets a certain standard. We provide an example of a hypothesis test about a single mean. The method can further extend to the hypothesis test about a one or more means/variances. Assume we have 50 random samples. We can calculate the sample mean as 875 by estimation while the sample standard derivation is 21. Assume the level of significance is 5% ($\alpha=5\%$), we would like to test whether the population mean is 885. The hypothesis is $H_0 : \mu=885$ and $H_1 : \mu<885$. Assume the population is a normal distribution. We have $z^*=\frac{\bar{x}-\mu_0}{\sqrt{\frac{\sigma^2}{n}}}=880+1.96*\frac{21}{\sqrt{50}}=-1.684$.

Because we have $|z^*|<z_{0.025}$, we do not reject H_0.

Correlation Analysis. Let x denote the advertisement expense for one product, and y denotes the number of this products sold. Ten data tuples are shown in Table 10. We would like to know how advertisement expense and sales are related. Values of the correlation coefficient are always between -1 and $+1$, and the larger the absolute value of the correlation coefficient is, the stronger the two variables are correlated. For Pearson product moment correlation coefficient, $\rho_{w_x,w_y}=\frac{\sigma_{x\wedge y}}{\sigma_{w_x}\rho_y}$, where $\rho_{x,y}=\sum_{i=1}^{N}(x_i-\bar{x})(y_i-\bar{y})/N$. We have $\bar{x}=183$, $\bar{y}=197.2$, $\sigma_{x,y}=1092.4$, $\sigma_x=30.676$ and $\sigma_y=36.589$. Therefore, $\rho_{x,y}=\frac{1092.4}{30.676\times36.589}=0.9733$.

Table 10. Example of Advertisement Expenses and Sales

Month	1	2	3	4	5	6	7	8	9	10
Expenses (x)	150	160	180	160	190	210	180	160	180	260
Sales (y)	156	180	190	170	198	250	189	168	191	280

Regression Analysis. Regression analysis is a statistical technique to estimate the relationships among variables. Linear regression uses linear prediction functions for the data modeling, in which finite unknown parameters are estimated from the data. We consider the linear regression by ordinary least squares method (OLS) involving two variables. For a data set of n points $\{(x_i, y_i)\}_{i=1}^n$, the regression function is a straight line $y = \alpha + \beta x$ such that $\sum_{i=1}^n (y_i - \alpha - \beta x_i)^2$ is minimized. The estimation $\hat{\beta}$ of β can be written as $\hat{\beta} = \frac{\sum_{i=1}^n (x_i - \overline{x})(y_i - \overline{y})}{\sum_{i=1}^n (x_i - \overline{x})^2}$, and $\hat{\alpha} = \overline{y} - \hat{\beta}\overline{x}$. The estimated regression function $y = \hat{\alpha} + \hat{\beta}x$ can be calculated by $\hat{\beta} = \frac{\sum_{i=1}^n (x_i - \overline{x})(y_i - \overline{y})}{\sum_{i=1}^n (x_i - \overline{x})^2} = \frac{1092.4}{941} = 1.1609$. $\hat{\alpha} = \overline{y} - \hat{\beta}\overline{x} = 197.2 - 1.1609 \times 183 = -15.2433$. The head office can estimate the sales when the budge is approved. Similarly, The head office can assess the budget of the advertisement expense in order to achieve the pre-set sales performance.

5   Security Analysis and Performance Evaluation

In this section, we present the security analysis of the proposed statistical framework by giving formal arguments. Moreover, we evaluate the storage, computation and communication costs for procedures and make comparison with the primitive approach to show its efficiency and practicality.

5.1   Security Analysis

On the one hand, the security of the content of the supplement data relies on the AES encryption scheme. The recovery of AES-128 key requires a computational complexity of $2^{126.1}$ which is computation-infeasible [25]. On the other hand, the security of the searchable keywords of a data is based on the security of the underlying HVE scheme. The HVE scheme in our framework achieves semantic security and attribute hiding property in the selective and honest-but-curious models [22]. For semantic security, the adversary cannot distinguish the ciphertext of a random string from the ciphertext of a user-defined string. For attribute hiding property, the adversary cannot distinguish the searchable keyword string x or an random string from the corresponding ciphertexts even when the adversary has access to the key generation procedure. Therefore, the CSP could not tell the content of the supplement data, the content of the ciphertext and the content of the searchable keyword. For the trapdoor privacy, the CSP only knows the set of files whose values of the non-'*' positions agree with the ones of the specified positions in the trapdoor. It is computation infeasible for the CSPs to obtain the content of the non-'*' positions in the trapdoor by guessing the meaning of each non-'*' position. Similarly, the search results (together the corresponding statistics) is meaningless for the CSP. Finally, to protect non-'*' positions in the trapdoor, inner product encryption (IPE) was proposed with doubled size of searchable keywords and much more processing time [26].

5.2   Performance Analysis

The cloud environment is simulated on the OpenStack platform at National Chiao-Tung University [27]. We solicit five virtual machines each of which has 2 QEMU Virtual CPU version 1.0 (2000.08-MHz K8-class CPU) running Freebsd 8.3-RELEASE-p3 to play the CSP role. On the other hand, we use local server with Intel Xeon processor E5620 at 2.40 GHz running Ubuntu 11.10 to play the CSC role. Both CSCs and CSPs use GMP [28] and PBC [29] libraries. We use a supersingular curve over one base field of size 512 bits and the embedding degree is 2. The size of an searchable keyword is equals to 2 elements in the field, which is 1026 bits in length. The AES-CTR ciphertext is as long as the plaintext.

For storing, the computation time is proportional to the length of the keyword vector x. In our example, the length is 13 (See Table 1), thus the time require to store each of the *str_rule* is in average 67.42 ms. For counting and sampling, the time is proportional to the number of non-asterisk values (which is 0 or 1) in the predicate vector. Computing each row of the frequency table in Table 6 takes $39.02+10.65N$ *ms* for 5 non-'*' keyword values, where N is the total number of stored transactions in the CSP. The time required for each trapdoor in Table 7 composes two parts: one is executed by the user to generate an encrypted trapdoor, while the latter is carried out by the CSP to search through the store transaction records. On the other hand, the first part of sampling algorithm works the same as the counting algorithm, while the second part involves the random sampling or systematic sampling. For obtaining statistics, descriptive and inferential statistics involves constructing the frequency table first and use the values from the table rows to obtain further statistics. The computation time is related to the respective counting and sampling time.

To make clear comparison between the primitive approach (including downloading, decrypting and searching all by the CSC) and our approach (generating the trapdoor, searching on encrypted data and downloading qualified data), we provide simulations of these two approaches. The computation time is proportional to the number of the transaction records. We assume the primitive approach uses Advanced Encryption Scheme (AES) (128-bit key) to protect the transaction records. Assume each record size is 1 KB, thus to download all the encrypted transaction records costs 1 MB to 1, 000 MB for the number of records ranging from 1, 000 to 1, 000, 000. Given the download bandwidth 10 MB/sec, it takes 0.1 to 100 seconds. Assume the CSP returns qualified data files which is 1% of the whole dataset, the time for download and decrypt qualified data files can be reduces accordingly. Our framework only submits the encrypted trapdoor to the CSP which is 2*13*1KB= 26 KB. For the storing in the primitive approach, encryption/decryption one 1 KB records takes 0.036 *ms*, while storing one encrypted records in our framework takes 67.421 *ms*. For the counting/statistical procedures, the time for the primitive approach is 100.024 *s*, while that for our statistical framework requires 403.171 *s* for 1 million transaction records in our Openstack platform with 10 processors conducting parallel processing. In addition, our statistical framework can recruit enough computing units to achieve quality of user experience (QoE). Given our framework can be further extended

Table 11. Performance of Obtaining Statistics (*time* is in *seconds.*)

framework	primitive (aes-ctr128)	proposed (10 cores)	proposed (1,000 cores)
number of transactions	*time*=(CSC *download*+ CSC *decrypt*+ CSC *search*)	*time*=(CSC *keygen* + CSP *search* + CSC *download*)	*time*=(CSC *keygen* + CSP *search* + CSC *download*)
1,000	0.100	0.408	0.010
5,000	0.500	2.020	0.030
10,000	1.000	4.036	0.055
50,000	5.002	20.163	0.256
100,000	10.004	40.321	0.507
500,000	50.018	201.588	2.516
1,000,000	100.036	403.171	5.027

to run with $1,000$ cores, we can greatly reduce the time required to retrieve the demanded records or statistics to 5.027 *s*. Our time is around 20 times faster than that for the primitive approach which takes 100.024 *s* (See Table 11).

6   Conclusion

In this paper, we present a statistical framework to securely and efficiently obtain the statistics on encrypted data collection in the cloud. We build our framework on top of the searchable public-key encryption called hidden vector encryption (HVE) and provide detailed transformation of the statistical procedures for the plain data to those for the encrypted data. We further give concrete examples of these secure statistical tools. Finally, we provide security analysis and performance evaluation of these procedures and demonstrate the effectiveness and efficiency of the proposed framework. For future work, we would like to consider more advanced statistical procedures and apply these transformations to various disciplines such as data mining and artificial intelligence.

Acknowledgments. This research is supported by National Science Council, Taiwan under contract No. 101-2221-E-009-138-, and Delta Electronics, Inc. under contract No. 102C003.

References

[1] Mell, P., Grance, T.: The nist definition of cloud computing (draft). NIST special publication 800-145 (2011)

[2] Armbrust, M., Fox, A., Griffith, R., Joseph, A.D., Katz, R., Konwinski, A., Lee, G., Patterson, D., Rabkin, A., Stoica, I., Zaharia, M.: A view of cloud computing. Commun. ACM 53(4), 50–58 (2010)

[3] Subashini, S., Kavitha, V.: Review: A survey on security issues in service delivery models of cloud computing. J. Netw. Comput. Appl. 34(1), 1–11 (2011)

[4] Virvilis, N., Dritsas, S., Gritzalis, D.: Secure cloud storage: Available infrastructures and architectures review and evaluation. In: Furnell, S., Lambrinoudakis, C., Pernul, G. (eds.) TrustBus 2011. LNCS, vol. 6863, pp. 74–85. Springer, Heidelberg (2011)

[5] NIST: Fips pub 197: Announcing the advanced encryption standard (aes). NIST (2001)

[6] Jonsson, J., Kaliski, B.: Public-Key Cryptography Standards (PKCS) #1: RSA Cryptography Specifications Version 2.1. (3) (February 2003)

[7] Smart, N.P., Vercauteren, F.: Fully homomorphic encryption with relatively small key and ciphertext sizes. In: Nguyen, P.Q., Pointcheval, D. (eds.) PKC 2010. LNCS, vol. 6056, pp. 420–443. Springer, Heidelberg (2010)

[8] Gentry, C., Halevi, S.: Implementing gentry's fully-homomorphic encryption scheme. In: Paterson, K.G. (ed.) EUROCRYPT 2011. LNCS, vol. 6632, pp. 129–148. Springer, Heidelberg (2011)

[9] Boneh, D., Di Crescenzo, G., Ostrovsky, R., Persiano, G.: Public key encryption with keyword search. In: Cachin, C., Camenisch, J.L. (eds.) EUROCRYPT 2004. LNCS, vol. 3027, pp. 506–522. Springer, Heidelberg (2004)

[10] Boneh, D., Waters, B.: Conjunctive, subset, and range queries on encrypted data. In: Vadhan, S.P. (ed.) TCC 2007. LNCS, vol. 4392, pp. 535–554. Springer, Heidelberg (2007)

[11] Lapin, L.L.: Probability and statistics for modern engineering (1990)

[12] Barnes, J.: Statistical analysis for engineers and scientists: a computer-based approach. McGraw-Hill, Inc. (1994)

[13] Walpole, R.E., Myers, R.H., Myers, S.L., Ye, K.: Probability and statistics for engineers and scientists, vol. 8. Prentice Hall, Upper Saddle River (1993)

[14] Kushilevitz, E., Ostrovsky, R.: Replication is not needed: single database, computationally-private information retrieval. In: Proceedings of the 38th Annual Symposium on Foundations of Computer Science, FOCS 1997, pp. 364–373. IEEE Computer Society, Washington, DC (1997)

[15] Chor, B., Kushilevitz, E., Goldreich, O., Sudan, M.: Private information retrieval. J. ACM 45(6), 965–981 (1998)

[16] Rabin, M.O.: How to exchange secrets with oblivious transfer

[17] Even, S., Goldreich, O., Lempel, A.: A randomized protocol for signing contracts. Commun. ACM 28(6), 637–647 (1985)

[18] Agrawal, R., Srikant, R.: Privacy-preserving data mining. In: Proceedings of the 2000 ACM SIGMOD International Conference on Management of Data, SIGMOD 2000, pp. 439–450. ACM, New York (2000)

[19] Goh, E.J.: Secure indexes. IACR Cryptology ePrint Archive (2003)

[20] Curtmola, R., Garay, J., Kamara, S., Ostrovsky, R.: Searchable symmetric encryption: improved definitions and efficient constructions. In: Proceedings of the 13th ACM Conference on Computer and Communications Security, CCS 2006, pp. 79–88. ACM, New York (2006)

[21] Tseng, F.K., Chen, R.J., Lin, B.S.P.: Toward authenticated and complete query results from cloud storages. In: 2013 IEEE 12th International Conference on Trust, Security and Privacy in Computing and Communications (TrustCom), pp. 452–458 (July 2013)

[22] Iovino, V., Persiano, G.: Hidden-vector encryption with groups of prime order. In: Galbraith, S.D., Paterson, K.G. (eds.) Pairing 2008. LNCS, vol. 5209, pp. 75–88. Springer, Heidelberg (2008)

[23] Tseng, F.K., Liu, Y.H., Chen, R.J.: Toward authenticated and complete query results from cloud storages. In: 2012 IEEE 11th International Conference on Trust, Security and Privacy in Computing and Communications (TrustCom), pp. 1204–1209 (June 2012)

[24] Tseng, F.K., Liu, Y.H., Chen, R.J.: Ensuring correctness of range searches on encrypted cloud data. In: 2012 IEEE 4th International Conference on Cloud Computing Technology and Science (CloudCom), pp. 570–573 (2012)

[25] Bogdanov, A., Khovratovich, D., Rechberger, C.: Biclique cryptanalysis of the full AES. In: Lee, D.H., Wang, X. (eds.) ASIACRYPT 2011. LNCS, vol. 7073, pp. 344–371. Springer, Heidelberg (2011)

[26] Blundo, C., Iovino, V., Persiano, G.: Private-key hidden vector encryption with key confidentiality. In: Garay, J.A., Miyaji, A., Otsuka, A. (eds.) CANS 2009. LNCS, vol. 5888, pp. 259–277. Springer, Heidelberg (2009)

[27] National Chiao-Tung University: NCTU Openstack Dashboard (2013), https://openstack.nctu.edu.tw/

[28] Free Software Foundation, Inc.: GMP: The GNU Multiple Precision Arithmetic Library (2006) http://gmplib.org/

[29] Lynn, B.: PBC: Pairing-Based Cryptography Library (2008), http://crypto.stanford.edu/pbc/

Toward Practical Searchable Symmetric Encryption

Wakaha Ogata[1], Keita Koiwa[2], Akira Kanaoka[3], and Shin'ichiro Matsuo[4]

[1] Tokyo Institute of Technology, Japan
ogata.w.aa@m.titech.ac.jp
[2] University of Tsukuba, Japan
koiwa@cipher.risk.tsukuba.ac.jp
[3] Toho University, Japan
akira.kanaoka@is.sci.toho-u.ac.jp
[4] National Institute of Information and Communications Technology, Japan
smatsuo@nict.go.jp

Abstract. Searchable symmetric encryption is a good building block toward ensuring privacy preserving keyword searches in a cloud computing environment. This area has recently attracted a great deal of attention and a large quantity of research has been conducted. A security protocol generally faces a trade-off between security/privacy requirements and efficiency. Existing works aim to achieve the highest levels of security requirements, so they also come with high overhead. In this paper, we reconsider the security/privacy requirements for searchable symmetric encryption and relax the requirements for practical use. Then, we propose schemes suitable for the new requirements. We also show experimental results of our schemes and comparison to existing schemes. The results show that the index sizes of our proposals are only a few times of that of a Lucene (without encryption). In document update, our proposal requests additional index which depends only on the size of new document.

1 Introduction

1.1 Background

In the last several years, the progress of network technology and computers, including broadband network and virtualization techniques, has made information technology (IT) environments more usable. The proliferation of cloud computing is a good example of this. Though cloud computing provides such a usable environment, its characteristics pose security issues since valuable information is stored and processed in uncontrollable locations for users, and this could be lead to information leakage by cloud operators.

Encrypting data stored in the cloud is considered to be a countermeasure for such threats, and a large number of studies have been conducted on this subject. In this research, data is encrypted in a manner that it can proceed in its encrypted form. Examples of such research are counting by using a homomorphic

K. Sakiyama and M. Terada (Eds.): IWSEC 2013, LNCS 8231, pp. 151–167, 2013.

encryption scheme, processing any calculation using a homomorphic encryption scheme, and searchable encryption schemes, and these fall within the scope of this paper.

In each scheme, security requirements are defined and a scheme with provable security is proposed. These security requirements include privacy of user requests for the cloud server as well as confidentiality of information. Since privacy requirements vary among entities, we have not provided effective and general security requirements for the scheme. However, security researchers generally tend to aim at stronger security requirements. Yet security requirements and efficiency have trade-off relationship in our hopes for strong security, and processing performance and communication efficiency decrease. We therefore have to find a good balance among usability in terms of cloud computing, security, privacy, and efficiency.

1.2   Related Works

In this research, we focus on security requirements for a searchable encryption scheme, which is a demanded service for cloud computing and has been extensively researched to date. In searchable encryption schemes, the data and keyword for the search are first encrypted. Ciphertext is stored in the server. Only a party possessing access right can produce valid information (trapdoor) for a keyword search. The server cannot know the keyword from the trapdoor. This characteristic protects the privacy of the keyword.

Searchable encryption based on a symmetric cipher was firstly proposed in [17], and then schemes with improved security definitions were proposed in [8,7]. In, [7] Curtmola et al. proposes two searchable symmetric encryption schemes. Then, Kamara and Roeder showed that the scheme can convert to secure against adaptive adversary in CCS 2012 [14]. To enhance efficiency for keyword search, researches on reducing cost for document update are conducted recently. In [10,11], new indexes are reconstructed based on a single private key. On the other hand, the scheme proposed in [14] does not construct additional indexes.

Boneh et al. first proposed searchable encryption based on a symmetric cipher [5] as an application of an identity-based encryption scheme. Following that, many schemes [1,4] have been proposed including those based on anonymous hierarchical identity-based encryption. There is also research on operation when a searchable encryption scheme is applied to cloud computing [12].

1.3   Our Contributions

In this paper, we focus on searchable symmetric encryption. We refine the security definitions that offer a good balance between privacy and efficiency.

We first show efficiency requirements for a practical searchable symmetric encryption scheme, and show that the existing scheme is not practical. After that, we reconsider the security requirement for searchable encryption. Next we propose new schemes that have smaller encrypted indexes and lower processing costs for adding documents. These schemes allow leakage of a part of privacy from search history, but this would not be a problem in most of practical usages. We also show experimental results of our schemes and existing schemes. The experimental results show that the original searchable symmetric encryption by Curtmola et al. has huge amount of index size against Lucene, and our proposal can reduce the index size to a few times of that of Lucene.

2   Definitions of Symmetric Searchable Encryption and Existing Schemes

2.1   System Model

We assume the following setting as in [7]: There is one user $\mathcal{U}$ and one server $\mathcal{S}$. $\mathcal{U}$ has a collection of documents $\mathcal{D} = \{D_1, \ldots, D_n\}$, each document D_j is stored on server $\mathcal{S}$ in an encrypted style. D_j is assigned a unique identifier $id(D_j)$ that does not reveal any confidential information, e.g., a sequential number ($id(D_j) = j$) or a ciphertext of the document name. We assume that the set of searchable keywords, $\Delta = \{w_1, \ldots, w_d\}$, is predetermined and is called a *dictionary*. An outcome of a search for $w \in \Delta$ is denoted by $\mathcal{D}(w) = \{id(D_j) \mid w \in D_j\}$.

In an ordinary file system (with no security or privacy), a database called an *index* is generated in advance for quick keyword searching. For example, $\{(w_i, \mathcal{D}(w_i))\}_{i=1,\ldots,d}$ is stored. When a user issues a search query to the file system, the file system searches $\mathcal{D}(w_i)$ in the database and returns it to the user. A symmetric searchable encryption system (SSE) is a system in which an encrypted index is built to prevent information leakage.

An SSE consists of four algorithms as follows.

Keygen(1^k): User $\mathcal{U}$ uses this algorithm to generate private key K based on security parameter k.

BuildIndex($K, \mathcal{D}, \Delta$): $\mathcal{U}$ uses this algorithm to build (encrypted) index $\mathcal{I}$ from document set $\mathcal{D}$. $\mathcal{I}$ is sent to server $\mathcal{S}$ along with encrypted documents $\zeta = (Enc(D_1), \ldots, Enc(D_n))$.

Trapdoor(K, w): $\mathcal{U}$ runs this algorithm when it searches in $\mathcal{D}$ for keyword w. The output $T_w = $ Trapdoor(K, w), called a *trapdoor*, is sent to $\mathcal{S}$.

Search($\mathcal{I}, T$): $\mathcal{S}$ uses this algorithm to search in encrypted documents. If $T = $ Trapdoor(K, w), then it is necessary that Search($\mathcal{I}, T$) $= \mathcal{D}(w)$. $\mathcal{S}$ returns the result $\mathcal{D}(w)$ to $\mathcal{U}$.

Although the search process in this model is a one-round protocol, it can generally be a multi-round protocol.

2.2   Security Requirement

Let $(w_1, \ldots, w_q)$ be a sequence of q keywords. A *history* is defined as

$$H_q = (\mathcal{D}, w_1, \ldots, w_q),$$

which determines an instantiation of an interaction between $\mathcal{U}$ and $\mathcal{S}$. A *partial history* of H_q is $H_q^t(\mathcal{D}, w_1, \ldots, w_t)$, where $t \leq q$. An adversary's *view* of H_q under secret key K is defined as

$$V_K(H_q) = (id(D_1), \ldots, id(D_n), \zeta, \mathcal{I}, T_1, \ldots, T_q),$$

where $\mathcal{I} = \mathsf{BuildIndex}(K, \mathcal{D}, \Delta)$ and $T_i = \mathsf{Trapdoor}(K, w_i)$. A *partial view* is

$$V_K^t(H_q) = (id(D_1), \ldots, id(D_n), \zeta, \mathcal{I}, T_1, \ldots, T_t),$$

where $t \leq q$.

Oblivious RAMs introduced in [9] realize secure searching in which $V_K(H_q)$ does not leak any information of H_q. However, this scheme is highly inefficient. It is not practical to require perfect secrecy such as with oblivious RAMs. Thus, some weak security definitions that allow leakage of partial information of the history to the server were defined in the literature.

Chang and Mitzenmacher [6] defined the security of SSEs. In [7], a vulnerability of the definition was pointed out, and the authors gave four new security definitions: semantic security against non-adaptive attack, semantic security against adaptive attack, indistinguishability against non-adaptive attack, and indistinguishability against adaptive attack. Since equivalence of semantic security and indistinguishability was shown [7], here we give only the definition of semantic security.

Definition 1 (Trace). *For a given history* $H_q = (\mathcal{D}, w_1, \ldots, w_q)$, *let* Π_q *be a* $q \times q$ *binary matrix where* $\Pi_q[i, j] = 1$ *if* $w_i = w_j$, $\Pi_q[i, j] = 0$ *otherwise. The trace of* H_q *is the sequence*

$$Tr(H_q) = (id(D_1), \ldots, id(D_n), |D_1|, \ldots, |D_n|, \mathcal{D}(w_1), \ldots, \mathcal{D}(w_q), \Pi_q).$$

Trace indicates information that we allow to leak to the server.

Definition 2 (Semantic security against non-adaptive attack). *We say that an SSE is non-adaptively semantically secure if all* $q \in N$ *and for any ppt adversary* $\mathcal{A}$, *there exists a ppt simulator Sim such that for all traces* Tr_q, *all polynomial samplable distributions* $\mathcal{H}_q$ *over* $\{H_q \in 2^{2^\Delta} \times \Delta^q : Tr(H_q) = Tr_q\}$, *all functions* f,

$$|\Pr[\mathcal{A}(V_K(H_q)) = f(H_q)] - \Pr[Sim(Tr(H_q)) = f(H_q)]|$$

is negligibly small, where $H_q \leftarrow \mathcal{H}_q$, $K \leftarrow \mathsf{Keygen}(1^k)$, *and the probabilities are taken over* $\mathcal{H}_q$ *and the internal coins of* Keygen, $\mathcal{A}$, Sim *and the underlying* $\mathsf{BuildIndex}$ *algorithm.*

Definition 3 (Semantic security against adaptive attack). *We say that SSE is adaptively semantically secure if all* $q \in N$ *and for all ppt adversaries* $\mathcal{A}$, *there exists a ppt simulator Sim such that for all traces* Tr_q, *all polynomial* *samplable distributions* $\mathcal{H}_q$ *over* $\{H_q \in 2^{2^\Delta} \times \Delta^q : Tr(H_q) = Tr_q\}$, *all functions* f, *all* $0 \le t \le q$:

$$| \Pr[\mathcal{A}(V_K^t(H_q)) = f(H_q^t)] - \Pr[Sim(Tr(H_q^t)) = f(H_q^t)]|$$

is negligibly small, where $H_q \leftarrow \mathcal{H}_q, K \leftarrow$ Keygen(1^k), *and the probabilities are taken over* $\mathcal{H}_q$ *and the internal coins of* Keygen, $\mathcal{A}$, Sim *and the underlying* BuildIndex *algorithm.*

2.3   Curtmola et al. Scheme (SSE-1)

Existing SSE schemes can be classified into two types. The first type uses a Boolean $n \times d$ matrix as an index (such as [6]), and the second type uses a list of $(w_i, \mathcal{D}(w_i))$ (such as [7]). We focus on the second type of scheme in this paper because computational complexity of searching in such schemes is $O(\log n)$, while it is $O(n)$ in the first type of scheme.

Curtmola et al. proposed two SSE schemes [7]. The first, called SSE-1, is secure against non-adaptive attacks, and is more efficient. The second, called SSE-2, is secure against adaptive attacks, but less efficient. On the other hand, in [14] it is shown that simple modification of SSE-1 can make it adaptively secure in the random oracle model. We review the SSE-1 scheme here.

Let k be a security parameter, p be the bit length of the longest keyword in Δ, *unit* be the bit length of the shortest keyword, and m be the total size of the plaintext documents $\mathcal{D}$ expressed in *unit*. Let $\mathcal{E}$ be a symmetric encryption function with key length ℓ. SSE-1 uses the following pseudo-random function f and pseudo-random permutations π, ψ.

- $f : \{0,1\}^k \times \{0,1\}^p \to \{0,1\}^{\ell + \log m}$
- $\pi : \{0,1\}^k \times \{0,1\}^p \to \{0,1\}^p$
- $\psi : \{0,1\}^k \times \{0,1\}^{\log m} \to \{0,1\}^{\log m}$

We give a list of parameters in Table 1 for convenience.

Keygen(1^k): Generate random keys $s, y, z \xleftarrow{R} \{0,1\}^k$ and output $K = (s, y, z, 1^\ell)$.
BuildIndex$(K, \mathcal{D}, \Delta)$:

Table 1. Parameters used in SSE-1

k	security parameter, key length of pseudo-random function/permutations
ℓ	key length of symmetric encryption
n	the number of documents in document collection $\mathcal{D}$
d	the number of keywords in dictionary Δ
p	bit length of the longest keyword
unit	bit length of the shortest keyword
m	total length of $\mathcal{D}$ expressed by *unit*

1. Scan $\mathcal{D}$ and build $\Delta'(\subseteq \Delta)$, which is the set of all keywords in $\mathcal{D}$. Build $\mathcal{D}(w)$ for each word $w \in \Delta'$.

2. Set up array A with m entries as follows. First, global counter ctr is set to 1. For each $w_i \in \Delta'$ choose a random ℓ-bit string $\kappa_{i,0}$, and for each $id_{i,j} \in \mathcal{D}(w_i)$ $(1 \leq j \leq |\mathcal{D}(w_i)|)$, set node $N_{i,j} = \langle id_{i,j}||\kappa_{i,j}||\psi_s(ctr+1)\rangle$, where $\kappa_{i,j}$ is a random ℓ-bit string.
   Compute $\mathcal{E}_{\kappa_{i,j-1}}(N_{i,j})$ and store it in $A[\psi_s(ctr)] = \mathcal{E}_{\kappa_{i,j-1}}(N_{i,j})$.
   Store a random string in all entries that are not used to store an encrypted node.

3. Build lookup table T with d entries as follows.
   For each $w_i \in \Delta'$, set $T[\pi_z(w_i)] = \langle \mathrm{addr}(A(N_{i,1}))||\kappa_{i,0}\rangle \oplus f_y(w_i)$, where $\mathrm{addr}(A(N_{i,1}))$ is the address add where $A[add] = \mathcal{E}_{\kappa_{i,0}}(N_{i,1})$.
   Store a random string in all $T[\pi_z(w_i)]$ s.t. $w_i \in \Delta \setminus \Delta'$.
   Output $\mathcal{I} = (A, T)$.

Trapdoor(K, w): Output $T_w = (\pi_z(w), f_y(w))$.

Search$(\mathcal{I}, T)$: Let $T = (\gamma, \eta)$. Retrieve $\theta = T[\gamma]$. Let $\langle \alpha||\kappa\rangle = \theta \oplus \eta$. Decrypt $A[\alpha]$ with κ to obtain $N_{i,1}$, which includes identifier $id_{i,1}$, the next random key $\kappa_{i,1}$, and the next address $\mathrm{addr}(A(N_{i,2}))$. Then decrypt $A[\mathrm{addr}(A(N_{i,2}))]$ with $\kappa_{i,1}$ to obtain $N_{i,2}$, which includes $id_{i,2}$. Iterating the same process to recover all $id_{i,j}$. Output all identifiers $\{id_{i,j}\}$.

It is shown that SSE-1 is semantically secure against non-adaptive attacks, if $\mathcal{E}$ is a secure symmetric encryption function, f is a pseudo-random function, and π, ψ are pseudo-random permutations.

2.4   Other Schemes Supporting Document Update

Consider the case that the user $\mathcal{U}$ keeps a set of documents $\mathcal{D}_1$ on a server $\mathcal{S}$ with an index $\mathcal{I} = \mathsf{BuildIndex}(K, \mathcal{D}_1, \Delta)$, and now $\mathcal{U}$ is going to store an additional set of documents $\mathcal{D}_2$. A simple way to add documents is that $\mathcal{U}$ builds a new index $\mathcal{I}' = \mathsf{BuildIndex}(K, \mathcal{D}_2, \Delta)$ and $\mathcal{S}$ replaces an old index $\mathcal{I}$ with $(\mathcal{I}, \mathcal{I}')$. In this case, however, $\mathcal{S}$ learns $\mathcal{D}_2(w)$ if $\mathcal{U}$ already made a search query for w in $\mathcal{D}_1$ (but not in $\mathcal{D}_2$) since $\mathcal{S}$ knows $T = \mathsf{Trapdoor}(K, w)$.

Accordingly, the following process is adopted in [6] and [7]. To add $\mathcal{D}_2$, $\mathcal{U}$ runs Keygen to generate a new key K', and builds a new index $\mathcal{I}' = \mathsf{BuildIndex}(K', \mathcal{D}_2, \Delta)$, which is sent to $\mathcal{S}$ with (encrypted) $\mathcal{D}_2$. When $\mathcal{U}$ wants to search for a keyword w, it sends two trapdoors $T = \mathsf{Trapdoor}(K, w)$ and $T' = \mathsf{Trapdoor}(K', w)$ to $\mathcal{S}$. $\mathcal{S}$ runs Search$(\mathcal{I}, T)$ and Search$(\mathcal{I}', T')$.

This process does not leak unnecessary information. However, if a few documents are added frequently, $\mathcal{U}$ has to keep many private keys and a set of many trapdoors has to be sent to search for a keyword.

Recently, some researchers have proposed SSE schemes in which the user can add document sets freely without increasing the size of the private key. SSE schemes proposed in [10] and [11] construct new index $\mathcal{I}'$ based on a single private key. On the other hand, the scheme proposed in [14] does not construct additional indexes but utilizes unused memory space of the original index.

3   What Is Practical SSE?

SSE schemes with high security are needed in special purposes. In most cases, however, we require practicality – reasonable index size and small communication/computational cost – rather than security, since we use storage services as a tool for improving convenience.

In this section, we first introduce requirements for practical SSE schemes. We then claim that existing schemes such as SSE-1 do not satisfy the requirements.

3.1   Requirements for Practicality

Here we introduce three requirements for practicality.

1. **Efficient search.** We perform keyword search repeatedly, so real-time response is required. We require that a much longer time than in an ordinary (unencrypted) system is not needed to search for a keyword.
2. **Reasonable index size.** In general, the size of an index depends on the total size of $\mathcal{D}$. We require that the size of index for $\mathcal{D}$ is not much larger than $\mathcal{D}$ itself.
3. **Scalability.** In most cases, new documents are added in storage one after another. On such occasions, the user must renew the index by performing an update protocol with the server. For scalability, it is desirable for an SSE scheme to have the following two properties.
   (R1) The size of secret key K and computation/communication cost for searching do not depend on the number of updates of the index.
   (R2) The computational cost to update the index depends on the additional document size, but not on the total document size.

3.2   Inefficiency of SSE-1

In SSE-1, index $\mathcal{I}$ consists of an array A and a lookup table T. A has m entries, where m is the total size of the plaintext documents $\mathcal{D}$ expressed in the shortest keyword length. Each entry consists of three parts: a document identifier, a random key, and the next address of A. Since the lengths of these parts are $\lceil \log n \rceil$, ℓ, $\lceil \log m \rceil$, respectively, the total size of A is $m(\lceil \log n \rceil + \ell + \lceil \log m \rceil)$ bits. T has d entries, each consisting of an address and a value, which are p bits and $(\ell + \lceil \log m \rceil)$ bits, respectively. Therefore, the size of T is $d(\ell + \lceil \log m \rceil + p)$. In total, the bit length of the index is $|\mathcal{I}| = m(\lceil \log n \rceil + \ell + \lceil \log m \rceil) + d(\ell + \lceil \log m \rceil + p)$.

Next, we estimate the sizes of $\mathcal{I}$ for concrete parameters. We assume that

- $\mathcal{D}$ consists of $n = 10^3$ documents, the size of each document is on average 10KB. The total size of $\mathcal{D}$ is 10MB.
- The dictionary includes 100,000 keywords, that is, $d = 10^5$. The length of the shortest keyword is 2 B (two letters) and that of the longest keyword is 20 B (20 letters, 10 units). Therefore, $p = 160$.

Table 2. Index size in SSE-1 and Lucene in a case[(*)]

	size of document set: $|\mathcal{D}|$	index size: $|\mathcal{I}|$	Ratio: $|\mathcal{I}|/|\mathcal{D}|$
SSE-1	10MB	836MB	83.6
Lucene	67MB	83MB	1.28

(*) $\mathcal{D}$ consists of $n = 10^3$ documents, the size of each document is on average 10KB.
The dictionary includes 100,000 keywords,
The shortest keyword is two letters, the longest keyword is 20 letters (10units).
The private key length is $\ell = 128$.

- The key length is $\ell = 128$ as in AES.

Under these parameters, $m = 5 \times 10^6$. We show the index size under these parameters in Table 2. For comparison, we also give the case of Lucene [2] as an example of systems that do not consider privacy at all.

From the table, we can see that the index is huge in SSE-1. Now consider the case in which we store documents using a free storage service. If the free space is 5 GB, we can store 2 GB (non-confidential) documents in total, along with a 2.6 BG index of Lucene. On the other hand, if we want to store them by using SSE-1, we cannot store 60 MB of documents in total since their index exceeds 5 GB.

3.3 Scalability of Existing SSE

As we mentioned before, SSE-1 does not have scalability.

In contrast, SSE schemes that support document updates have scalability. However, they require huge indexes as well as SSE-1.

4 Relaxation of Security

As we show in section 3.2, a serious disadvantage of SSE-1 (and its variations) is index size, especially the size of array A. A has m entries, but only $m'(= \sum_{w \in \Delta'} |\mathcal{D}(w)|)$ entries are used to store meaningful values. The remaining entries are prepared to hide the number m'. This means that if the user does not mind the server knowing the number m', the number of entries of A can be reduced to $m'(\ll m)$. Similarly, there is a possibility that a rather efficient SSE scheme can be constructed if the user does not mind leakage of some additional information.

In this section, we discuss the need for adaptive indistinguishability and define several levels of security.

4.1 Adaptive Attack

An adversary that mounts a chosen-keyword attack (cka) has the ability to obtain trapdoors corresponding to the keywords. We discuss the feasibility of cka.

The general attack scenario of active attacks such as chosen-keyword attacks and chosen-ciphertext attacks is a lunchtime attack. That is, an adversary illegitimately accesses a computer that is used to make trapdoors. Do we possess other means against the attack other than cryptographical control? Yes, we will be able to avoid such illegitimate use by adequately managing a private key K.

Another attack scenario of a chosen-keyword attack is a social attack, as follows.

- An adversary popularizes a target keyword w. Accordingly, the user would search for w in his documents by sending a trapdoor $T = \mathsf{Trapdoor}(K, w)$.
- A malicious administrator of the server tells the user a forged notification that word w is not allowed to be stored in storage (e.g., for certain political reasons). Accordingly, the user searches for w in his documents.

Although it is difficult to avoid such social attacks, we think that an adversary cannot frequently succeed in obtaining desirable trapdoors. It seems particularly hard to adaptively obtain desirable trapdoors.

From the above discussion, if adaptively indistinguishable SSE schemes are much more inefficient than non-adaptively indistinguishable ones, one practical choice is to use an efficient non-adaptively indistinguishable one together with appropriate key management and other controls against social attacks.

4.2 Relaxed Security Definitions

In [7], a trace of history H_q is defined as

$$Tr(H_q) = (id(D_1), \ldots, id(D_n), |D_1|, \ldots, |D_n|, \mathcal{D}(w_1), \ldots, \mathcal{D}(w_q), \Pi_q).$$

As mentioned before, $Tr(H_q)$ indicates partial information of H_q that we allow to leak to the server. Below, we define some variations of trace.

For given dictionary $\Delta = \{w_1, \ldots, w_d\}$ with d words and document set $\mathcal{D} = \{D_1, \ldots, D_n\}$, we define *index matrix P* which is expressed by a binary matrix:

$$P = \begin{bmatrix} p_{1,1} & p_{1,2} & \cdots & p_{1,n} \\ p_{2,1} & p_{2,2} & \cdots & p_{2,n} \\ \vdots & \vdots & \ddots & \vdots \\ p_{d,1} & p_{d,2} & \cdots & p_{d,n} \end{bmatrix},$$

$$p_{i,j} = \begin{cases} 1 & \text{if } w_i \in D_j, \\ 0 & \text{otherwise.} \end{cases}$$

Let

$$W_H(w_i) = \sum_{j=1}^{n} p_{i,j},$$

$$W_H(D_j) = \sum_{i=1}^{d} p_{i,j},$$

$$W_H(P) = \sum_{i=1}^{d} \sum_{j=1}^{n} p_{i,j}.$$

$W_H(w_i)$ is the number of documents in $\mathcal{D}$ that include keyword $w_i(\in \Delta)$, that is, $W_H(w_i) = |\mathcal{D}(w_i)|$. $W_H(D_j)$ is the number of keywords in document D_j. For randomly chosen permutations π_d (over $\{1, \ldots, d\}$), let $\hat{P}$ be a binary matrix such that rows of P are permuted by π_d. We call $\hat{P}$ a randomized index matrix.[1]

Our new security definitions are as follows.

Definition 4. *For a given history $H_q = (\mathcal{D}, w_1, \ldots, w_q)$, define*

$$Tr^{(0)}(H_q) = Tr(H_q),$$
$$Tr^{(1)}(H_q) = (Tr(H_q), W_H(P)),$$
$$Tr^{(2)}(H_q) = (Tr(H_q), W_H(D_1), \ldots, W_H(D_n)),$$
$$Tr^{(3)}(H_q) = (Tr(H_q), W_H(w_1), \ldots, W_H(w_d)),$$
$$Tr^{(4)}(H_q) = (Tr(H_q), \hat{P}).$$

(In this definition, we only consider non-adaptive semantic security.) For $k \in \{0, 1, 2, 3, 4\}$, $Tr^{(k)}$*-security is defined the same way as in Def. 2 except that* $Tr(H_q)$ *is replaced with* $Tr^{(k)}(H_q)$.

From the definition, $Tr^{(0)}$-security is equivalent to the original semantic security. Clearly,

$$Tr^{(0)}\text{-secure} \Rightarrow Tr^{(1)}\text{-secure} \Rightarrow Tr^{(2)}\text{-secure and } Tr^{(3)}\text{-secure},$$

$$Tr^{(2)}\text{-secure or } Tr^{(3)}\text{-secure} \Rightarrow Tr^{(4)}\text{-secure}$$

hold.

When a user searches for a keyword, the server learns a vector in the randomized index matrix $\hat{P}$ even if the scheme has $Tr^{(0)}$-semantic security. Therefore, after the user searches for all keywords in the dictionary, the server learns the entire $\hat{P}$. This means that $Tr^{(0)}$-semantic security gets closer to $Tr^{(4)}$-security the more keywords are searched.

In the next subsection, we further discuss the relation among the security notions focusing on document update.

4.3 Relations among the Security Notions

We assume that an SSE scheme has $Tr^{(3)}$-semantic security, that is, $\mathcal{I}$ leaks $W_H(w_i) = \sum_{j=1}^{n} p_{i,j}$ for all $w_i \in \Delta$. Consider the case that the user add a new document D_{n+1} and the index is replaced with $\mathcal{I}'$ that has information about D_{n+1}. At this moment, the server learns $(p_{1,n+1}, \ldots, p_{d,n+1})$ since

$$\sum_{j=1}^{n+1} p_{i,j} - \sum_{j=1}^{n} p_{i,j} = p_{i,n+1}$$

---

[1] Note that the order of documents is not randomized, since it does not have any confidential information.

holds. If documents are added one by one, the server learns the entire $\hat{P}$ (even if no keyword is queried). This situation happens independently of the scheme and the way of index update. Therefore, in the case documents are added one by one (or only a few at a time), $Tr^{(3)}$-semantic security is very close to $Tr^{(4)}$-semantic security. With the same argument, $Tr^{(1)}$-semantic security and $Tr^{(2)}$-semantic security is also very close in such a situation.

5   Practical SSE Schemes

In this section, we show how we can improve the efficiency of SSE schemes by relaxing the security requirement. For this purpose, two efficient SSE schemes are given.

5.1   Simplest Scheme (Simple-SSE)

Before showing SSE schemes, we describe a search scheme with no security measure— SEARCH. In SEARCH, an index is built as $\mathcal{I}_0 = \{(w_i, \mathcal{D}(w_i))\}_{i=1,...,d}$ from document set $\mathcal{D}$ beforehand. (We assume that the entries in $\mathcal{I}_0$ are sorted in alphabetical order.) When a user requests a search for $w \in \Delta$, the server finds an entry $(w_i = w, \mathcal{D}(w_i))$ in $\mathcal{I}_0$ (with $O(\log d)$ computational cost), and answers $\mathcal{D}(w_i)$ to the user.

Needless to say, SEARCH is absolutely insecure since the server knows which keywords are included with which documents, and also learns which keywords the user searched for. By replacing all keywords with random strings we can obtain an SSE scheme which we call Simple-SSE. The description is as follows. In this scheme, $H : \{0,1\}^* \rightarrow \{0,1\}^{\ell_H}$ is a collision resistance hash function.

Keygen(1^k): Choose $K \xleftarrow{R} \{0,1\}^k$ and output K.
BuildIndex($K, \mathcal{D}, \Delta$): Build $\mathcal{I}_0 = \{(w_i, \mathcal{D}(w_i))\}_{i=1,...,d}$. For each $w_i \in \Delta$ compute
   $\hat{w}_i = H(K\|w_i)$. Replace each entry $(w_i, \mathcal{D}(w_i))$ of $\mathcal{I}_0$ with $(\hat{w}_i, \mathcal{D}(w_i))$, and
   then sort the entries in alphabetical order of $\hat{w}_i$. The result is $\mathcal{I}$.
Trapdoor(K, w): Output $\hat{w} = H(K\|w)$.
Search($\mathcal{I}, T$): Search $(\hat{w} = T, \mathcal{D}(w))$ in $\mathcal{I}$ and output $\mathcal{D}(w)$.

Theorem 1. *If a pseudo-random encryption function Enc is used to encrypt each document, Simple-SSE has $Tr^{(4)}$-semantic security in the random oracle model. More precisely,*

$$| \Pr(A(V_K(H_q)) = f(H_q)) - \Pr(Sim(Tr^{(4)}(H_q)) = f(H_q))| \leq q_H/2^k + Adv_{Enc}$$

holds, where q_H is the number of oracle queries, k is the private key length, and Adv_{Enc} is an advantage of pseudo-randomness of Enc.

Proof. We consider Sim as follows. The input of Sim is

$$Tr^{(4)}(H_q) = (id(D_1), ..., id(D_n), |D_1|, ..., |D_n|, \mathcal{D}(w_1), ..., \mathcal{D}(w_q), \Pi_q, \hat{P}),$$

where $\hat{P} = \{p_{ij}\}$. Sim computes $\mathcal{I}$ as follows.

1) For all $i(1 \leq i \leq d)$,
   1a) choose a random string $\hat{w}$ with length ℓ_H bits and set $List \leftarrow \{\}$;
   1b) for all $j(1 \leq j \leq n)$, if $p_{ij} = 1$, add $id(D_j)$ to $List$;
   1c) set $Entry_i = (\hat{w}, List)$;
2) Sort d entries in alphabetical order of $\hat{w}$ to obtain $\mathcal{I}$.

Next, Sim computes the list of trapdoors as follows.
   For all $i(1 \leq i \leq q)$,

1) search in $\mathcal{I}$ and find an entry $(\hat{w}_j, List_j)$ such that $List_j = \mathcal{D}(w_i)$;
2) set $T_i \leftarrow \hat{w}_j$.

Then, Sim runs adversary A as a subroutine with input

$$view = (id(D_1), \ldots, id(D_n), \zeta, \mathcal{I}, T_1, \ldots, T_q),$$

where ζ is a set of random strings and each length is determined by $|D_i|$.

The adversary A may issue random oracle queries. To answer them, Sim chooses random key K^* at first, and initializes a list $L_H = \emptyset$. When A makes query $(K||w)$, Sim first checks if $K = K^*$. If so, Sim aborts. Otherwise, Sim searches $K||w$ in L_H. If there exists $\langle K||w, \hat{w}\rangle$, then Sim returns $\hat{w}$ to A. Otherwise, chooses ℓ_H-bit random string $\hat{w}$, adds $\langle K||w, \hat{w}\rangle$ in L_H, and returns $\hat{w}$ to A.

When A outputs $f(H_q)$, Sim outputs it as own result.

If Sim does not abort, A's view is the same as the real attack scenario except the distribution of ζ; it consists of random strings in the above simulation, while real ciphertexts in the real attack scenario. Therefore,

$$|\Pr(A(V_K(H_q)) = f(H_q)) - \Pr(Sim(Tr^{(4)}(H_q)) = f(H_q))| \leq \Pr(Sim \text{ aborts}) + \text{Adv}_{Enc}.$$

Sim aborts only if A queries K^*. So, $\Pr(Sim \text{ aborts}) \leq q_H/2^k$, where q_H is the number of oracle queries. That is, Simple-SSE holds $Tr^{(4)}$-semantic security.   □

The computational costs for searching in Simple-SSE are almost the same as those in SEARCH. The computational costs of BuildIndex are d hashes and sorting, which are very lightweight.

We can therefore say that the extra computational cost needed to guarantee $Tr^{(4)}$-semantic security is very small.

5.2   Lightened SSE-1 (SSE-1′)

We consider a lightened version of SSE-1, called SSE-1′, in which A has only $m'(= W_H(P))$ entries, that is, we eliminate all entries that store random strings.

Theorem 2. *SSE-1′ has $Tr^{(1)}$-semantic security, if $\mathcal{E}$ is a secure symmetric encryption function, f is a pseudo-random function, and π, ψ are pseudo-random permutations.*

Proof. SSE-1′ is the same as SSE-1 except the size of array A. In the security proof of SSE-1, Sim simulates adversary's view, which includes encrypted data, index (T, A), and trapdoors. (A has m entries, and m is determined by the sizes of documents.)

We consider a simulator Sim' that operates in the same way to Sim except that A has only m' entries. Note that Sim' knows $m' = W_H(P)$ because it is included in $Tr^{(1)}(H_q)$ (but not in $Tr(H_q)$). Then, the simulated view by Sim' and the real view are indistinguishable from the same reason in the proof of original SSE-1. □

5.3 Index Size of Proposed Schemes

To show the efficiency of our schemes, we first compare the size of index in SEARCH, Simple-SSE, and SSE-1′. We denote them with $\mathcal{I}_0$ and $\mathcal{I}_{\text{Simple}}$ and $\mathcal{I}_{\text{SSE1'}}$. In the following discussion, we use $\rho = W_H(P)/dn$, which is the average hit rate.

Since $\mathcal{I}_0 = \{(w_i, \mathcal{D}(w_i))\}_{i=1,\dots,d}$,

$$|\mathcal{I}_0| = \sum_{i=1}^{d} |w_i| + W_H(P)\lceil \log n \rceil = d(ave(|w_i|) + n\rho\lceil \log n \rceil)$$

where d and n are the number of keywords in Δ and the number of documents in $\mathcal{D}$, respectively.

$|\mathcal{I}_{\text{Simple}}|$ is estimated as

$$|\mathcal{I}_{\text{Simple}}| = d\ell_H + W_H(P)\lceil \log n \rceil = d(\ell_H + n\rho\lceil \log n \rceil).$$

$|\mathcal{I}_{\text{SSE1'}}|$ is estimated as

$$|\mathcal{I}_{\text{SSE1'}}| = W_H(P)(\lceil \log n \rceil + \ell + \lceil \log W_H(P) \rceil) + d(\ell + \lceil \log W_H(P) \rceil + p)$$

$$= d\left((\ell + p) + n\rho(\lceil \log n \rceil + \ell + \frac{n\rho + 1}{n\rho}\lceil \log dn\rho \rceil)\right)$$

If $\ell_H = 160$ (as in SHA-1) and it is longer than the average length of keywords, $\mathcal{I}$ in Simple-SSE is larger than $\mathcal{I}_0$. However, the difference between them is not so large.

Next, we compare $|\mathcal{I}_{\text{SSE1'}}|$ and $|\mathcal{I}_{\text{Simple}}|$. Assuming that $\ell_h = 160, \ell = 128, p = 160$, the first term of $|\mathcal{I}_{\text{SSE1'}}|$ is not as large as twice the first term of $|\mathcal{I}_{\text{Simple}}|$. The ratio of the second terms is $1 + (\log n)^{-1}(\ell + \frac{n\rho+1}{n\rho} \log dn\rho)$, which is $1 + (\log n)^{-1}(\ell + 2\log d)$ when $n\rho \approx 1$ and $2 + (\log n)^{-1}(\ell + \log d\rho)$ when $n\rho \gg 1$. When $n = 2^{10} \sim 2^{20}$, $\rho = 2^{-4} \sim 2^{-10}$, $d = 2^{10} \sim 2^{20}$, and $\ell = 128$, it is estimated between 8 and 18.

5.4 Scalability of Simple-SSE and SSE-1′

In Simple-SSE, we can update the index as follows.

- Let $\mathcal{D}'$ be the additional documents. $\mathcal{U}$ first builds $\mathcal{I}_0' = \{(w_i, \mathcal{D}'(w_i))\}_{i=1,\ldots,d}$. For each $w_i \in \Delta$, computes $\hat{w}_i = H(K\|w_i)$ as in BuildIndex, replaces w_i in $\mathcal{I}_0'$ which $\hat{w}_i$, and then sorts $(\hat{w}_i, \mathcal{D}'(w_i))$ to obtain $\mathcal{I}'$. $\mathcal{U}$ sends $\mathcal{I}'$ to $\mathcal{S}$ along with ciphertext of $\mathcal{D}'$.
- Upon receiving $\mathcal{I}'$, $\mathcal{S}$ updates $\mathcal{I}$ as follows. For all $(\hat{w}_i, \mathcal{D}'(w_i)) \in \mathcal{I}'$, replaces $(\hat{w}_i, \mathcal{D}(w_i))$ in $\mathcal{I}$ with $(\hat{w}_i, \mathcal{D}(w_i) \cup \mathcal{D}'(w_i))$.

After an update of the index, an original private key K can be used and the user can search as in the same process as before. Therefore, the update satisfies (R1). Computation and communication costs for an update of the index are also proportional to the size of $\mathcal{I}'$. The size of $\mathcal{I}'$ depends on the bit length of new documents $\mathcal{D}'$, but not on the existing document set. So, the update satisfies (R2). That is, we can say that Simple-SSE has scalability.

On the other hand, updating of index in SSE-1' can be done similar way to [11] to satisfy scalability.[2] (Unfortunately, Hirano et al.'s technique [10] does not satisfy (R1); the technique introduced in [14] leaks additional information and degrades security.)

6   Implementation and Evaluation of SSE

To confirm that the new schemes are practical, we implement SSE-1, Simple-SSE, and SSE-1', and evaluate the index size and execution time of the search for each scheme. We use Java for implementation of each program.

6.1   Preparation of Implementation

Document set $\mathcal{D}$: We use the following presented papers (total of 974) as documents of the targeted search.

- USENIX Security Symposium (2002–2011)
- IEEE Symposium on Security and Privacy (2003–2012)
- ACM Conference on Computer and Communications Security (2002–2011)

Since these papers are published on the Web by the PDF file, we convert them into text files.[3] The sum total of the size of converted documents is 65,011,003 B.

Dictionary Δ: The dictionary in our implementation is $\Delta = \Delta_1 \cup \Delta_2$, where Δ_1 is SINGLE.TXT on Moby Word Lists[16], and Δ_2 was produced by Lucene

---

[2] In [11], the following techniques are used: (a) To keep the key size to be constant, a secret key used to make each index is generated from a unique master secret key. (b) To keep the trapdoor size to be constant, all indexes are linked by putting a trapdoor in the next index.

[3] We use the pdftotext command of Xpdf 3.03 for conversion to text files.

from $\mathcal{D}$.[4] This dictionary has 514,045 words, that is, $d = 514,045$. The longest words in Δ is 248 letters[5], i.e., the bit length is 1,984 bit.

Other Parameters: For implementing a pseudo-random function f, we use HMAC which is in the javax.crypto package and javax.crypto.spec package. More precisely, $f_k(w)$ is computed by

$$f_k(w) = \mathrm{HMAC}(w\|0)\|\mathrm{HMAC}(w\|1)\| \cdots \|\mathrm{HMAC}(w\|(s-1)),$$

where $s = \lceil n/160 \rceil$ and n is the output length of f.

We implement pseudo-random permutations ψ and π by using AES[3]. The input length of ψ is $\log m$ in SSE-1 and $\log m'$ in SSE-1', which are less than the block length of AES, 128. However, the input length of π, 1,984 bit, is much longer than 128. Therefore, we adopt the ECB-mode[6], considering a word as a 16-block plaintext.

We also adopt AES as the symmetric encryption $\mathcal{E}$.

In π, ψ and $\mathcal{E}$, the shortest key length, $\ell = 128$, is used.

6.2 Execution Environment

We measure execution time via a machine with the following specifications.

- OS: Linux 2.6.35 x86_64, Ubuntu server 10.10
- CPU: Intel Core i7 2600
- Memory: DDR3-1333 SDRAM 4GB × 2
- Software: JRE 1.6.0_29

6.3 Numerical Results

Index Size: Table 3 shows the index sizes of SSE-1, SSE-1', and Simple-SSE. As a comparison with the case in which privacy protection is not taken into consideration, we also measured the size of the index using StandardAnalyzer in Lucene[2].This table also shows the size comparison.

Table 3 shows that the index becomes large as compared with original documents or the index of Lucene. However, Simple-SSE and SSE-1' have succeeded in drastic reduction of the size of the index as compared with SSE-1.

---

[4] Here we use Lucene only to create a set of words from a targeted file set.

[5] Such a long word is because the documents include numerical data and binary data. If we exclude such long (pseudo)words, the index sizes in SSE-1 and SSE-1' would become 100MB smaller than our results.

[6] AES-ECB is a permutation but not pseudo-random. Therefore, we have to adopt other implementation to satisfy the security definition. Though evaluation time of π increases by this change, it is thought that the increment does not affect searching time so much since π is evaluated only once in a search.

Table 3. Comparison of index size

	Index size: $|\mathcal{I}|$	$|\mathcal{I}|/|\mathcal{D}|$	Ratio to SSE-1	Ratio to Lucene
SSE-1	1,836MB	28.24	(1.00)	22.12
SSE-1$'$	397MB	6.10	0.22	4.78
Simple-SSE	275MB	4.23	0.15	3.32
Lucene	83MB	1.28	–	(1.00)

Table 4. Execution time of Search

	Execution time of Search (msec)		
	words in $\Delta - \Delta'$	words in Δ'	random character string
SSE-1	0.0941	0.8383	0.0817
Simple-SSE	0.0603	0.6406	0.0602
SSE-1$'$	0.0603	0.7534	0.0609

Search Time: We measured each execution time of Search in order to evaluate the performance of SSE. Since the execution time of Search may depend on the number of search results, we measured it by classifying keywords into the following three cases.

- Words in $\Delta - \Delta'$: Words that can be searched although not contained in documents of a targeted search. The number of search results is zero.
- Words in Δ': Words contained in documents of the targeted search. The number of search results changes in accordance with words.
- Random character string: Words that cannot be searched. The number of search results is zero. Here, we make 1,000 random character strings of 16 characters.

Table 4 shows the results in execution time very small in all schemes. This means that the measures for privacy protection do not have a bad influence on efficiency.

From the evaluation results concerning SSE-1, Simple-SSE, and SSE-1$'$, we can say that both Simple-SSE and SSE-1$'$ satisfy the objectives of "efficient search" and "reasonable index size" mentioned in section 3.

7   Conclusion

In this paper, we reconsidered the balance between efficiency and security/privacy of a searchable symmetric encryption scheme.

By excluding consideration of active attacks, we proposed light searchable symmetric encryption schemes. We showed that they have some leakage, but this would pose no problems in most of practical cases.

We also showed experimental results of our scheme and comparison with existing schemes. The result showed that the index sizes in our schemes are only a few times of that of a general search engine (without encryption). Thus, our schemes are sufficiently secure and efficient enough for practical use.

References

1. Abdalla, M., et al.: Searchable Encryption Revisited: Consistency Properties, Relation to Anonymous IBE, and Extensions. In: Shoup, V. (ed.) CRYPTO 2005. LNCS, vol. 3621, pp. 205–222. Springer, Heidelberg (2005)
2. Apache Lucene - Welcome to Apache Lucene, http://lucene.apache.org/
3. Announcing the ADVANCED ENCRYPTION STANDARD (AES). Federal Information Processing Standards Publication 197 (November 2001)
4. Bellare, M., Boldyreva, A., O'Neill, A.: Deterministic and Efficiently Searchable Encryption. In: Menezes, A. (ed.) CRYPTO 2007. LNCS, vol. 4622, pp. 535–552. Springer, Heidelberg (2007)
5. Boneh, D., Di Crescenzo, G., Ostrovsky, R., Persiano, G.: Public key encryption with keyword search. In: Cachin, C., Camenisch, J.L. (eds.) EUROCRYPT 2004. LNCS, vol. 3027, pp. 506–522. Springer, Heidelberg (2004)
6. Chang, Y.-C., Mitzenmacher, M.: Privacy Preserving Keyword Searches on Remote Encrypted Data. In: Ioannidis, J., Keromytis, A.D., Yung, M. (eds.) ACNS 2005. LNCS, vol. 3531, pp. 442–455. Springer, Heidelberg (2005)
7. Curtmola, R., Garay, J., Kamara, S., Ostrovsky, R.: Searchable Symmetric Encryption: Improved Definitions and Efficient constructions. In: ACM Conference on Computer and Communications Security (CCS 2006), pp. 79–88. ACM, New York (2006)
8. Goh, E.-J.: Secure indexes. Technical Report 2003/216, IACR ePrint Cryptography Archive (2003)
9. Goldreich, O., Ostrovsky, R.: Software protection and simulation on oblivious RAMs. J. of the ACM 43(3), 431–473 (1996)
10. Hirano, T., Mori, T., Hattori, M., Ito, T., Matsuda, N., Kawai, Y., Sakai, Y., Ohta, K.: Security Notions for Searchable Symmetric Encryption with Extra Multiple Documents. In: The 29th Symposium on Cryptography and Information Security, SCIS 2012, 2B3-1 (2012) (in Japanese)
11. Iwanami, J., Ogata, W.: Secure and Efficient Searchable Symmetric Encryption with Document Addition. In: The 30th Symposium on Cryptography and Information Security, SCIS 2013, 3A3-1 (2013) (in Japanese)
12. Kamara, S., Lauter, K.: Cryptographic Cloud Storage. In: Sion, R., Curtmola, R., Dietrich, S., Kiayias, A., Miret, J.M., Sako, K., Sebé, F. (eds.) FC 2010 Workshops. LNCS, vol. 6054, pp. 136–149. Springer, Heidelberg (2010)
13. Kamara, S., Papamanthou, C., Roeder, T.: CS2: A searchable cryptographic cloud storage system. MSR Tech Report no. MSR-TR-2011-58. Microsoft, Redmond (2011)
14. Kamara, S., Roeder, T.: Dynamic Searchable Symmetric Encryption. In: Proc. of the 2012 ACM Conference on Computer and Communications Security, pp. 965–976. ACM, New York (2012)
15. van Liesdonk, P., Sedghi, S., Doumen, J., Hartel, P., Jonker, W.: Computationally efficient searchable symmetric encryption. In: Jonker, W., Petković, M. (eds.) SDM 2010. LNCS, vol. 6358, pp. 87–100. Springer, Heidelberg (2010)
16. Moby Word Lists, The Institute for Language, Speech and Hearing, http://icon.shef.ac.uk/Moby/
17. Song, D., Wagner, D., Perrig, A.: Practical techniques for searching on encrypted data. In: Proc. of 2000 IEEE Symposium on Security and Privacy, pp. 44–55 (2000)

Unconditionally Secure Oblivious Transfer from Real Network Behavior

Paolo Palmieri[1,*] and Olivier Pereira[2]

[1] Delft University of Technology, Parallel and Distributed Systems Group
Mekelweg 4, 2628 CD Delft, The Netherlands
`p.palmieri@tudelft.nl`
[2] Université catholique de Louvain, UCL Crypto Group
Place du Levant 3, B-1348 Louvain-la-Neuve, Belgium
`olivier.pereira@uclouvain.be`

Abstract. Secure multi-party computation (MPC) deals with the problem of shared computation between parties that do not trust each other: they are interested in performing a joint task, but they also want to keep their respective inputs private. In a world where an ever-increasing amount of computation is outsourced, for example to the cloud, MPC is a subject of crucial importance. However, unconditionally secure MPC protocols have never found practical application: the lack of realistic noisy channel models, that are required to achieve security against computationally unbounded adversaries, prevents implementation over real-world, standard communication protocols.

In this paper we show for the first time that the inherent noise of wireless communication can be used to build multi-party protocols that are secure in the information-theoretic setting. In order to do so, we propose a new noisy channel, the Delaying-Erasing Channel (DEC), that models network communication in both wired and wireless contexts. This channel integrates erasures and delays as sources of noise, and models reordered, lost and corrupt packets. We provide a protocol that uses the properties of the DEC to achieve Oblivious Transfer (OT), a fundamental primitive in cryptography that implies any secure computation. In order to show that the DEC reflects the behavior of wireless communication, we run an experiment over a 802.11n wireless link, and gather extensive experimental evidence supporting our claim. We also analyze the collected data in order to estimate the level of security that such a network can provide in our model. We show the flexibility of our construction by choosing for our implementation of OT a standard communication protocol, the Real-time Transport Protocol (RTP). Since the RTP is used in a number of multimedia streaming and teleconference applications, we can imagine a wide variety of practical uses and application settings for our construction.

---

* This work was accomplished while the author was at the Crypto Group of the Université catholique de Louvain.

K. Sakiyama and M. Terada (Eds.): IWSEC 2013, LNCS 8231, pp. 168–182, 2013.

1   Introduction

Multi-party computation protocols that are secure against computationally unbounded adversaries have seen, up until now, little or no practical use. This is
mainly due to the strong assumptions that need to be satisfied for them to work.
In particular, they require the availability of a noisy channel, the theoretical ab-
straction of an error-prone communication medium, since security can not be
achieved over a clear channel. The aim of this paper is to show that, through
the use of realistic channel models and efficient constructions, we can achieve secure multi-party computation over standard, commonly used network protocols
today.

In a 1-out-of-2 oblivious transfer protocol, Rachel (the receiver) wants to
learn one of the two secret bits b_0, b_1 that Sam (the sender) knows, but without
revealing to him her selection s. Sam, on the other hand, wants to make sure
that Rachel will not get any information about the other bit in the process.
The first protocol to achieve this over a noisy channel was designed by Crépeau
and Kilian, and used the Binary Symmetric Channel (BSC) [4]. The BSC is
a simple channel model where each binary input has a probability p of being
"flipped" when output: a 0 flipped becomes a 1 and vice versa. Since the BSC
does not provide a realistic model of communication, new channel models have
been subsequently proposed. Most of these models are modifications of the BSC
itself, that introduce more freedom for the attacker in order to increase the
generality of the construction. In particular, the Unfair Noisy Channel (UNC),
proposed by Damgard et al. in 1999 [6] and later improved in 2004 [5], lets
the adversary choose the error probability within a specific (narrow) range. The
Weak Binary Symmetric Channel (WBSC), designed by Wullschleger in [22], lets
a dishonest player know with a certain probability if a bit was received correctly.

While these constructions ease the assumptions needed to build OT from a
theoretical point of view, they hardly make the channel models closer to any
real communication channel. To address this problem, recent constructions use
noisy channels that try to model common transmission errors occurring in actual
networks. In particular, the use of transmission delays as source of noise has been
proposed in [12], where Palmieri and Pereira provide a protocol for achieving
oblivious transfer over the Binary Discrete-time Delaying Channel (BDDC). A
modified version of the protocol, secure against malicious players, has later been
introduced by Cheong and Miyaji [1].

The suitability of the BDDC to model packet reordering over IP networks has
been shown in [13]. However, the BDDC does not take into account the possibility of packets being lost, which is a common occurrence in real communication
settings. Moreover, it does not limit the number of times a packet can be delayed: however unlikely, it is possible for a packet to be delayed indefinitely. The
behavior of a real packet-switching network would be instead to drop a packet
after a certain time, usually called *time to live* (TTL).

1.1   Contribution

In this paper we propose a new noisy channel, the Delaying-Erasing Channel (DEC). The DEC integrates delays and erasures (lost packets) and introduces a limit to the number of possible delays. This channel, while being based on discrete times like its predecessors, addresses the lacks of the BDDC, and provides a realistic model for network communication, in both wireless and wired settings. We propose a protocol for achieving oblivious transfer over the DEC, and we study the security of the construction against both semi-honest and malicious adversaries.

   The main goal of the DEC is to finally provide a realistic noisy channel model for network communication. In order to show that the DEC achieves this goal, we conduct an experiment simulating our OT protocol over a wireless network, and we collect extensive statistical evidence that supports our claims of security and flexibility for the construction. We analyze the collected data using several standard tools for entropy estimation, whose results confirm the suitability of the wireless medium to be used as a noisy channel. Our implementation of OT is based on the Real-time Transport Protocol (RTP), an application layer protocol frequently used for the streaming of multimedia content.

1.2   Outline of the Paper

In section 2 we give a security definition of oblivious transfer In section 3 we introduce the Delaying-Erasing Channel (DEC), and we provide a protocol implementing oblivious transfer over it. In section 3.2 we prove the security of the construction in the semi-honest setting, while in 3.3 we discuss the case of malicious adversaries. In section 4 we show that packets transmitted over a 802.11n wireless link show a behavior consistent with the channel definition. We analyze the experimental results and measure the entropy of the network errors in section 4.4.

2   Preliminaries

For a protocol to successfully implement oblivious transfer, three conditions must be satisfied after an execution: the receiver, Rachel, learns the value of the chosen bit b_s (correctness); the sender, Sam, learns nothing about the value of the selection bit s (security for Rachel); the receiver learns no further information about the value of the other bit b_{1-s} (security for Sam) [4]. When proving the security of our construction, we use the security definition of oblivious transfer provided in [12]. The definition uses the concept of *prediction advantage*, a measure of the advantage that an adversary has in guessing a secret bit by using all the information available to him. We use the notation found in [21].

Definition 1. ([21]) *Let P_{XY} be a distribution over $\{0,1\} \times \mathcal{Y}$. The* maximal bit prediction advantage *of X from Y is*

$$\mathrm{PredAdv}\,(X \mid Y) = 2 \cdot \max_{f} \mathrm{Pr}\,[f\,(Y) = X] - 1 \ . \tag{1}$$

The *view* of a player consists of all the information that the player learns during the protocol execution. The sender, the receiver and the potential adversary all have different views. The security definition for OT follows.

Definition 2. [12] *A protocol Π between a sender and a receiver, where the sender inputs $(b_0, b_1) \in \{0, 1\}$ and outputs nothing, and the receiver inputs $s \in \{0, 1\}$ and outputs S, securely computes 1-2 oblivious transfer with an error of at most ε, assuming that U and V represent the sender and receiver views respectively, if the following conditions are satisfied:*

- (Correctness) *If both players are honest, we have*

$$Pr\left[S = b_s\right] \geq 1 - \varepsilon . \tag{2}$$

- (Security for Sam) *For an honest sender and an honest (but curious) receiver we have*

$$\mathrm{PredAdv}\left(b_{1-s} \mid V, s\right) \leq \varepsilon . \tag{3}$$

- (Security for Rachel) *For an honest receiver and an honest (but curious) sender we have*

$$\mathrm{PredAdv}\left(s \mid U, b_0, b_1\right) \leq \varepsilon . \tag{4}$$

3   Delaying-Erasing Channel

The channel model we propose combines the erasure and delaying channels. It takes into account the possibility for an input string to be delayed or to be lost (that is, erased). The channel also sets a limit to the number of delays that an input can suffer, and considers lost (erased) any string delayed a number of times equal or higher than that. We call p the delaying probability, r the maximum number of delays and q the erasing probability. Consequently, the probability that a string will be considered lost by a receiver is $(q + p^r)$, consisting of the erasing probability plus the probability for the string to be delayed r times.

Definition 3. *A Delaying-Erasing Channel (DEC) with delaying probability p, erasing probability q and maximum number of delays per single input r accepts as input a sequence $T = \langle t_1, t_2, \ldots \rangle$ of sets of strings $t_i \in (\{0, 1\}^n)^*$, called input times, and outputs a sequence $U = \langle u_1, u_2, \ldots \rangle$ of sets of strings $u_i \in (\{0, 1\}^n)^*$ called output times. Each string X admitted into the channel at input time $t_i \in T$ is output at most once by the channel, with probability of being output at time $u_j \in U$*

$$\Pr\left[X \in u_j \mid X \in t_i\right] = \begin{cases} (1 - q) \cdot p^{j-i} \cdot (1 - p) \\ \qquad\qquad with\ 0 \leq j - i < r, \\ 0 \qquad\qquad otherwise. \end{cases} \tag{5}$$

In practice, the channel works as follows. An input string X that enters the channel at time t_i is due to be output by the channel at time u_i with probability $(1 - p - q)$. The channel has probability q of erasing the string (we call this event *impromptu loss*), in which case the string is not output by the channel. If an impromptu loss does not occur, the string has a probability p of being delayed until the next output time. The delay event can happen multiple times: once a string is delayed, it can be delayed again. Therefore, the string has probability p^d of being delayed d times and being output at u_{i+d}, as long as $d < r$. Finally, the string is not output by the channel if it is delayed r times or more, which happens with probability p^r.

3.1   OT Protocol

Our oblivious transfer protocol follows the general scheme proposed by Crépeau and Kilian in [4], which has been at the base of every following OT construction. The idea behind the scheme is to generate, through a specific transmission strategy, a simple erasure channel over the available noisy channel (in our case the DEC), and then use it a number of times in order to realize OT and achieve security by privacy amplification. The protocol we propose is also similar to the one proposed in [12] for the Binary Discrete-time Delaying Channel, in the sense that it uses a precomputation phase during which two sets of packets are created. Contrary to the case of the BDDC, our protocol streams the two sets of packets by interleaving them: we send the first packet of the first set at t_1, then the second packet of the first set and the first of the second set at t_2 and so on. This allows us to exploit the uncertainty caused by the lost and delayed packets.

The protocol works as follows. First the sender, Sam, precomputes a sequence of packets. For simplicity, we can assume that the packets only contain their sequence number i. Then, he starts sending the packets over the DEC to the receiver, Rachel. Each packet is sent twice: the first packet at times t_1 and t_2, the second one at t_2 and t_3 and so on. Each of the two times the same packet is sent, Sam also attaches to it a unique identifier (e_i for the first transmission and e_i' for the second one), so that he will be able to tell them apart. However, he will not reveal to Rachel which identifier is used for which transmission of the packet. Rachel keeps track of the packets lost on the way and of the arrival times of those she receives. However, the channel does not give her any feedback on the delays that occur during transmission. Therefore, in the case of a packet received for the first time later than the expected time u_i, she is not able to tell which of the two copies of the packet was sent with the first transmission, and which with the second. The same is true in case only one copy arrives and it is received after the expected time u_i, or in case both copies are lost. At the same time, Sam does not know the arrival time and order of the packets. We use this uncertainty to build oblivious transfer. Rachel assigns to her selection bit s the packets for which she knows with certainty the identifier e, and to the other bit $(1 - s)$ the other packets. Then, she sends her two selections of packets to Sam. Sam encodes the secret bits b_0 and b_1 using the identifiers e attached to the packets during the first transmission, according to the selection operated by Rachel. Using the same identifiers, Rachel is able to decode b_s, but not b_{1-s}.

Protocol 1. The parties have a clear channel and a p-q-r-DEC with $0 < (p+q) < \frac{1}{2}$, $r > 1$ available for communication. Sam selects two disjoint sets E and E', each composed of n distinct binary strings of length l: $e_1, \ldots, e_n \in E$ and $e'_1, \ldots, e'_n \in E'$. From E and E' Sam builds the sets $C = \{c_1, \ldots, c_n\}$ and $C' = \{c'_1, \ldots, c'_n\}$, according to the following rules: $c_i := e_i \| i$ and $c'_i := e'_i \| i$. Then the parties communicate as follows:

1. Sam sends the set C to Rachel over the DEC, one string at each input time, starting at t_1. At t_2 he starts sending C' as well. This way, at each t_i, c_i and c'_{i-1} are sent.
2. Rachel receives over the DEC the strings in $\{C \cup C'\}$ that have not been erased, in the order produced by the channel.
3. Rachel selects the set I_s, where $s \in \{0, 1\}$ is her selection bit, such that $|I_s| = \frac{n}{2}$ and so that $i \in I_s$ only if she is able to distinguish $c_i \in C$ from $c'_i \in C'$. This happens in two cases: c_i has been received at u_i; or c_i, c'_i have not been erased and c'_i has been received at u_{i+r}. If less than $\frac{n}{2}$ strings can be placed in I_s, Rachel instructs Sam to abort the communication. Otherwise she selects $I_{1-s} = \{1, \ldots, n\} \setminus I_s$ and sends I_0 and I_1 to Sam over the clear channel. [1]
4. Sam receives the sets I_0 and I_1. Then, he chooses two universal hash functions f^0, f^1, whose output is 1-bit long for any input. Let $E_j \subset E$ be the set containing every $e_i \in E$ corresponding to an $i \in I_j$, such that

$$e_i \in E_j \Leftrightarrow i \in I_j \ . \tag{6}$$

For each set I_j, Sam computes the string g_j by concatenating each $e^j_k \in E_j$, ordering them for increasing binary value, so that

$$g_j = \left(e^j_1 \| \ldots \| e^j_{\frac{n}{2}} \right) \quad \text{with } e^j_1, \ldots, e^j_{\frac{n}{2}} \in E_j \ . \tag{7}$$

Sam computes $h_0 = f^0(g_0)$, $h_1 = f^1(g_1)$ and sends to Rachel over the clear channel the functions f^0, f^1 and the two values

$$k_0 = (h_0 \oplus b_0) \ , \qquad k_1 = (h_1 \oplus b_1) \ . \tag{8}$$

5. Rachel computes her guess for b_s

$$b_s = f^s(g_s) \oplus k_s \ . \tag{9}$$

3.2   Security: Honest-But-Curious Adversaries

In the *semi-honest* setting, the players follow the protocol, but try to use any information available to them in order to guess the other player's secret. We prove the security of our construction by proving each of the three conditions of the security definition of oblivious transfer (Definition 2).

---

[1] In order to improve the efficiency of the protocol in a real setting, the receiver can send just one of these two sets, for example always I_0, as the sender can easily reconstruct the other.

Correctness. The first condition of Definition 2 states that, if both players behave in an honest way, the secret bit must be correctly received and decoded by the receiver party. In practice, the protocol succeeds when Rachel is able to identify with certainty at least $\frac{n}{2}$ strings from C among all the strings she receives. As stated in step 3 of the protocol, a string c_i is known by Rachel to be $\in C$ with certainty either when it is received at u_i; or when c_i is not erased and the corresponding string $c'_i \in C'$ is received at u_{i+r}. Therefore, the probability that a string c_i will not be identifiable as being part of C is upper-bounded by the probability $(p + q)$ that c_i is erased, or delayed at least once. Let us denote by X the random variable counting the number of strings not affected by the noise (that is, erased or delayed) out of the n total strings in C. We have that $\Pr\left[X \le \frac{n}{2}\right]$, the probability that not enough strings in C are received correctly and on time for the protocol to succeed, follows the cumulative distribution function of the binomial distribution. For Hoeffding's inequality we have that

$$\Pr\left[X \le \frac{n}{2}\right] \le \exp\left(-2n\left(p + q - \frac{1}{2}\right)^2\right). \tag{10}$$

Therefore, the correctness condition is satisfied with overwhelming probability in n as soon as $p + q < \frac{1}{2}$, as per the protocol definition.

Security for Sam. A curious Rachel is interested in learning b_{1-s}. She has two ways of obtaining the value: either by decoding k_{1-s} on the correct g_{1-s}, or by trying to guess it on a (partially) incorrect g_{1-s}. In the latter case, the probability of a correct guess is upper-bounded by $\frac{1}{2}$, for the properties of a universal hash function. In the following we evaluate the probability of the former.

For each pair of strings $(c_i \in C, c'_i \in C')$, Rachel receives two or less strings, in the order produced by the channel. She is interested in determining c_i, in order to learn e_i. We analyze in the following her ability of doing so, based on the different events that can happen after the transmission of the strings through the delaying-erasing channel. We suppose that, in case only one string is received, Rachel assumes to have received c_i.[2] For each (c_i, c'_i) we can have that:

- c_i is neither erased nor delayed. Independently of what happens to c'_i, Rachel learns e_i. This happens with probability $(1 - p - q)$.
- c_i is erased. Independently of what happens to c'_i, Rachel is not able to recover the identifier e_i. This happens with probability q.
- c_i is delayed. This happens with probability p. In this case, Rachel's probability to learn e_i depends on c'_i. We can have that:
  - c'_i is erased. Following the strategy of using the identifier she possesses, Rachel succeeds in guessing the right identifier. This happens with probability $p \cdot q$.

---

[2] This is always the best strategy, since a wrong assumption does not lower her probability of learning e_i: we assume that guessing e_i with no information has a negligible probability of succeeding.

- c_i' is not erased. This happens with probability $p(1-q)$. If c_i' is delayed $r-1$ times, Rachel learns the right identifier. This happens with probability $p^r(1-q)$. Otherwise, Rachel guesses the right identifier with probability $\frac{1}{2}$. In fact, the probability for the strings to arrive in the same order in which they are sent is equal to the probability for them to arrive in the reverse order ($\frac{p^2}{1+p}$). Therefore she does not have any strategy better than tossing a coin in both cases, as well as when the strings arrive at the same time.

Therefore, for each pair of strings (c_i, c_i'), Rachel does not learn e_i with probability

$$\Pr\left[\neg e_i\right] = q + \frac{p(1-q) - p^r(1-q)}{2} \ , \tag{11}$$

which is > 0 as soon as $0 < (p+q) < \frac{1}{2}$ and $r > 1$ as per the protocol definition. Therefore, Rachel's probability of building the correct g_{1-s} by learning the correct e_i for every $i \in I_{1-s}$ is

$$\Pr\left[g_{1-s}\right] = (1 - \Pr\left[\neg e_i\right])^n \ , \tag{12}$$

which is negligible in n.

Security for Rachel. Since the delaying-erasing channel does not give any feedback to the sender on the state of transmitted strings, Sam ignores whether a string has been correctly received or not, and if it has been delayed during transmission. Therefore, from the point of view of a curious sender the distribution of (I_0, I_1) is independent of s.

3.3   Security: Malicious Adversaries

We observe that the semi-honest assumption of our construction is only required for the sender, but not for the receiver. This is also the case for the oblivious transfer protocol proposed for the BDDC [12]. In fact, a malicious Rachel can either send to Sam a malformed set I_{1-s}, where she puts only indices of strings not affected by the noise (for instance, by sending less i's than required or by including i's already in I_s), or swap strings affected by the noise with non-affected ones between the sets I_s and I_{1-s}. If Rachel chooses the former strategy, Sam can detect her malicious behavior by implementing a simple additional check on I_{1-s}, and abort the protocol in case the behavior of the receiver deviates from the protocol. The latter strategy, instead, increases Rachel's probability to learn the other bit b_{1-s}, by moving delayed or erased strings from I_{1-s} to I_s, but only at the cost of lowering her probability to learn the selected bit b_s. In fact, the number of strings that have been delayed or erased by the channel, which is also the number of guesses that Rachel needs to make, remains the same. Therefore the probability for Rachel to decode both b_s and b_{1-s} is the same whether she acts honestly or in a malicious way.

As already noted in [10], we can use an oblivious transfer protocol secure against a malicious receiver and a semi-honest sender to obtain a protocol secure against a semi-honest receiver and a malicious sender. This is possible thanks to the symmetry property of oblivious transfer, proved for the first time in [20]. A black-box combiner for this reversal operation has been proposed in [8], where a compiler that combines the two protocols into one that is secure against generic malicious adversaries, originally designed for the case of OT based on trapdoor functions, is also presented.

4   From Noisy Channel to Real Network Behavior

The aim of this section is to show that the DEC realistically models actual network behavior. In order to do so, we simulate the OT protocol over a wireless point-to-point connection between two hosts, and we study the amount of errors that occur during the transmission and the predictability of such errors. We show the flexibility of our construction by implementing our OT protocol over a standard Internet protocol, the Real-time Transport Protocol.

4.1   Real-Time Transport Protocol (RTP)

The Real-time Transport Protocol (RTP) [7,15] is an application layer protocol designed for the delivery of real-time information. Its typical use is the delivery of real-time audio and video, as in the case of multimedia streaming or teleconferencing. It is often used in conjunction with the Real Time Streaming Protocol (RTSP) [16], that provides a framework for controlling the data flow. RTP typically runs on top of the User Datagram Protocol (UDP). Both protocols are particularly suited to be used in our construction: they do not guarantee reliable transmission or quality-of-service and they do not support error correction and lost packet resending. The protocol specification for RTP expressly states that it "does not guarantee delivery or prevent out-of-order delivery, nor does it assume that the underlying network is reliable and delivers packets in sequence" [15].

4.2   OT over RTP

Taking advantage of the fact that RTP does not prevent packet loss or reoredering, we can use it as the base for our oblivious transfer construction. In particular, the parties use RTP at step 1 of the protocol, while communicating over a wireless (or wired) link, that acts as the noisy channel. The sender sends two distinct RTP streams composed of the same number of packets. The content of each packet can be arbitrarily selected by the sender, as long as it is unique with respect to both streams, since it is to be used as the packet identifier. Some of the fields of a standard RTP packet header (see RFC 3550 [15] for reference) require special care in our application. The Sequence Number value will be used, with the same meaning, also in the OT protocol. Packets sharing the same position in the two streams will be forged by the sender in order to have identical headers. In particular, this has to be enforced for the `Timestamp` field. Similarly, the identifier of the synchronization source (that is, the sender) has to be replicated in both streams. The `Payload Type` field, which indicates the encoded format of the data sent with RTP, can be chosen arbitrarily. The underlying protocols (UDP or IP) do not add information that could make the streams distinguishable, so no specific intervention is needed at levels lower than the application layer, other than selecting the desired time-to-live at the IP level.

The contemporary transmission of multiple RTP streams from the same source does not reveal the specific use we make of the protocol. In fact, it is a common occurrence: for instance, in the transmission of multimedia content, audio and video streams usually have separate RTP sessions, enabling a receiver to deselect a particular stream.

The following steps of the protocol remain unchanged, as they are performed over a clear channel.

4.3 Experiment

The aim of this experiment is double: to show that the DEC realistically models the noise introduced during network communication, and to analyze the unpredictability, and therefore the suitability for secure computation of that noise. We conduct the experiment as follows.

The party acting as sender is simulated by a wireless router running the open source and Linux-based custom firmware OpenWRT. This particular configuration let us use the RTP/RTSP streaming server Live 555 directly on the device. The receiver party, a notebook computer, connects to the router before starting the OT protocol using the IEEE 802.11n-2009 wireless transmission method [9], and receives an IP address through a DHCP request. This way, the receiver and sender parties are directly connected by a wireless link.

The notebook computer simulating the receiver party is placed at about 12 meters of distance from the router. No physical obstacles block the line of sight between the two devices. Both parties are not engaged in any network communication other than the RTP streaming. The streaming session, initiated by the sender, runs for 158.28 seconds, with a total of 5629 packets sent. The packets reaching the receiver party are collected in the order they are received using the open source sniffing tool WireShark. [3] Steps 3 to 5 of the protocol (encoding and decoding of the secret bits and communication over a clear channel) are not simulated during the experiment.

The results of the experiment are shown in Table 1, and appear to be consistent with relevant literature (see, for instance, [14]). The number of lost packets (erasures) and sequence errors (delays) has been obtained using the RTP Stream Analysis tool provided with WireShark. In the following we analyze the results from the security point of view.

---

[3] The sample data transmitted, and the dump of the packets received is available at the URL: `http://www.uclouvain.be/crypto/ot-wireless-tests.tar.gz`

Table 1. Average lost (erased) and displaced (delayed) packets during video streaming using the RTP protocol over a wireless link

Total RTP packets:	5629	
Erasures:	65	1.15%
Delays:	109	1.94%
Total errors:	174	3.09%

4.4   Analysis

The amount of noise that we observed during the experiment indicates that both lost packets and sequence errors are relatively common occurrences, as shown in Table 1.

The security of our construction, however, also depends on the (im)possibility, for an attacker, of being able to predict errors. In other words, we want the distribution of the displaced and lost packets into the sequence to be as uniform as possible. In order to evaluate how much this assumption reflects the reality of wireless communication, we convert the sequence of packets generated during the experiment into a binary string, using the following strategy: the packets affected by the noise are represented by a bit of value 1, those not affected by a bit of value 0. Then, we estimate the entropy of the generated binary string using a set of standard test suites, in particular: ent [17], Maurer's test including Coron's modification [11,3,2] and the Context-Tree Weighting (CTW) method [19,18]. The main idea behind these tools for entropy estimation is to compare the length of an input sequence with its output after compression. Since the probability of errors (and therefore of 1's) is lower than 0.5, we compare it to the Shannon entropy normalized to the actual probability, calculated using the standard definition

$$H_b(p) = -p \log_2 p - (1-p) \log_2(1-p) \tag{13}$$

and the amount of noise observed during our experiment. Since we fix the probability p to the observed value, $H_b(p)$ is the maximum possible entropy, and not an upper-bound. This does not affect the reliability of the results, since our goal is to detect the presence of any pattern in the error distribution that might lead to predictability, and not to evaluate the error probability itself. In the case of packet delays, we have $p = 0.0194$, and therefore $H_b(p) = 0.1392$. Entropy estimations calculated by the three tests mentioned above are shown in Table 2: the closer to the maximum entropy $H_b(p)$ the estimated values are, the less likely we are to find any pattern in the sequence.

While in the case of the ent test the entropy estimation is virtually identical to the maximum value, the context-tree weighting method is able to compress to a higher ratio. In fact, the CTW algorithm produces an output whose size is 82% of the one that would be obtained compressing an input where errors are uniformly distributed. The Maurer-Coron test is the most effective, reaching a compression ratio of 71%. However, this is partly due to a requirement in the

Table 2. Entropy estimation for one bit, given $p = 0.0196$, as observed during the experiment

Max. normalized entropy $H_b (p)$: 0.1392	
Ent	0.1392
Maurer*	0.0994
CTW	0.1144

algorithm that imposes a minimum input length higher than the size of our test string. Therefore, during the test execution, about 800 bits of the input string are read twice, since the test loops the input in case of an insufficient amount of data to elaborate. Overall, these results confirm that, even in a setting where a low amount of noise can be expected, errors are both enough frequent and randomly distributed to allow for a significant security margin to be achieved.

5   Conclusion

In this paper we propose a noisy channel model that reflects, for the first time, the behavior of real networks. We present experimental evidence collected during an experiment over wireless communication supporting this claim, and we show the flexibility of the model by running the experiment using a commonly used Internet streaming protocol, the Real-time Transport Protocol.

Analysis of the noise introduced by the wireless medium during the experiment supports the assumptions that the channel makes in terms of unpredictability of that noise. In fact, using standard entropy estimation tools, we estimate the normalized entropy to be between 71% and 100% of the theoretical maximum, depending on the test, even for a relatively clean channel where the amount of noise observed is, on average, 3.09%. This allows us to construct, for the first time, an oblivious transfer protocol secure against computationally unbounded adversaries over a real network. We believe that the flexibility of our model and construction will help open the way to widespread implementation of secure multi-party computation.

Acknowledgments. This research work was supported by the SCOOP Action de Recherche Concertées. Olivier Pereira is a Research Associate of the F.R.S.-FNRS.

References

1. Cheong, K.-Y., Miyaji, A.: Unconditionally secure oblivious transfer based on channel delays. In: Qing, S., Susilo, W., Wang, G., Liu, D. (eds.) ICICS 2011. LNCS, vol. 7043, pp. 112–120. Springer, Heidelberg (2011)
2. Coron, J.-S.: On the security of random sources. In: Imai, H., Zheng, Y. (eds.) PKC 1999. LNCS, vol. 1560, pp. 29–42. Springer, Heidelberg (1999)

3. Coron, J.-S., Naccache, D.: An accurate evaluation of maurer's universal test. In: Tavares, S., Meijer, H. (eds.) SAC 1998. LNCS, vol. 1556, pp. 57–71. Springer, Heidelberg (1999)
4. Crépeau, C., Kilian, J.: Achieving oblivious transfer using weakened security assumptions (extended abstract). In: FOCS, pp. 42–52. IEEE (1988)
5. Damgård, I., Fehr, S., Morozov, K., Salvail, L.: Unfair noisy channels and oblivious transfer. In: Naor, M. (ed.) TCC 2004. LNCS, vol. 2951, pp. 355–373. Springer, Heidelberg (2004)
6. Damgård, I., Kilian, J., Salvail, L.: On the (Im)possibility of basing oblivious transfer and bit commitment on weakened security assumptions. In: Stern, J. (ed.) EUROCRYPT 1999. LNCS, vol. 1592, pp. 56–73. Springer, Heidelberg (1999)
7. Group, A.V.T.W., Schulzrinne, H., Casner, S., Frederick, R., Jacobson, V.: RTP: A Transport Protocol for Real-Time Applications. RFC 1889 (Proposed Standard) (January 1996), http://www.ietf.org/rfc/rfc1889.txt, obsoleted by RFC 3550
8. Haitner, I.: Semi-honest to malicious oblivious transfer—The black-box way. In: Canetti, R. (ed.) TCC 2008. LNCS, vol. 4948, pp. 412–426. Springer, Heidelberg (2008)
9. IEEE-SA: Ieee 802.11n-2009 amendment 5: Enhancements for higher throughput (October 2009)
10. Ishai, Y., Kushilevitz, E., Lindell, Y., Petrank, E.: Black-box constructions for secure computation. In: Kleinberg, J.M. (ed.) STOC, pp. 99–108. ACM (2006)
11. Menezes, A.J., Vanstone, S.A., Oorschot, P.C.V.: Handbook of Applied Cryptography, 1st edn. CRC Press, Inc., Boca Raton (1996)
12. Palmieri, P., Pereira, O.: Building oblivious transfer on channel delays. In: Lai, X., Yung, M., Lin, D. (eds.) Inscrypt 2010. LNCS, vol. 6584, pp. 125–138. Springer, Heidelberg (2011)
13. Palmieri, P., Pereira, O.: Implementing information-theoretically secure oblivious transfer from packet reordering. In: Kim, H. (ed.) ICISC 2011. LNCS, vol. 7259, pp. 332–345. Springer, Heidelberg (2012)
14. Salyers, D., Striegel, A., Poellabauer, C.: Wireless reliability: Rethinking 802.11 packet loss. In: WOWMOM, pp. 1–4. IEEE (2008)
15. Schulzrinne, H., Casner, S., Frederick, R., Jacobson, V.: RTP: A Transport Protocol for Real-Time Applications. RFC 3550 (Standard) (July 2003), http://www.ietf.org/rfc/rfc3550.txt
16. Schulzrinne, H., Rao, A., Lanphier, R.: Real Time Streaming Protocol (RTSP). RFC 2326 (Proposed Standard) (April 1998), http://www.ietf.org/rfc/rfc2326.txt
17. Walker, J.: Ent: A pseudorandom number sequence test program, http://www.fourmilab.ch/random/
18. Willems, F.M.J.: The context-tree weighting method: Extensions. IEEE Transactions on Information Theory 44(2), 792–798 (1998)
19. Willems, F.M.J., Shtarkov, Y.M., Tjalkens, T.J.: The context-tree weighting method: basic properties. IEEE Transactions on Information Theory 41(3), 653–664 (1995)
20. Wolf, S., Wullschleger, J.: Oblivious transfer is symmetric. In: Vaudenay, S. (ed.) EUROCRYPT 2006. LNCS, vol. 4004, pp. 222–232. Springer, Heidelberg (2006)
21. Wullschleger, J.: Oblivious-transfer amplification. In: Naor, M. (ed.) EUROCRYPT 2007. LNCS, vol. 4515, pp. 555–572. Springer, Heidelberg (2007)
22. Wullschleger, J.: Oblivious transfer from weak noisy channels. In: Reingold, O. (ed.) TCC 2009. LNCS, vol. 5444, pp. 332–349. Springer, Heidelberg (2009)

A   Equipment and Configuration

The wireless router used for the purpose of the experiment is a Netgear N600 (WNDR3800). It is a dual band (2.4 or 5.0 GHz), 802.11a/b/g/n capable device. It is powered by an Atheros AR7161 rev. 2 680 MHz CPU, and has 128MiB of RAM and 16MiB of flash memory. [4]

The OpenWRT version installed on the router is 10.03.1, the latest at the time of writing. The open source LIVE555[TM] Media Server (updated to version 2012.05.17) was installed, and used for streaming packets with the RTP/RTSP protocol.

The USB Wireless LAN adapter used during the experiment is a Linksys AE2500 (branded Cisco). This adapter is capable of working according to the latest WIEEE 802.11n standard (but can also work in 802.11b or 802.11g compatible modes). It supports dual band communication (2.4 GHz or 5 GHz).[5]

On the client side, the stream was displayed using the open source media player VLC (version 2.0.2 "Twoflower"), and packets were dumped using the WireShark open source sniffing tool.

The wireless configuration used for the router/access point (AP) during the experiment is shown in table 3.

Table 3. Configuration of the Wireless router-AP for the experiment

W-LAN:	IEEE 802.11n
	(2.4 GHz band)
AP Channel:	6 (2437 MHz)
AP Security:	WPA-CCMP(AES)
	Pre-Shared Key (PSK)
Active STA's:	1

B   RTP Packet Header

The header of an RTP packet is shown in Figure 1, as described in [15]. The first field specifies the protocol revision used (the current version is 2). The padding (P) field indicates if there are extra padding bytes at the end of the packet. X, extension, indicates the presence of application or profile specific headers between the standard header and the payload data. Extensions of the protocol can also use the marker (M) field, to indicate that the current packet has some special relevance for the application. The Real-time Transport Protocol allows the transmitted information to be generated by multiple sources. In this

---

[4] Full specifications are available at the manufacturer's website:
http://www.netgear.com/home/products/wirelessrouters/high-performance/
WNDR3700.aspx

[5] Full specifications are available at the manufacturer's website:
http://home.cisco.com/en-eu/products/adapters/AE2500

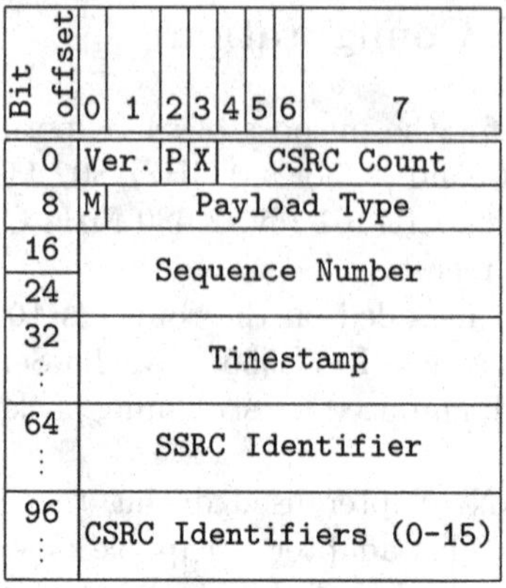

Bit offset	0	1	2	3	4	5	6	7
0	Ver.		P	X				CSRC Count
8	M				Payload Type			
16					Sequence Number			
24								
32 ⋮					Timestamp			
64 ⋮					SSRC Identifier			
96 ⋮	CSRC Identifiers (0-15)							

Fig. 1. RTP Packet Header

case, the packet flow will be synchronized by a unique *synchronization source* (SSRC), while any additional source will act as *contributing source* (CSRC). Both SSRC and CSRC's have unique identifiers, whose value is contained in the SSRC Identifier and CSRC Identifiers fields respectively. The maximum number of CSRC's is 16, and the actual number for a specific stream is defined in the CSRC Count field. For the purpose of our oblivious transfer protocol, only the synchronization source is used. The RTP header also contains information about the format used for the payload data (Payload Type) and specifies for each packet a Sequence Number and a Timestamp.

Cryptographically-Secure and Efficient Remote Cancelable Biometrics Based on Public-Key Homomorphic Encryption

Takato Hirano, Mitsuhiro Hattori, Takashi Ito, and Nori Matsuda

Information Technology R&D Center, Mitsubishi Electric Corporation, Japan
{Hirano.Takato@ay, Hattori.Mitsuhiro@eb, Ito.Takashi@aj,
Matsuda.Nori@ea}.MitsubishiElectric.co.jp

Abstract. Cancelable biometrics is known as a template protection approach, and concrete protocols with high accuracy and efficiency have been proposed. Nevertheless, most known protocols, including the Hattori et al. protocol (Journal of Information Processing, 2012), pay little attention to security against the replay attack, which leads to severe authenticity violation in the remote authentication setting. In this paper, we revisit the Hattori et al. protocol based on the Boneh-Goh-Nissim encryption scheme, and propose a secure variant while keeping user-friendliness of the original protocol. Our protocol uses the revocation method of the original protocol in a proactive manner, i.e., in our protocol, the public key assigned to a user is randomly re-generated in every authentication process. We define a general and formal security game that covers the replay attack and considers fuzziness of biometric feature extraction, and show that our protocol is secure in that model. The computation and communication costs of our protocol are more efficient than those of similar protocols.

Keywords: Cancelable biometrics, remote authentication, replay, security game for biometrics, homomorphic encryption.

1 Introduction

Biometric authentication is based on physical characteristics of claimants such as fingerprints, facial features, iris, retina, DNA, vein, and their combinations. Since these characteristics have adequate amount of information for user authentication and are easily captured from the human body when needed, claimants do not require to remember long and complex information nor tokens. In other words, biometric authentication has advantage in user-friendliness and applicability over traditional methods such as knowledge-based authentication or possession-based authentication. However, (ordinary) biometric authentication has crucial issues, *privacy* and *revocability*.

As for *privacy*, biometric information such as face, fingerprints, DNA, etc. can be regarded as sensitive information because it represents physical characteristics of individuals. For example, it is known that retinal patterns might provide

K. Sakiyama and M. Terada (Eds.): IWSEC 2013, LNCS 8231, pp. 183–200, 2013.

medical information on diabetes. Therefore, biometric information should not be disclosed to others, even to verifier (e.g. authentication servers).

As for *revocability*, it is infeasible to revoke biometric information because only a limited number of biometric information can be used for substitution. Furthermore, once templates of biometric information stored in the authentication server are leaked out, they might be used in another authentication system for disguise. This situation would allow illegitimate users to gain illegal access to services.

Therefore, template protection techniques have been studied, in order to address these issues.

1.1 Related Works

Roughly speaking, the template protection techniques can be classified into three approaches: Cancelable Biometrics [15,4,6,19,17,1,7,18,8], Biometric Cryptosystem [11,10], and ZeroBio [16,12,13].

Cancelable Biometrics: The first approach, Cancelable Biometrics, is that by using a randomization key, user's biometric feature is transformed to random data [15]. Up to now, concrete protocols with high accuracy and efficiency have been proposed. Since cancelable biometrics uses encryption-like techniques, there are several protocols which can be proved secure template protection under some cryptographic assumptions. For example, as symmetric-key encryption approach, informationally secure protocols based on the correlation invariant random filtering have been proposed (e.g. [17,18]), and as public-key encryption approach, computationally secure protocols based on (additively or somewhat) homomorphic encryption have been proposed (e.g. [4,7]).

Especially, Hattori, Ito, Matsuda, Shibata, Takashima, and Yoneda proposed two cancelable biometric protocols which are cryptographically-secure template protection against passive adversaries in a semi-honest model, by using public-key homomorphic encryption which can evaluate 2-DNF (disjunctive normal form) on ciphertexts [7]. Concretely, one is based on the Boneh-Goh-Nissim (BGN) encryption scheme [2] and the other is based on the Okamoto-Takashima encryption scheme [14]. In addition to template protection, their protocols are efficient and "user-friendly". Here, "user-friendly" means that each user has only to access or possess public information such as his public key and ID, and thus no direct access to (the corresponding) secret key is needed.

However, most known cancelable biometric protocols, including the Hattori et al. one, have a crucial security issue that the replay attack is applicable in the remote authentication setting (i.e. authentication is proceeded through public networks). That is, cancelable biometrics focuses on security for template protection and pays little attention to security in the remote authentication setting, although biometrics is one of the authentication protocols.

There are a small number of cancelable biometric protocols which also consider the cryptographic security for (remote) authentication [6,19,1,8].

The protocols proposed in [6,19,1] are based on public-key homomorphic encryption and have resistance against the replay attack and other impersonation attacks by interacting between the users and the authentication server in challenge-response manners. However, security arguments for the protocols are intuitive (i.e. detailed cryptographic security proofs are not given), and the user-friendliness is lacked because the user must have the secret key in addition to his biometric information (i.e. their protocols can be regarded as token-based authentication).

Recently, Hirano, Hattori, Ito, Matsuda, and Mori proposed a variant of the Hattori et al. protocols [7], by combining a challenge-response approach with additive homomorphicity [8]. Although their protocol has the user-friendliness and the cryptographic security for both template protection and authentication against passive adversaries in a semi-honest model, the computational and communication costs are huge because a large number of random numbers are used in the authentication phase.

Biometric Cryptosystem: The second approach, Biometric Cryptosystem, is based on error-correcting codes, and its main focus is on generating a secret key which can be applied to various cryptosystems such as authentication protocols. Biometric Cryptosystem includes fuzzy commitment [11], fuzzy vault [10], etc. Since concrete schemes of Biometric Cryptosystem and authentication protocols can be chosen independently, this approach has high flexibility and can guarantee the security for authentication by using "secure" authentication protocols with the generated secret key. However, it is known that this approach has some practical issues such as a trade-off between the security and the accuracy.

ZeroBio: The third approach, ZeroBio, is to apply zero-knowledge interactive proof to biometrics. Therefore, this approach can consider the security for both template protection and authentication. The protocol proposed by Nishigaki, Watanabe, Oda, Yoneyama, Yamamoto, Takahashi, Ogata, and Kikuchi has the user-friendliness and the cryptographic security against passive adversaries in a semi-honest model [13]. However, their protocol and other protocols (e.g. [16,12]) have huge computational and communication costs because a zero-knowledge protocol runs multiple times in the authentication phase.

1.2   Our Contribution

In this paper, we study remote cancelable biometrics secure against impersonation attacks. Especially, we focus on the Hattori et al. cancelable biometric protocols [7] which are user-friendly and cryptographically-secure template protection. Note that their protocols are vulnerable to the replay attack, similarly to known ones [15,4,17,7,18].

As a simple way, we observe that cancelable biometric protocols using secure communication techniques (e.g. SSL/TLS [5]) would resist against the replay attack and several impersonation attacks. However, the approach cannot address the situation that the templates stored in the authentication server are stolen.

That is, it is impossible to prevent the attacker who uses the stolen templates and impersonates a legitimate user even if we use such secure communication techniques. Additionally, the revocation method of cancelable biometrics might be impractical and insufficient, although it is a solution for the attack. This is because, when the templates are stolen from the authentication server, can we notice the fact and run the revocation method immediately? Furthermore, there would be some impersonation attacks which cannot be avoided even if we use the revocation method.

In this paper, we construct a variant of the Hattori et al. protocol based on the BGN encryption scheme without losing its user-friendliness. Roughly speaking, our idea is to use their revocation method in a proactive manner, i.e., in our protocol, the public key assigned to the user is randomly re-generated in every authentication process. Fortunately, it is easy to introduce their revocation method into the authentication process in a straightforward way, since the revocation method is simple and constructed by using additive homomorphicity of the BGN encryption scheme. Note that this approach is not always succeed in general because of depending the structure of the revocation method. In our protocol, it is not required to re-generate users' secret keys although their public keys are changed in every authentication process. Moreover, in our protocol, there exists a unique secret key which is common among all users.

We also define a general and formal security game for biometric authentication against passive adversaries with a semi-honest manner. Our security definition covers the replay attack and considers fuzziness of biometric feature extraction. Then, we can show that our variant is secure in that definition, by assuming the hardness of a decisional bilinear Diffie-Hellman type problem on composite order group and high entropy of user's biometric features. This entropic approach was also used in [13,8], in order to prove the cryptographic security for authentication. In addition to the authentication security, we show that our variant has the cryptographic security for template protection in the same security model formalized by Hattori et al. under the subgroup decision assumption, without the high entropic assumption on user's biometric features.

In contrast to [15,4,6,19,17,1,7,18], our variant can show the cryptographic security for authentication. Furthermore, the computational and communication costs of our protocol is more efficient than those of the protocols [13,8] which have the user-friendliness and the cryptographically security for authentication. Additionally, it is fair to say that our protocol is more secure than the protocols [13,8], since our security definition for biometric authentication is more general and formal than those of them.

Note that we can also construct a variant based on the Okamoto-Takashima encryption scheme [14], by combining our construction idea with the Hattori et al. protocol based on the Okamoto-Takashima one.

1.3   Organization

The rest of the paper is as follows. In Section 2, we give some definitions and notions used throughout the paper. In Section 3, we review the Hattori et al.

protocol based on the BGN encryption scheme. In Section 4, we propose a cancelable biometric protocol based on the Hattori et al. one, and show that our protocol is cryptographically secure in the sense of template protection and authentication. In Section 5, we conclude.

2 Preliminaries

In this section, we give some notations and definitions, recall the BGN encryption scheme [2] and its related assumptions, and review desirable properties for cancelable biometrics formalized by Jain, Nandakumar, and Nagar [9].

2.1 Notations and Definitions

Let N be a positive integer. We denote $\{0, 1, \ldots, N-1\}$ by $\mathbb{Z}/N\mathbb{Z}$, and its reduced residue class group by $(\mathbb{Z}/N\mathbb{Z})^\times$, namely, $(\mathbb{Z}/N\mathbb{Z})^\times = \{x \in \mathbb{Z}/N\mathbb{Z} \mid \gcd(x, N) = 1\}$.

We denote the set of positive real numbers by $\mathbb{R}^+$. We say that a function $\epsilon : \mathbb{N} \to \mathbb{R}^+$ is negligible if for any (positive) polynomial $poly$, there exists $n_0 \in \mathbb{N}$ such that for all $n \geq n_0$, it holds $\epsilon(n) < 1/poly(n)$.

If A is a randomized algorithm, $y \leftarrow A(x)$ denotes running A on input x with a uniformly-chosen random tape and assigning the output to y. If S is a finite set, $s \xleftarrow{u} S$ denotes that s is uniformly chosen from S.

2.2 The BGN Encryption Scheme and Its Related Assumptions

Let $\mathbb{G}$ and $\mathbb{G}_T$ be multiplicative cyclic groups of order N, and e be a non-degenerate (symmetric) bilinear map from $\mathbb{G} \times \mathbb{G}$ to $\mathbb{G}_T$. That is, for a generator g of $\mathbb{G}$, it holds (1) $e(g^a, g^b) = e(g, g)^{ab}$ for any $a, b \in \mathbb{Z}/N\mathbb{Z}$, and (2) $e(g, g) \neq 1_{\mathbb{G}_T}$ and $e(g, g)$ is a generator of $\mathbb{G}_T$, where $1_{\mathbb{G}_T}$ is the identity element of $\mathbb{G}_T$. In this paper, we assume that e is polynomial-time computable.

Let p and q be prime, and $N = pq$. Let $\mathbb{G}_p$ and $\mathbb{G}_q$ be subgroups of order p and q of $\mathbb{G}$, respectively. Then, for all $g_p \in \mathbb{G}_p$ and for all $g_q \in \mathbb{G}_q$, it holds $e(g_p, g_q) = 1_{\mathbb{G}_T}$.

The BGN encryption scheme consists of the following algorithms ($\mathrm{Gen_{BGN}}$, $\mathrm{Enc_{BGN}}$, $\mathrm{Dec_{BGN}}$):

$\mathrm{Gen_{BGN}}$: Take as an input a security parameter $1^\lambda \in \mathbb{N}$, output a pair of a public key $PK = (g, h, N, e)$ and a secret key $SK = (p, q)$, where g and h are randomly chosen generators of $\mathbb{G}$ and $\mathbb{G}_q$, respectively.

$\mathrm{Enc_{BGN}}$: Take as inputs the public key PK, a message $m \in \{0, 1\}^M$, and a randomness $r \in \mathbb{Z}/N\mathbb{Z}$, output a ciphertext $c = g^m h^r$, where M is much smaller than λ.[1]

---

[1] Hereafter, we denote $\mathrm{Enc_{BGN}}(PK, m, r)$ by $E_{PK}(m)$, shortly, where E is regarded as a probabilistic algorithm.

Dec$_{\mathbf{BGN}}$: Take as inputs the public key PK, the secret key SK, and a ciphertext c, output a message $m = \mathsf{DLog}_{g^q} c^q$ (we note that m is the discrete logarithm of c^q to the base g^q over $\mathbb{G}$).

From a security point of view, the BGN encryption scheme is secure in the sense of IND-CPA under the following subgroup decision (SD) assumption [2].

Definition 1 (SD Assumption). *Let* $g, x \xleftarrow{u} \mathbb{G}$, $h, y \xleftarrow{u} \mathbb{G}_p$, $S = (N, g, h, e)$. *We say that the SD assumption holds if for any PPT algorithm* $\mathcal{A}$, $|\Pr[\mathcal{A}(S, x) = 1] - \Pr[\mathcal{A}(S, y) = 1]|$ *is negligible in the security parameter* λ.

Additionally, encryption schemes with bilinear maps on composite order groups rely its indistinguishability on the following decisional bilinear Diffie-Hellman on composite order group (DBDH on COG) assumption which can be regarded as a decisional assumption of the computational assumption defined in [3, Section 4.2].

Definition 2 (DBDH on COG Assumption). *Let* $g_p \xleftarrow{u} \mathbb{G}_p$, $g_q \xleftarrow{u} \mathbb{G}_q$, $a, b, c, z \xleftarrow{u} \mathbb{Z}/N\mathbb{Z}$, *and* $S = (N, \mathbb{G}, \mathbb{G}_T, e, g_p, g_q, g_p^a, g_p^b, g_p^c)$. *We say that the DBDH assumption holds if for any PPT algorithm* $\mathcal{A}$, $|\Pr[\mathcal{A}(S, e(g_p, g_p)^{abc}) = 1] - \Pr[\mathcal{A}(S, e(g_p, g_p)^z) = 1]|$ *is negligible in the security parameter* λ.

2.3 Desirable Properties for Cancelable Biometrics

Jain et al. stated that cancelable biometrics should satisfy the following properties [9].

Accuracy: In general, an error might occur in evaluating the similarity of the biometric features in the cancelable biometric system, and the accuracy (i.e. false acceptance rate and false rejection rate) may be degraded from that of the original biometric system. It is important that the degree of accuracy degradation is small enough.

Diversity: It should be possible to produce a very large number of cancelable templates (to be used in different applications) from the same biometric feature. Furthermore, it should be impossible to match cancelable templates from different applications.

Revocability: It should be straightforward to revoke a compromised template and reissue a new one based on the same biometric feature.

Security: It should be infeasible to obtain any partial information on users' feature vectors from the data that appears in the protocol.

3   The Hattori et al. Protocol

In this section, we describe the Hattori et al. protocol [7]. Concretely, we give the system model, the construction based on the BGN encryption scheme, and the security definition. Further, we point out that the protocol is not secure against the replay attack.

3.1   System Model

Let us denote the system model, biometric features, and metrics used in the Hattori et al. protocol.

Entities: In their system model, there are three kinds of entities: users $\mathcal{U}_1, \ldots, \mathcal{U}_n$, an authentication server $\mathcal{S}$, and a decryptor $\mathcal{D}$. It is assumed that $\mathcal{S}$ and $\mathcal{D}$ do not collude.

Biometric Features and Metrics: Their protocol employs generic feature vectors of fixed length as a biometric feature and the squared Euclidean distance of two vectors as a similarity metrics. Concretely, each feature vector consists of D elements of integers; e.g. $\boldsymbol{x} = (x_1, \ldots, x_D) \in \mathbb{Z}^D$, and the squared Euclidean distance of two vectors $\boldsymbol{x}$ and $\boldsymbol{y}$ is defined by $d_{E^2}(\boldsymbol{x}, \boldsymbol{y}) = \sum_{i=1}^{D} (x_i - y_i)^2$. The distance will be compared with $\theta \in \mathbb{Z}$ which is the pre-defined threshold in their system.

3.2   Construction Based on the BGN Encryption Scheme

Let us describe the construction of the Hattori et al. protocol based on the BGN encryption scheme, as follows.

- The setup process is as follows.
    1. $\mathcal{D}$ invokes $(p, q, \mathbb{G}, \mathbb{G}_T, e) \leftarrow \mathrm{Gen}_{\mathrm{BGN}}(1^\lambda)$.
    2. $\mathcal{D}$ picks $g_k, u_k \xleftarrow{u} \mathbb{G}$ for $k = 1, \ldots, n$ where n is the number of users, and sets $\{h_k = u_k^p, \ g_{T,k} = e(g_k, g_k)\}_{k=1}^n$.
    3. $\mathcal{D}$ sets $PK = (N, \mathbb{G}, \mathbb{G}_T, e)$, $SK = q$, and $\{PK_k = (g_k, h_k, g_{T,k})\}_{k=1}^n$, and makes PK and $\{PK_k\}_{k=1}^n$ publicly available, where PK is the public parameter of the system, PK_k is the public key for the user $\mathcal{U}_k$, and SK is the secret key of the system and stored secretly in $\mathcal{D}$.
- The enrollment process of $\mathcal{U}_k$ is as follows.
    1. $\mathcal{U}_k$ picks $r_1, \ldots, r_D \xleftarrow{u} \mathbb{Z}/N\mathbb{Z}$ and encrypts his biometric feature vector $\boldsymbol{x} = (x_1, \ldots, x_D) \in (\mathbb{Z}/N\mathbb{Z})^D$ under PK_k as $c_{k_E,i} = g_k^{x_i} h_k^{r_i}$ for $i = 1, \ldots, D$.
    2. $\mathcal{U}_k$ sends $E_{PK_k}(\boldsymbol{x}) = (c_{k_E,1}, \ldots, c_{k_E,D})$ to $\mathcal{S}$.
    3. $\mathcal{S}$ stores a tuple $(k, PK_k, E_{PK_k}(\boldsymbol{x}))$ as a template.
- The authentication process of $\mathcal{U}_k$ is as follows.
    1. $\mathcal{U}_k$ picks $s_1, \ldots, s_D \xleftarrow{u} \mathbb{Z}/N\mathbb{Z}$ and encrypts his biometric feature vector $\boldsymbol{y} = (y_1, \ldots, y_D) \in (\mathbb{Z}/N\mathbb{Z})^D$ under PK_k as $c_{k_A,i} = g_k^{y_i} h_k^{s_i}$ for $i = 1, \ldots, D$.
    2. $\mathcal{U}_k$ sends $E_{PK_k}(\boldsymbol{y}) = (c_{k_A,1}, \ldots, c_{k_A,D})$ to $\mathcal{S}$.
    3. $\mathcal{S}$ picks $t_1, t_2 \xleftarrow{u} \mathbb{Z}/N\mathbb{Z}$, takes $(k, PK_k, E_{PK_k}(\boldsymbol{x}))$, and computes the encrypted distance $\Delta_{PK_k} = \Delta_{PK_k}(E_{PK_k}(\boldsymbol{x}), E_{PK_k}(\boldsymbol{y}))$ by $\Delta_{PK_k} = e(g_k, h_k)^{t_1} \cdot e(h_k, h_k)^{t_2} \cdot \prod_{i=1}^{D} \left(e(c_{k_E,i}, c_{k_E,i}) \cdot e(c_{k_A,i}, c_{k_A,i}) \cdot e(c_{k_E,i}, c_{k_A,i})^{-2} \right)$.
    4. $\mathcal{S}$ sends the encrypted distance Δ_{PK_k} to $\mathcal{D}$.

5. $\mathcal{D}$ computes the discrete logarithm of $\Delta_{PK_k}^q$ to the base $g_{T,k}^q$, that is, $d_{E^2}(\boldsymbol{x}, \boldsymbol{y}) = \mathsf{DLog}_{g_{T,k}^q} \Delta_{PK_k}^q$, where $d_{E^2}(\cdot, \cdot)$ is the squared Euclidean distance.

6. $\mathcal{D}$ compares $d_{E^2}(\boldsymbol{x}, \boldsymbol{y})$ to the pre-defined threshold θ. If $d_{E^2}(\boldsymbol{x}, \boldsymbol{y}) \leq \theta$, $\mathcal{D}$ returns accept to $\mathcal{S}$; otherwise, $\mathcal{D}$ returns reject to $\mathcal{S}$.

– The revocation process of $\mathcal{U}_k$ is as follows.

1. $\mathcal{S}$ picks $\delta \xleftarrow{u} \mathbb{Z}/N\mathbb{Z}$ and computes $g_k' = g_k^\delta, h_k' = h_k^\delta, g_{T,k}' = g_{T,k}^{\delta^2}, c_{k_E,i}' = c_{k_E,i}^\delta$ for $i = 1, \ldots, D$.

2. $\mathcal{S}$ stores a new tuple $(k, PK_k' = (g_k', h_k', g_{T,k}'), E_{PK_k'}(\boldsymbol{x}) = (c_{k_E,1}', \ldots, c_{k_E,D}'))$.

3. $\mathcal{S}$ makes PK_k' publicly available.

We note that, at Step 5 of the authentication process, we must solve discrete logarithms in $\mathbb{G}_T$, which is known as an expensive computation. It is known that, by using the Pollard lambda method, discrete logarithms laid in an interval $[0, \tau]$ can be solved in $O(\sqrt{\tau})$-time.

In the protocol, $\mathcal{D}$ decides reject if the distance is over θ. Also, from the fact above, $\mathcal{D}$ can conclude that the distance exceeds θ (i.e. reject), if $\mathcal{D}$ could not obtain the discrete logarithms within $O(\sqrt{\theta})$-time. Therefore, θ and D should be much smaller than N (for example, $\theta = c$ or $\ln^c N$, where c is constant).

On the other hand, x_i and y_i in the protocol can be chosen from $\mathbb{Z}/N\mathbb{Z}$ rather than $\{0, 1\}^M$, although the message space of the BGN encryption scheme (i.e. $\{0, 1\}^M$) is much smaller than $\mathbb{Z}/N\mathbb{Z}$. This is because, in the protocol, it is solely checked whether the distance between two feature vectors is small.

3.3 Security

Hattori et al. stated that the general security requirement in their system can further be classified into the following three requirements:

Security against the Authentication Server $\mathcal{S}$: It is required that $\mathcal{S}$ cannot obtain extra information other than the binary result (accept or reject) of authentication.

Security against the Decryptor $\mathcal{D}$: It is required that $\mathcal{D}$ cannot obtain extra information other than the squared Euclidean distance of two feature vectors.

Security against Eavesdroppers $\mathcal{E}$: It is required that eavesdroppers $\mathcal{E}$ cannot obtain extra information other than the binary result (accept or reject) of authentication.

Security against $\mathcal{S}$: Hattori et al. considered the security in the semi-honest model and only passive eavesdroppers. Then, they defined the security game between $\mathcal{S}$ and the challenger $\mathcal{C}$ and its advantage of $\mathcal{S}$ as follows.

Setup: $\mathcal{C}$ runs the setup process and gives $PK, \{PK_k\}_{k=1}^n$ to $\mathcal{S}$. Then, $\mathcal{C}$ picks $\beta \xleftarrow{u} \{0, 1\}$ and uses the same value throughout the game.

Query: S adaptively makes two types of queries in an arbitrary order:

- *Enrollment query.* S sends $(i, \boldsymbol{x}_0^{(i)}, \boldsymbol{x}_1^{(i)})$ to $\mathcal{C}$, where $1 \leq i \leq n$. $\mathcal{C}$ returns $E_{PK_i}(\boldsymbol{x}_\beta^{(i)})$.
- *Authentication query.* This query consists of two consecutive procedures:
  1. S sends $(j, \boldsymbol{y}_0^{(j)}, \boldsymbol{y}_1^{(j)})$ to $\mathcal{C}$, where $1 \leq j \leq n$. $\mathcal{C}$ returns $E_{PK_j}(\boldsymbol{y}_\beta^{(j)})$.
  2. After receiving $E_{PK_j}(\boldsymbol{y}_\beta^{(j)})$, S computes $\Delta_{PK_j}(E_{PK_j}(\boldsymbol{x}_\beta^{(j)}),$ $E_{PK_j}(\boldsymbol{y}_\beta^{(j)}))$ by using $E_{PK_j}(\boldsymbol{x}_\beta^{(j)})$ which must have been queried in the enrollment query, and sends $\Delta_{PK_j}(E_{PK_j}(\boldsymbol{x}_\beta^{(j)}), E_{PK_j}(\boldsymbol{y}_\beta^{(j)}))$ to $\mathcal{C}$. $\mathcal{C}$ returns the authentication result (accept or reject).

  Here, we allow S to make a polynomial number of authentication queries. The restriction of the authentication query is that if $\boldsymbol{x}_0^{(j)}$ and $\boldsymbol{y}_0^{(j)}$ are accepted vectors, then $\boldsymbol{x}_1^{(j)}$ and $\boldsymbol{y}_1^{(j)}$ must also be accepted ones, and vice versa.

Guess: S outputs a guess β' of β.

The advantage of S in the above game is defined as $\mathbf{Adv}_S(\lambda) = \left| \Pr[\beta' = \beta] - \frac{1}{2} \right|$.

Definition 3. *We say that a biometric protocol is secure in the sense of template protection against S if for all PPT servers S, $\mathbf{Adv}_S(\lambda)$ is negligible in λ.*

Their protocol satisfies the above security under the SD assumption.

Security against $\mathcal{D}$: In addition to the security against the authentication server S, they defined the security against the decryptor $\mathcal{D}$, as follows.

Definition 4. *We say that a biometric protocol is secure against $\mathcal{D}$ if for all unconditionally computable adversarial decryptors $\mathcal{D}$, it is impossible to obtain any partial information on $\boldsymbol{x}$ and $\boldsymbol{y}$, except for $d_{E^2}(\boldsymbol{x}, \boldsymbol{y})$.*

Their protocol satisfies the above security with no assumption.

Security against $\mathcal{E}$: We need to further consider the security against eavesdroppers $\mathcal{E}$. Fortunately, it can be reduced to that of the authentication server S, since all the transmissions $\mathcal{E}$ can observe are equally observed by S. That is, if it is secure against S, it holds that it is also secure against $\mathcal{E}$. We note that this claim is only valid for passive adversaries.

3.4   Security Issues

As described in Section 3.3, Hattori et al. solely focused on privacy aspects of biometrics; that is, authenticity aspects were outside of their scope, similarly to most cancelable biometric protocols (e.g. [15,4,17,7,18]). Therefore, they did not consider attacks on authenticity, such as impersonation attacks. Unfortunately, the Hattori et al. protocol (similarly to [15,4,17,7,18]) is not secure against the replay attack. Indeed, if $\mathcal{E}$ can capture an encrypted feature vector $E_{PK_k}(\boldsymbol{y})$ sent through a public network, $\mathcal{E}$ can impersonate as $\mathcal{U}_k$ by directly sending $E_{PK_k}(\boldsymbol{y})$ to S, without knowing his plain biometric information or feature vectors.

4 Our Variant

In this section, we firstly propose a cancelable biometric protocol which is a variant of the Hattori et al. protocol. Secondly, we compare efficiency among our protocol and other biometric protocols [7,13,8]. Thirdly, we consider the security for template protection, replay, and authentication of our protocol.

4.1 Construction

As pointed out in Section 3.4, the attacker for the Hattori et al. protocol can impersonate a legitimate user by directly sending eavesdropped encrypted feature vectors of legitimate users to S. On the other hand, we see that, after running their revocation process, it would be infeasible to use a former templates $(k, PK_k, E_{PK_k}(x))$ for impersonation, since user's public key is changed.

From the fact, we observe an idea that their revocation method would be effective for preventing several impersonation attacks, including the replay attack. Concretely, we randomly re-generate user's public key (equivalently, generate a one-time public key) via the revocation process in the authentication phase. Fortunately, it is easy to introduce their revocation method into the authentication process in a straightforward way because the revocation method is simple. Then, we construct a cancelable biometric protocol with the above idea, as follow:

- The setup process is as follows.
    1. D invokes $(p, q, \mathbb{G}, \mathbb{G}_T, e) \leftarrow \mathrm{Gen}_{\mathrm{BGN}}(1^\lambda)$.
    2. D picks $g_k, u_k \xleftarrow{u} \mathbb{G}$ for $k = 1, \ldots, n$ where n is the number of users, and sets $\{h_k = u_k^p, \ g_{T,k} = e(g_k, g_k)\}_{k=1}^n$.
    3. D sets $PK = (N, \mathbb{G}, \mathbb{G}_T, e)$, $SK = q$, and $\{PK_k = (g_k, h_k, g_{T,k})\}_{k=1}^n$, and makes PK and $\{PK_k\}_{k=1}^n$ publicly available, where PK is the public parameter of the system, PK_k is the public key for the user $\mathcal{U}_k$, and SK is the secret key of the system and stored secretly in D.

- The enrollment process of $\mathcal{U}_k$ is as follows.
    1. $\mathcal{U}_k$ picks $r_1, \ldots, r_D \xleftarrow{u} \mathbb{Z}/N\mathbb{Z}$ and encrypts his biometric feature vector $x = (x_1, \ldots, x_D) \in (\mathbb{Z}/N\mathbb{Z})^D$ under PK_k as $c_{k_E,i} = g_k^{x_i} h_k^{r_i}$ for $i = 1, \ldots, D$.
    2. $\mathcal{U}_k$ sends $E_{PK_k}(x) = (c_{k_E,1}, \ldots, c_{k_E,D})$ to S.
    3. S stores a tuple $(k, PK_k, E_{PK_k}(x))$ as a template.

- The authentication process of $\mathcal{U}_k$ is as follows.
    1. S picks $\gamma \xleftarrow{u} \mathbb{Z}/N\mathbb{Z}$ and computes $g_{OT_k} = g_k^\gamma, h_{OT_k} = h_k^\gamma, g_{T,OT_k} = g_{T,k}^{\gamma^2}$ from PK_k.
    2. S sends $PK_{OT_k} = (g_{OT_k}, h_{OT_k}, g_{T,OT_k})$ to $\mathcal{U}_k$.
    3. $\mathcal{U}_k$ picks $s_1, \ldots, s_D \xleftarrow{u} \mathbb{Z}/N\mathbb{Z}$ and encrypts his biometric feature vector $y = (y_1, \ldots, y_D) \in (\mathbb{Z}/N\mathbb{Z})^D$ under PK_{OT_k} as $c_{k_A,i} = g_{OT_k}^{y_i} h_{OT_k}^{s_i}$ for $i = 1, \ldots, D$.
    4. $\mathcal{U}_k$ sends $E_{PK_{OT_k}}(y) = (c_{k_A,1}, \ldots, c_{k_A,D})$ to S.

5. $\mathcal{S}$ picks $t_1, t_2 \xleftarrow{u} \mathbb{Z}/N\mathbb{Z}$, takes $(k, PK_k, E_{PK_k}(\boldsymbol{x}))$, and computes the encrypted distance $\Delta_{PK_{OT_k}} = \Delta_{PK_{OT_k}}(E_{PK_k}(\boldsymbol{x}), E_{PK_{OT_k}}(\boldsymbol{y}), \gamma)$ by $\Delta_{PK_{OT_k}} = e(g_{OT_k}, h_{OT_k})^{t_1} \cdot e(h_{OT_k}, h_{OT_k})^{t_2} \cdot \prod_{i=1}^{D} \left(e(c_{k_E,i}^{\gamma}, c_{k_E,i}^{\gamma}) \cdot e(c_{k_A,i}, c_{k_A,i}) \cdot e(c_{k_E,i}^{\gamma}, c_{k_A,i})^{-2} \right)$.

6. $\mathcal{S}$ sends the encrypted distance $\Delta_{PK_{OT_k}}$ and g_{T,OT_k} to $\mathcal{D}$.

7. $\mathcal{D}$ computes the discrete logarithm of $\Delta_{PK_{OT_k}}^q$ to the base g_{T,OT_k}^q, that is, $d_{E^2}(\boldsymbol{x}, \boldsymbol{y}) = \mathsf{DLog}_{g_{T,OT_k}^q} \Delta_{PK_{OT_k}}^q$, where $d_{E^2}(\cdot, \cdot)$ is the squared Euclidean distance.

8. $\mathcal{D}$ compares $d_{E^2}(\boldsymbol{x}, \boldsymbol{y})$ to the pre-defined threshold θ. If $d_{E^2}(\boldsymbol{x}, \boldsymbol{y}) \leq \theta$, $\mathcal{D}$ returns accept to $\mathcal{S}$; otherwise, $\mathcal{D}$ returns reject to $\mathcal{S}$.

– The revocation process of $\mathcal{U}_k$ is identical to the Hattori et al. one.

We note that $\mathcal{D}$ can also conclude reject if $\mathcal{D}$ could not compute the discrete logarithm within $O(\sqrt{\theta})$ at Step 7 of the authentication phase.

4.2   Correctness

We show that $\mathsf{DLog}_{g_{T,OT_k}^q} \Delta_{PK_{OT_k}}^q$ is exactly the squared Euclidean distance of two vectors $\boldsymbol{x}$ and $\boldsymbol{y}$.

For $i = 1, \ldots, D$, it holds $c_{k_E,i}^{\gamma} = g_k^{x_i\gamma} h_k^{r_i\gamma} = g_{OT_k}^{x_i} h_{OT_k}^{r_i}$. This means that $(c_{k_E,1}^{\gamma}, \ldots, c_{k_E,D}^{\gamma})$ can be regarded as $E_{PK_{OT_k}}(\boldsymbol{x})$. Since $e(g_{OT_k}, h_{OT_k})^q = e(g_{OT_k}, u_k^{\gamma})^{pq} = 1_{\mathbb{G}_T}$ and $e(h_{OT_k}, h_{OT_k})^q = e(h_{OT_k}, u_k^{\gamma})^{pq} = 1_{\mathbb{G}_T}$, we have

$$\Delta_{PK_{OT_k}}^q = \prod_{i=1}^{D} \left(e(c_{k_E,i}^{\gamma}, c_{k_E,i}^{\gamma}) \cdot e(c_{k_A,i}, c_{k_A,i}) \cdot e(c_{k_E,i}^{\gamma}, c_{k_A,i})^{-2} \right)^q.$$

From bilinearity of e, we have

$$\begin{aligned} e(c_{k_E,i}^{\gamma}, c_{k_E,i}^{\gamma})^q &= e(g_{OT_k}^{x_i} h_{OT_k}^{r_i}, g_{OT_k}^{x_i} h_{OT_k}^{r_i})^q \\ &= \left(e(g_{OT_k}^{x_i}, g_{OT_k}^{x_i}) \cdot e(h_{OT_k}^{r_i}, g_{OT_k}^{x_i}) \cdot e(g_{OT_k}^{x_i}, h_{OT_k}^{r_i}) \cdot e(h_{OT_k}^{r_i}, h_{OT_k}^{r_i}) \right)^q \\ &= g_{T,OT_k}^{x_i^2 q} \end{aligned}$$

for $i = 1, \ldots, D$. Similarly, we have $e(c_{k_A,i}, c_{k_A,i})^q = g_{T,OT_k}^{y_i^2 q}$ and $e(c_{k_E,i}^{\gamma}, c_{k_A,i})^{-2q} = g_{T,OT_k}^{-2x_i y_i q}$ for $i = 1, \ldots, D$. Then, we obtain

$$\Delta_{PK_{OT_k}}^q = \prod_{i=1}^{D} g_{T,OT_k}^{x_i^2 q} \cdot g_{T,OT_k}^{y_i^2 q} \cdot g_{T,OT_k}^{-2x_i y_i q} = g_{T,OT_k}^{q \sum_{i=1}^{D}(x_i - y_i)^2} = (g_{T,OT_k}^q)^{d_{E^2}(\boldsymbol{x}, \boldsymbol{y})}.$$

Therefore, it holds $\mathsf{DLog}_{g_{T,OT_k}^q} \Delta_{PK_{OT_k}}^q = d_{E^2}(\boldsymbol{x}, \boldsymbol{y})$.

Table 1. Comparison of efficiency among our protocol and other biometric protocols

Protocol	Entity	Setup & Enrollment Phases		Authentication Phase	
		Computation	Communication	Computation	Communication
[13]	$\mathcal{U}_k$	$(2\theta D+2)e_p$	$(4\theta D+4)\ell_p$	$(4\theta D+4)e_p'$	$(4\theta D+4)\ell_p$
	$\mathcal{A}$	-	-	$(10\theta D+8)e_p$	$(6\theta D+6)\ell_p$
[7]	$\mathcal{U}_k$	$2De_G$	$D\ell_G$	$2De_G$	$D\ell_G$
	$\mathcal{A}$	-	-	$(D+2)e_{G_T}+(3D+2)p$	ℓ_{G_T}
	$\mathcal{D}$	ne_G+np	$2n\ell_G+n\ell_{G_T}+\lg N$	$2e_{G_T}+O(\sqrt{\theta})$	1
[8]	$\mathcal{U}_k$	$2De_G$	$D\ell_G$	$2De_G$	$D\ell_G$
	$\mathcal{A}$	-	-	$3De_G+(D+2)e_{G_T}+(3D+2)p$	$D\ell_G+\ell_{G_T}$
	$\mathcal{D}$	ne_G+np	$2n\ell_G+n\ell_{G_T}+\lg N$	$2e_{G_T}+O(\sqrt{\theta})$	1
Ours	$\mathcal{U}_k$	$2De_G$	$D\ell_G$	$2De_G$	$D\ell_G$
	$\mathcal{A}$	-	-	$De_G+(D+2)e_{G_T}+(3D+2)p$	$2\ell_{G_T}$
	$\mathcal{D}$	ne_G+np	$2n\ell_G+n\ell_{G_T}+\lg N$	$2e_{G_T}+O(\sqrt{\theta})$	1

4.3 Comparison of Efficiency

Here we compare efficiency among our protocol and other protocols [7,13,8].

Table 1 summarizes a result of the comparison of efficiency among our protocol and other biometric protocols [7,13,8]. Concretely, Table 1 gives the computation and communication costs of their protocols, where the computation costs are focused on exponentiation and pairing operation. Note that the Hattori et al. protocol [7] is not secure against the replay attack in contrast to [13,8], and that there is no decryptor $\mathcal{D}$ in the system model of [13].

The notations used in the table are defined as follows. ℓ_p, ℓ_G, and ℓ_{G_T} mean the data size of an element of $\mathbb{Z}/p\mathbb{Z}$, $\mathbb{G}$, and $\mathbb{G}_T$, respectively. e_p, e_G, and e_{G_T} mean an exponentiation cost in $\mathbb{Z}/p\mathbb{Z}$, $\mathbb{G}$, and $\mathbb{G}_T$, respectively. p means a pairing operation cost in $\mathbb{G}$.

From Table 1, we see that the authentication process of our protocol is more efficient than that of the Hirano et al. protocol [8] which is also a variant of the Hattori et al. protocol [7]. Further, it is fair to say that our protocol can be considered as more efficient than the Nishigaki et al. ZeroBio protocol [13] (at least, the communication cost), although the groups used in our protocol (i.e. $\mathbb{G}_N$ and $\mathbb{G}_T$) are different from that in [13] (i.e. $\mathbb{Z}/p\mathbb{Z}$).

4.4 Security

Now, let us show the cryptographic security of our protocol. Firstly, we show the cryptographic security for template protection under our security model which is slightly modified the original one formalized in [7]. Secondly, we check that our protocol is secure against the replay attack. After that, in order to show that our protocol is secure against other impersonate attacks, we define a general and formal security model for biometric authentication, and show that our protocol is secure in that model.

The Cryptographic Security for Template Protection: In order to show that our protocol has the cryptographic security for template protection, we slightly modify the security game against S, described in Section 3.3:

The security game between S and the challenger C, and the advantage of S, in our protocol is defined as follows.

Setup: C runs the setup process and gives $PK, \{PK_k\}_{k=1}^n$ to S. Then, C picks $\beta \xleftarrow{u} \{0,1\}$ and uses the same value throughout the game.

Query: S adaptively makes two types of queries in an arbitrary order:

- *Enrollment query.* S sends $(i, x_0^{(i)}, x_1^{(i)})$ to C, where $1 \leq i \leq n$. C returns $E_{PK_i}(x_\beta^{(i)})$.

- *Authentication query.* This query consists of two consecutive procedures:
  1. S picks $\gamma_j \xleftarrow{u} \mathcal{R}_{PK}$, computes PK_{OT_j} from PK_j and γ_j, and sends $(j, PK_{OT_j}, y_0^{(j)}, y_1^{(j)})$ to C, where $1 \leq j \leq n$ and $\mathcal{R}_{PK}$ is a set of random numbers depending on PK. C returns $E_{PK_{OT_j}}(y_\beta^{(j)})$.
  2. After receiving $E_{PK_{OT_j}}(y_\beta^{(j)})$, S computes $\Delta_{PK_{OT_j}}(E_{PK_j}(x_\beta^{(j)})$, $E_{PK_{OT_j}}(y_\beta^{(j)}))$ from γ_j, where $E_{PK_j}(x_\beta^{(j)})$ must have been queried in the enrollment query, and sends $\Delta_{PK_{OT_j}}(E_{PK_j}(x_\beta^{(j)}), E_{PK_{OT_j}}(y_\beta^{(j)}))$ to C. C returns the authentication result (accept or reject).

  Here, we allow S to make a polynomial number of authentication queries. The restriction of the authentication query is that if $x_0^{(j)}$ and $y_0^{(j)}$ are accepted vectors, then $x_1^{(j)}$ and $y_1^{(j)}$ must also be accepted ones, and vice versa.

Guess: S outputs a guess β' of β.

The advantage of S in the above game is defined as $\mathbf{Adv}'_S(\lambda) = \left| \Pr[\beta' = \beta] - \frac{1}{2} \right|$.

Definition 5. *We say that a biometric protocol is secure in the sense of template protection against S if for all PPT servers S, $\mathbf{Adv}'_S(\lambda)$ is negligible in λ.*

Our protocol is slightly modified from the Hattori et al. one, and any adversary in the security game above would have no information on β since one-time public keys are generated correctly and uniformly in the authentication query. That is, we can easily construct a PPT algorithm which breaks the template security of the Hattori et al. protocol with non-negligible probability, if there exists a PPT adversary which breaks that of our protocol with non-negligible probability. Therefore, in a similar fashion to the security proof of the Hattori et al. protocol, we can directly show that our protocol is secure in our model under the SD assumption defined in Definition 1.

Theorem 1. *Our protocol is secure in the sense of template protection under the SD assumption.*

Further, we can show that our protocol is secure against D with no assumption, in a similar fashion to the security proof of the Hattori et al. protocol.

The Cryptographic Security for Authentication: Firstly, let us consider the replay attack by the eavesdropper $\mathcal{E}$.

Let $(k, PK_k, E_{PK_k}(\boldsymbol{x}))$ be a template of the user $\mathcal{U}_k$ and $(PK_{OT_k}, E_{PK_{OT_k}}(\boldsymbol{y}))$ be eavesdropped authentication data of $\mathcal{U}_k$, where $PK_{OT_k} = (g_k^\gamma, h_k^\gamma, g_{T,k}^\gamma)$, $\gamma \xleftarrow{u} \mathbb{Z}/N\mathbb{Z}$, and $d_{E^2}(\boldsymbol{x}, \boldsymbol{y}) \leq \theta$. For a one-time public key $PK_{OT_k'} = (g_{OT_k'}, h_{OT_k'}, g_{T,OT_k'}) = (g_k^{\gamma'}, h_k^{\gamma'}, g_{T,k}^{\gamma'^2})$ generated from $\gamma' \xleftarrow{u} \mathbb{Z}/N\mathbb{Z}$ in the authentication phase, we have

$$
\Delta_{PK_{OT_k'}}^q = \prod_{i=1}^{D} \left(e(c_{k_E,i}^{\gamma'}, c_{k_E,i}^{\gamma'}) \cdot e(c_{k_A,i}, c_{k_A,i}) \cdot e(c_{k_E,i}^{\gamma'}, c_{k_A,i})^{-2} \right)^q
$$
$$
= \prod_{i=1}^{D} e(g_k^{\gamma'}, g_k^{\gamma'})^{x_i^2 q} \cdot e(g_k^{\gamma'}, g_k^{\gamma'})^{y_i^2 \alpha^2 q} \cdot e(g_k^{\gamma'}, g_k^{\gamma'})^{-2x_i y_i \alpha q}
$$
$$
= (g_{T,OT_k'}^q)^{d_{E^2}(\boldsymbol{x}, \alpha\boldsymbol{y})},
$$

where $\alpha \equiv \gamma/\gamma' \pmod{N}$. Obviously, the discrete logarithm of $\Delta_{PK_{OT_k'}}^q$ to the base $g_{T,OT_k'}^q$, that is, $d_{E^2}(\boldsymbol{x}, \alpha\boldsymbol{y})$ is not constant with overwhelming probability in λ. Therefore, we see that the replay attack is not applicable to our protocol.

Theorem 2. *Our protocol is secure against the replay attack.*

Secondly, in addition to the security against the replay attack, let us consider the security for (remote) biometric authentication against general impersonation attacks. Here, we define the security game against passive eavesdroppers $\mathcal{E}$ in a semi-honest manner. Our security model covers the replay attack and considers fuzziness of biometric feature extraction. Therefore, it is more general and formal than those of the protocols [13,8].

The security game between $\mathcal{E}$ and the challenger $\mathcal{C}$, and the advantage of $\mathcal{E}$ in our protocol are defined as follows.

Setup: $\mathcal{C}$ runs the setup process and gives $PK, \{PK_k\}_{k=1}^n$ to $\mathcal{E}$.

Query: $\mathcal{E}$ adaptively makes two types of queries in an arbitrary order:
- *Enrollment query.* $\mathcal{E}$ sends i to $\mathcal{C}$, where $1 \leq i \leq n$. $\mathcal{C}$ picks $\boldsymbol{x}_i \in \mathcal{B}_i$, and returns $E_{PK_i}(\boldsymbol{x}_i)$, where $\mathcal{B}_i$ is a set of feature vectors of $\mathcal{U}_i$.
- *Authentication query.* $\mathcal{E}$ sends j to $\mathcal{C}$, where $1 \leq j \leq n$ and j must have been queried in the enrollment query. $\mathcal{C}$ picks $\gamma_j \xleftarrow{u} \mathcal{R}_{PK}$ and $\boldsymbol{y}_j \in \mathcal{B}_j$, and computes PK_{OT_j} and $E_{PK_{OT_j}}(\boldsymbol{y}_j)$ from PK_j, where $\mathcal{R}_{PK}$ is a set of random numbers depending on PK. $\mathcal{C}$ checks $d_{E^2}(\boldsymbol{x}_j, \boldsymbol{y}_j) \leq \theta$, decides result $=$ accept or reject, and returns $(PK_{OT_j}, E_{PK_{OT_j}}(\boldsymbol{y}_j), \text{result})$ to $\mathcal{E}$. Here, we allow $\mathcal{E}$ to make a polynomial number of authentication queries.

Challenge: $\mathcal{E}$ picks ℓ, and sends to $\mathcal{C}$, where $1 \leq \ell \leq n$ and ℓ must have been queried in the enrollment query. $\mathcal{C}$ picks $\gamma_\ell \xleftarrow{u} \mathcal{R}_{PK}$, computes PK_{OT_ℓ}, and returns PK_{OT_ℓ}.

Compute: $\mathcal{E}$ outputs $E_{PK_{OT_\ell}}(\boldsymbol{x}_\ell')$.

The advantage of $\mathcal{E}$ in the above is defined as $\mathbf{Adv}'_{\mathcal{E}}(\lambda) = \Pr[d_{E^2}(\boldsymbol{x}_\ell, \boldsymbol{x}_\ell') \leq \theta]$.

Definition 6. *We say that a biometric protocol is secure in the sense of authentication against $\mathcal{E}$ if for all PPT eavesdroppers $\mathcal{E}$, $\mathbf{Adv}'_{\mathcal{E}}(\lambda)$ is negligible in λ.*

In order to show that our protocol is secure under the above definition, we use the following three heuristic assumptions, in addition to the DBDH on COG assumption formalized in Definition 2.

The first one is an assumption that entropy of user's biometric feature is very high. In other words, we assume that each a_i of $\mathcal{U}_k$'s biometric feature $\boldsymbol{a} = (a_1, \ldots, a_D) \in \mathcal{B}_k$ is uniformly distributed over $\mathbb{Z}/N\mathbb{Z}$. Moreover, for feature vectors $\boldsymbol{v}_{k_1}$ and $\boldsymbol{v}_{k_2}$ extracted from $\mathcal{U}_{k_1}$ and $\mathcal{U}_{k_2}$, respectively, $d_{E^2}(\boldsymbol{v}_{k_1}, \boldsymbol{v}_{k_2})$ is exponentially large with overwhelming probability in λ. This assumption were also used in [13,8] in order to show the authentication security of their protocols.

The second one is that for any enrollment vector $\boldsymbol{x}$ and any authentication vector $\boldsymbol{y}$ extracted from $\mathcal{U}_k$, it will hold $|x_i - y_i| \leq d_{E^2}(\boldsymbol{x}, \boldsymbol{y}) \leq \mu$ for $1 \leq j \leq D$, where μ is positive constant. That is, $\boldsymbol{x} - \boldsymbol{y} \in Err(\mu, D) := \{(e_1, \ldots, e_D) \in \mathbb{Z}^D \mid \sum_{i=1}^{D} e_i^2 \leq \mu\}$. Note that $e \xleftarrow{u} Err(\mu, D)$ is efficiently samplable if μ and D are very small. Although this assumption gives a restriction for user's biometric feature, it is a plausible model in practical biometrics.

The third one is that for the plain feature vector $\boldsymbol{x}'_{\ell} = (x'_1, \ldots, x'_D)$ of $E_{PK_{OT_{\ell}}}(\boldsymbol{x}'_{\ell})$ outputted by $\mathcal{E}$, it will hold $|x_j - x'_j| \leq \nu$ for $1 \leq j \leq D$, in addition to $d_{E^2}(\boldsymbol{x}_{\ell}, \boldsymbol{x}'_{\ell}) \leq \theta$, where ν is positive constant. This assumption is closely related to the second one and might be considered as a practical condition.

Then, we can show the security of our protocol under the DBDH on COG assumption and the heuristic assumptions.

Theorem 3. *Our protocol is secure in the sense of authentication against $\mathcal{E}$ under the DBDH on COG assumption and the heuristic assumptions.*

Proof. We will construct a PPT algorithm $\mathcal{A}$ that breaks the DBDH on COG assumption with non-negligible probability by using a PPT adversary $\mathcal{E}$ that breaks our authentication security with non-negligible probability $\varepsilon(\lambda)$.

Let $(N, \mathbb{G}, \mathbb{G}_T, e, g_p, g_q, g_p^a, g_p^b, g_p^c, e(g_p, g_p)^z)$ be a DBDH on COG instance. For the instance, $\mathcal{A}$ picks $X, Y, Z \xleftarrow{u} \mathbb{Z}/N\mathbb{Z}$ and sets $g = g_p^X g_q^Y$ and $h = g_q^Z$. Here, we observe that g and h are random generators of $\mathbb{G}$ and $\mathbb{G}_q$, respectively, since $\mathbb{Z}/N\mathbb{Z} \simeq \mathbb{Z}/p\mathbb{Z} \times \mathbb{Z}/q\mathbb{Z}$ by the Chinese remainder theorem. $\mathcal{A}$ takes $X_k, Y_k, Z_k \xleftarrow{u} \mathbb{Z}/N\mathbb{Z}$ and sets $g_k = g_p^{X_k} g_q^{Y_k}$ and $h_k = g_q^{Z_k}$ for $1 \leq k \leq n$ where n is the number of users. Then, $\mathcal{A}$ sets

$$PK = (N, \langle g \rangle, \langle e(g, g) \rangle, e) \text{ and } \{PK_k = (g_k, h_k, e(g_k, g_k))\}_{k=1}^{n},$$

and gives them to $\mathcal{E}$ as the setup phase. Note that for any $\alpha, \beta \in \mathbb{Z}/N\mathbb{Z}$, we have $g_k^{\alpha} h_k^{\beta} = g_p^{\alpha X_k} g_q^{\alpha Y_k + \beta Z_k}$. Intuitively, the information on α appears only on the exponent of g_p but not g_q when $\beta \xleftarrow{u} \mathbb{Z}/N\mathbb{Z}$.

In the enrollment query phase, $\mathcal{A}$ picks $r_{i,1}, \ldots, r_{i,D}, \xi_{i,1}, \ldots, \xi_{i,D} \xleftarrow{u} \mathbb{Z}/N\mathbb{Z}$, gives

$$c_{k_E}^{(i)} = ((g_p^a g_p^{\xi_{i,1}})^{X_i} g_q^{r_{i,1}}, \ldots, (g_p^a g_p^{\xi_{i,D}})^{X_i} g_q^{r_{i,D}})$$

to $\mathcal{E}$, and keeps $\{\xi_{i,m}\}_{m=1}^{D}$, when $\mathcal{E}$ queries on i. Note that $(g_p^a g_p^{\xi_{i,m}})^{X_i} g_q^{r_{i,m}}$ for $1 \leq m \leq D$ can be regarded as a valid ciphertext $E_{PK_i}(a + \xi_{i,m})$ since it holds $g_i^q = g_p^{X_i q}$ and $((g_p^a g_p^{\xi_{i,m}})^{X_i} g_q^{r_{i,m}})^q = (g_p^{X_i q})^{(a+\xi_{i,m})}$.

In the authentication query phase, $\mathcal{A}$ picks $e = (e_1, \ldots, e_D) \xleftarrow{u} Err(\mu, D)$ and $\gamma_j, s_{j,1}, \ldots, s_{j,D} \xleftarrow{u} \mathbb{Z}/N\mathbb{Z}$, and computes $PK_{OT_j} = (g_j^{\gamma_j}, h_j^{\gamma_j}, e(g_j, g_j)^{\gamma_j^2})$ and

$$c_{k_A}^{(j)} = ((g_p^a g_p^{\xi_{j,1}+e_1})^{X_j \gamma_j} g_q^{s_{j,1}}, \ldots, (g_p^a g_p^{\xi_{j,D}+e_D})^{X_j \gamma_j} g_q^{s_{j,D}})$$

from $\{\xi_{j,m}\}_{m=1}^{D}$, when $\mathcal{E}$ queries on j, where μ is constant. If $\sum_{m=1}^{D} e_m^2 \leq \theta$, then $\mathcal{A}$ sets result = accept, otherwise, result = reject. Then, $\mathcal{A}$ gives $(PK_{OT_j}, c_{k_A}^{(j)}, \text{result})$ to $\mathcal{E}$. Note that $(g_p^a g_p^{\xi_{j,m}+e_m})^{X_m \gamma_j} g_q^{s_{j,m}}$ for $1 \leq m \leq D$ can be regarded as a valid ciphertext $E_{PK_{OT_j}}(a + \xi_{j,m} + e_m)$ since it holds $g_j^{\gamma_j q} = g_p^{X_j \gamma_j q}$ and $((g_p^a g_p^{\xi_{j,m}+e_m})^{X_j \gamma_j} g_q^{s_{j,m}})^q = (g_p^{X_j \gamma_j q})^{a+\xi_{j,m}+e_m}$. Additionally, $E_{PK_{OT_j}}(a + \xi_{j,m} + e_m)$ can also be regarded as $E_{PK_j}((a + \xi_{j,m} + e_m)\gamma_j)$.

In the challenge phase, $\mathcal{A}$ picks $Y_\ell', Z_\ell' \xleftarrow{u} \mathbb{Z}/N\mathbb{Z}$, compute

$$PK_{OT_\ell} = (g_{OT_\ell}, h_{OT_\ell}, g_{T,OT_\ell}) = (g_p^b g_q^{Y_\ell'}, g_q^{Z_\ell'}, e(g_p^b g_q^{Y_\ell'}, g_p^b g_q^{Y_\ell'})),$$

and returns PK_{OT_ℓ} as a one-time public key to $\mathcal{E}$ when $\mathcal{E}$ sends $\ell \in \{1, \ldots, n\}$.

Then, $\mathcal{E}$ outputs $(E_{PK_{OT_\ell}}(a + \xi_{\ell,1} + e_1'), \ldots, E_{PK_{OT_\ell}}(a + \xi_{\ell,D} + e_D'))$ with non-negligible probability ε, where $|e_m'| \leq \nu$ for $1 \leq m \leq D$ and ν is constant. Here, we observe

$$E_{PK_{OT_\ell}}(a + \xi_{\ell,m} + e_m') = (g_p^b g_q^{Y_\ell'})^{a+\xi_{\ell,m}+e_m'}(g_q^{Z_\ell'})^{R_m}$$
$$= g_p^{b(a+\xi_{\ell,m}+e_m')} g_q^{Y_\ell'(a+\xi_{\ell,m}+e_m')+Z_\ell' R_m},$$

for $1 \leq m \leq D$, where $R_m \in \mathbb{Z}/N\mathbb{Z}$. Therefore, we see that for $1 \leq m \leq D$, it holds

$$e(E_{PK_{OT_\ell}}(a + \xi_{\ell,m} + e_m'), g_p^c) = e(g_p, g_p)^{bc(a+\xi_{\ell,m}+e_m')}$$
$$= e(g_p, g_p)^{abc} \cdot e(g_p, g_p)^{bc(\xi_{\ell,m}+e_m')}$$

since $e(g_p, g_q) = 1_{\mathbb{G}_T}$.

From the above fact, $\mathcal{A}$ fixes $m \in \{1, \ldots, D\}$, computes a set W as

$$W = \{e(g_p^b, g_p^c)^{\xi_{\ell,m}}, e(g_p^b, g_p^c)^{\xi_{\ell,m}\pm 1}, \ldots, e(g_p^b, g_p^c)^{\xi_{\ell,m}\pm\nu}\},$$

by using $\xi_{\ell,m}$ and the DBDH on COG instance, and checks whether there exists an element $w \in W$ such that

$$e(E_{PK_{OT_\ell}}(a + \xi_{\ell,m} + e_m'), g_p^c)/w = e(g_p, g_p)^z.$$

If there exists such w, then $\mathcal{A}$ decides $e(g_p, g_p)^z = e(g_p, g_p)^{abc}$, otherwise, $e(g_p, g_p)^z \neq e(g_p, g_p)^{abc}$. In the both cases, $\mathcal{A}$ can break the DBDH on COG assumption with non-negligible probability $\varepsilon(\lambda)$. Thus, we conclude that our protocol is secure in the sense of authentication against $\mathcal{E}$.   □

4.5   Desirable Four Properties

We briefly mention that our protocol satisfies the desired four properties.

Accuracy: Our protocol does not affect the accuracy, because in the authentication process, the decryptor $\mathcal{D}$ can recover the squared Euclidean distance which is also used in the ordinary (non-cryptographic) biometrics.

Diversity: A very large number of cancelable templates can be produced in our protocol by changing random numbers such as $r_1, \ldots, r_D \xleftarrow{u} \mathbb{Z}/N\mathbb{Z}$. Also, it is impossible to cross-match templates even within a single database. In our protocol, templates for users i and j are $E_{PK_i}(\boldsymbol{x}) = (g_i^{x_1} h_i^{r_1}, \ldots, g_i^{x_D} h_i^{r_D})$ and $E_{PK_j}(\boldsymbol{x}) = (g_j^{x_1} h_j^{r'_1}, \ldots, g_j^{x_D} h_j^{r'_D})$, respectively, where $g_i, g_j \xleftarrow{u} \mathbb{G}$ and $h_i, h_j \xleftarrow{u} \mathbb{G}_q$. Therefore, two templates are totally independent. Moreover, since there exists $\zeta \in \mathbb{Z}/N\mathbb{Z}$ such that $g_i = g_j^\zeta$, $E_{PK_i}(\boldsymbol{x})$ can be regarded as $E_{PK_j}(\zeta\boldsymbol{x})$. Then, $d_{E^2}(\zeta\boldsymbol{x}, \boldsymbol{x})$ is not constant with overwhelming probability in λ. Thus the cross-matching is impossible.

Revocability: An attacker who obtained the former template $E_{PK_k}(\boldsymbol{x})$ and the new public key PK'_k cannot update his template into $E_{PK'_k}(\boldsymbol{x})$. This is because computing $E_{PK'_k}(\boldsymbol{x}) = (g_k^{\delta x_1} h_k^{\delta r_1}, \ldots, g_k^{\delta x_D} h_k^{\delta r_D})$ from $E_{PK_k}(\boldsymbol{x}) = (g_k^{x_1} h_k^{r_1}, \ldots, g_k^{x_D} h_k^{r_D})$ and $PK'_k = (g_k^\delta, h_k^\delta, g_{T,k}^{\delta^2})$ is exactly the computational Diffie-Hellman problem.

Security: Security properties have been proved in Section 4.4.

5   Conclusion

We have proposed an efficient and secure variant of the Hattori et al. cancelable biometric protocols [7], and have shown that our protocol is cryptographically secure for both template protection and authentication under the DBDH on COG assumption and the heuristic assumptions. Our security definition for authentication is more general and formal than those formalized in [13,8]. Note that we can also construct a variant based on the Okamoto-Takashima encryption scheme [14], by combining the construction idea of our BGN-based protocol with the Hattori et al. protocol based on the Okamoto-Takashima one.

Acknowledgments. The authors would like to thank Kazuo Ohta, Yusuke Sakai, and Takumi Yamamoto for helpful discussions, and the anonymous reviewers of IWSEC'13 for valuable comments.

References

1. Blanton, M., Gasti, P.: Secure and efficient protocols for iris and fingerprint identification. In: Atluri, V., Diaz, C. (eds.) ESORICS 2011. LNCS, vol. 6879, pp. 190–209. Springer, Heidelberg (2011)
2. Boneh, D., Goh, E.-J., Nissim, K.: Evaluating 2-DNF formulas on ciphertexts. In: Kilian, J. (ed.) TCC 2005. LNCS, vol. 3378, pp. 325–341. Springer, Heidelberg (2005)

3. Boneh, D., Waters, B.: Conjunctive, subset, and range queries on encrypted data. In: Vadhan, S.P. (ed.) TCC 2007. LNCS, vol. 4392, pp. 535–554. Springer, Heidelberg (2007)

4. Bringer, J., Chabanne, H.: An authentication protocol with encrypted biometric data. In: Vaudenay, S. (ed.) AFRICACRYPT 2008. LNCS, vol. 5023, pp. 109–124. Springer, Heidelberg (2008)

5. Dierks, T., Rescorla, E.: The transport layer security (TLS) protocol version 1.2. RFC 5246 (2008)

6. Erkin, Z., Franz, M., Guajardo, J., Katzenbeisser, S., Lagendijk, I., Toft, T.: Privacy-preserving face recognition. In: Goldberg, I., Atallah, M.J. (eds.) PETS 2009. LNCS, vol. 5672, pp. 235–253. Springer, Heidelberg (2009)

7. Hattori, M., Shibata, Y., Ito, T., Matsuda, N., Takashima, K., Yoneda, T.: Provably-secure cancelable biometrics using 2-DNF evaluation. Journal of Information Processing 20(2), 496–507 (2012)

8. Hirano, T., Hattori, M., Ito, T., Matsuda, N., Mori, T.: Homomorphic encryption based cancelable biometrics secure against replay and its related attack. In: ISITA 2012, pp. 421–425 (2012)

9. Jain, A.K., Nandakumar, K., Nagar, A.: Biometric template security. EURASIP Journal on Advances in Signal Processing 2008 (2008)

10. Juels, A., Sudan, M.: A fuzzy vault scheme. Designs, Codes and Cryptography 38(2), 237–257 (2006)

11. Juels, A., Wattenberg, M.: A fuzzy commitment scheme. In: ACM CCS 1999, pp. 28–36 (1999)

12. Kikuchi, H., Nagai, K., Ogata, W., Nishigaki, M.: Privacy-preserving similarity evaluation and application to remote biometrics authentication. Soft Computing 14(5), 529–536 (2010)

13. Nishigaki, M., Watanabe, Y., Oda, M., Yoneyama, Y., Yamamoto, T., Takahashi, K., Ogata, W., Kikuchi, H.: Template-protecting biometrics authentication using oblivious evaluation of feature value function with fuzzy polynomial. IPSJ Journal 53(9), 2254–2266 (2012) (in Japanese)

14. Okamoto, T., Takashima, K.: Homomorphic encryption and signatures from vector decomposition. In: Galbraith, S.D., Paterson, K.G. (eds.) Pairing 2008. LNCS, vol. 5209, pp. 57–74. Springer, Heidelberg (2008)

15. Ratha, N.K., Connell, J.H., Bolle, R.M.: Enhancing security and privacy in biometrics-based authentication systems. IBM Systems Journal 40(3), 614–634 (2001)

16. Sakashita, T., Shibata, Y., Yamamoto, T., Takahashi, K., Ogata, W., Kikuchi, H., Nishigaki, M.: A proposal of efficient remote biometric authentication protocol. In: Takagi, T., Mambo, M. (eds.) IWSEC 2009. LNCS, vol. 5824, pp. 212–227. Springer, Heidelberg (2009)

17. Takahashi, K., Hirata, S.: Cancelable biometrics with provable security and its application to fingerprint verification. IEICE Transactions on Fundamentals of Electronics, Communications and Computer Sciences E94-A(1), 233–244 (2011)

18. Takahashi, K., Naganuma, K.: Unconditionally provably secure cancellable biometrics based on a quotient polynomial ring. IET Biometrics 1(1), 63–71 (2012)

19. Upmanyu, M., Namboodiri, A.M., Srinathan, K., Jawahar, C.V.: Blind authentication: A secure crypto-biometric verification protocol. IEEE Transactions on Information Forensics and Security 5(2), 255–268 (2010)

Efficient Algorithm for Tate Pairing
of Composite Order

Yutaro Kiyomura[1] and Tsuyoshi Takagi[2]

[1] Graduate School of Mathematics, Kyushu University,
744, Motooka, Nishi-ku, Fukuoka 819-0395, Japan
[2] Institute of Mathematics for Industry, Kyushu University,
744, Motooka, Nishi-ku, Fukuoka 819-0395, Japan

Abstract. A lot of important cryptographic schemes such as fully secure leakage-resilient encryption and keyword searchable encryption are based on pairings of composite order. Miller's algorithm is used to compute pairings, and the time taken to compute the pairings depends on the cost of calculating the Miller loop. As a way of speeding up calculations of the parings of prime order, the number of iterations of the Miller loop can be reduced by choosing a prime order of low hamming weight. However, it is difficult to choose a particular composite order that can speed up the pairings of composite order. Kobayashi et al. proposed an efficient algorithm for computing Miller's algorithm by using a window method, called Window Miller's algorithm. We can compute scalar multiplication of points on elliptic curves by using a window hybrid binary-ternary form (w-HBTF). In this paper, we propose a Miller's algorithm that uses w-HBTF to compute Tate pairing efficiently. This algorithm needs a precomputation of the points on an elliptic curve and rational functions. The proposed algorithm was implemented in Java on a PC and compared with Window Miller's Algorithm in terms of the time and memory needed to make their precomputed tables. We used the supersingular elliptic curve $y^2 = x^3 + x$ of embedding degree 2 and a composite order of size of 2048 bits. The proposed algorithm with $w = 6 = 2 \cdot 3$ was about 12% faster than Window Miller's Algorithm with $w = 2$ given smallest precomputed tables of the same memory size. Moreover, the proposed algorithm with $w = 162 = 2 \cdot 3^4$ was about 8.5% faster than Window Miller's algorithm with $w = 7$ on each fastest algorithm.

Keywords: Composite order pairing, Miller's Algorithm, NAF, w-HBTF.

1   Introduction

A lot of cryptographic schemes, such as ID based encryption [5] and keyword searchable encryption [6], are constructed on pairings over elliptic curves. New cryptographic schemes that have high functionality such as fully secure leakage-resilient encryption and keyword searchable encryption use pairings of composite order [4][23][24]. Recently, Guillevic reported the timing of pairings of composite

K. Sakiyama and M. Terada (Eds.): IWSEC 2013, LNCS 8231, pp. 201–216, 2013.

order which has some prime factors $(2, 3, \ldots)$ [13]. Miller's Algorithm [22] is used to compute the pairings, and the time taken to compute the pairings depends on the cost of calculating the Miller loop. To speed up the parings of prime order, the number of iterations of the Miller loop can be reduced by choosing a prime order of low hamming weight. There are efficient algorithms of pairings such as η_T pairing [3], Ate pairing [15], R-ate pairing [20], and Optimal ate pairing [26] to compute pairings of prime order. Recently, Le et al. analyzed techniques to speed up Ate pairing computation in affine coordinates using 4-ary Miller algorithm for elliptic curve of embedding degrees $k = 9, 15$ [19]. In the case of the pairings of composite order, the number of iterations of the Miller loop can be reduced by constructing a composite order of low hamming weight. However, we should consider the security of lattice attack when we construct such composite order [16]. In this paper, we choose a random composite order which has 2 prime factors for our algorithm. On the other hand, there are some conversion techniques that translate cryptographic schemes constructed over composite order groups to similar ones in the prime order setting, but Meiklejohn et al. showed some limitations on such transformations [21]. If we know the factors of composite order N, we can calculate pairings of composite order efficiently by using Chinese reminder theorem[21]. However, we assume that we don't know the factors of N in this paper.

Kobayashi et al. proposed an efficient algorithm for computing Miller's algorithm using a window method, called Window Miller's algorithm [17]. By using the width-w non-adjacent form $(w\text{NAF})$[14] or double-base chains [9][10][11] or window hybrid binary-ternary form $(w\text{-HBTF})$ [1], we are able to compute scalar multiplications on points of elliptic curves efficiently. Note that w-HBTF is a special case of double-base chains [1].

Let p be a prime number satisfying $N < p$. Let k be a embedding degree. The security of pairings of composite order is based on the difficulty of solving several problems such as the factorization problem of N, the discrete logarithm problem on a finite field $\mathbb{F}_{p^k}$, and the elliptic curve discrete logarithm problem. The difficulty of solving the discrete logarithm problem on a finite field $\mathbb{F}_{p^k}$ and the elliptic curve discrete logarithm problem is satisfied if the difficulty of solving the factorization problem of N is satisfied. Let us define $\rho = \lfloor \log_2 p \rfloor / \lfloor \log_2 N \rfloor$. To compute pairings efficiently, we need to choose an elliptic curve whose ρ is small. The elliptic curves $y^2 = x^3 - x$ and $y^2 = x^3 - 4x$ are elliptic curves of embedding degree 1 when $p \equiv 1 \mod 4$. In this case, we can choose p and N satisfying $\rho = 2$ [18]. On the other hand, $y^2 = x^3 + x$ is a supersingular elliptic curve of embedding degree 2 when $p \equiv 3 \mod 4$. In this case, we can choose p and N satisfying $\rho = 1$. The supersingular elliptic curve is suitable for efficiently calculating pairings of composite order, because $k = 2$ is the smallest embedding degree among pairings satisfying $\rho = 1$.

In this paper, we propose an efficient Miller's Algorithm using w-HBTF that computes the Tate pairing. The proposed algorithm needs a precomputation of points on an elliptic curve and rational functions to speed up the pairings computation. We also propose a computation of f^3 for $f \in \mathbb{F}_{p^2}$ of the TTRL algorithm. We implemented the proposed algorithm and the Window Miller's Algorithm in Java on a PC and compared them in terms of the time and the memory size of the precomputed table. In this implementation, we used the supersingular elliptic curve $y^2 = x^3 + x$ of $k = 2$ and a composite order of size of 2048 bits. We will explain how to construct composite integer N and prime p in the Section 4.2. The proposed algorithm with $w = 6 = 2 \cdot 3$ turned out to be about 12% faster than Window Miller's Algorithm with $w = 2$ on the smallest precomputed tables of the same memory size. Moreover, the proposed algorithm with $w = 162 = 2 \cdot 3^4$ was about 8.5% faster than Window Miller's algorithm with $w = 7$ on each fastest algorithm.

2   Mathematical Preparations

Here, we explain the necessary mathematical concepts, i.e., elliptic curves and the extension field for Tate pairing, Miller's algorithm, the width-w non-adjacent form (wNAF), and the window hybrid binary-ternary form (w-HBTF).

2.1   Elliptic Curve

Let p be a prime number such that $p \equiv 3 \mod 4$. A supersingular elliptic curve of $k = 2$ is defined over $\mathbb{F}_p$ as follows:

$$E(\mathbb{F}_p) = \{(x, y) \in \mathbb{F}_p \times \mathbb{F}_p \mid y^2 = x^3 + x\} \cup \{O\} \tag{1}$$

where O is a point at infinity of $E(\mathbb{F}_p)$.

We can explain the elliptic curve operations using affine coordinates. Given points $P_1 = (x_1, y_1), P_2 = (x_2, y_2) \in E(\mathbb{F}_p)$, we define $P_1 + P_2 = (x_3, y_3)$ as follows if $P_1 \neq \pm P_2$

$$x_3 = \left(\frac{y_2 - y_1}{x_2 - x_1}\right)^2 - x_1 - x_2, \quad y_3 = \left(\frac{y_2 - y_1}{x_2 - x_1}\right)(x_1 - x_3) - y_1$$

and define $P_1 + P_2 = O$ if $P_1 = -P_2$. Given a point $P_1 = (x_1, y_1) \in E(\mathbb{F}_p)$, we define the double $2P_1 = (x_4, y_4)$ as follows if $P_1 \neq -P_1$,

$$x_4 = \left(\frac{3x_1^2 + 1}{2y_1}\right)^2 - 2x_1, \quad y_4 = \left(\frac{3x_1^2 + 1}{2y_1}\right)(x_1 - x_4) - y_1$$

and define $2P_1 = O$ if $P_1 = -P_1$.

The elliptic curve $E(\mathbb{F}_p)$ forms an abelian group: (1) O is the zero element. (2) Define the negative $-P_1 = (x_1, -y_1)$ for $P_1 = (x_1, y_1) \in E(\mathbb{F}_p)$. (3) The sums $P_1 + P_2$ and $2P_1$ are defined according to the above operations. The group order of $E(\mathbb{F}_p)$ is $p + 1$. Let N be a composite number such that $N \mid p + 1$. Define $E(\mathbb{F}_p)[N]$ to be a subgroup of $E(\mathbb{F}_p)$ that has group order N.

2.2 Extension Field $\mathbb{F}_{p^2}$

The quadratic extension field is represented as follows:

$$\mathbb{F}_{p^2} = \mathbb{F}_p[i]/(i^2 + 1)$$

where $i^2 = -1$. If $p \equiv 3 \mod 4$, -1 is a quadratic non-residue over $\mathbb{F}_p$ and $i^2 + 1$ is irreducible over $\mathbb{F}_p[i]$. Now we will explain the extension field $\mathbb{F}_{p^2}$ operations. Let $A = a_0 + a_1 i, B = b_0 + b_1 i \in \mathbb{F}_{p^2}$ ($a_0, a_1, b_0, b_1 \in \mathbb{F}_p$). The Karatsuba method [8] computes the product $A \cdot B$ in $\mathbb{F}_{p^2}$ as

$$A \cdot B = (a_0 b_0 - a_1 b_1) + \{(a_0 + a_1)(b_0 + b_1) - a_0 b_0 - a_1 b_1\}i$$

which needs three multiplications and five additions of $\mathbb{F}_p$. The Complex method [8] computes the square A^2 in $\mathbb{F}_{p^2}$ as

$$A^2 = \{(a_0 + a_1)(a_0 - a_1)\} + 2a_0 a_1 i$$

which needs two multiplications and two additions of $\mathbb{F}_p$.

2.3 Tate Pairing

Here, we define the Tate pairing. The embedding degree of the elliptic curve $E(\mathbb{F}_p)$ is 2. So there exists a distortion map $\phi : E(\mathbb{F}_p) \to E(\mathbb{F}_{p^2})$ defined by $(x, y) \mapsto (-x, yi)$. Define $Q = \phi(Q') \in E(\mathbb{F}_{p^2})$ for $Q' \in E(\mathbb{F}_p)[N]$. Let f_i be a rational function with the following divisor :

$$\mathrm{div}(f_i) = i(P) - (iP) - (i - 1)(O) \tag{2}$$

where $i \in \mathbb{Z}$, $P \in E(\mathbb{F}_p)[N]$. Let $l_{R,S}$ be the equation of a line through $R, S \in E(\mathbb{F}_p)$, and v_T be the vertical line through $T \in E(\mathbb{F}_p)$. For a given $Q \in E(\mathbb{F}_{p^2})$, we have

$$f_{i+j}(Q) = f_i(Q) \cdot f_j(Q) \cdot \frac{l_{iP,jP}(Q)}{v_{(i+j)P}(Q)} \tag{3}$$

where $i, j \in \mathbb{Z}$. In the case of embedding degree 2, we can redefine equation (3) as follows [2]:

$$f_{i+j}(Q) = f_i(Q) \cdot f_j(Q) \cdot l_{iP,jP}(Q) \tag{4}$$

Finally, the Tate pairing $e(P, Q) \in \mathbb{F}_{p^2}^{\times}/(\mathbb{F}_{p^2}^{\times})^N$ is defined by

$$e(P, Q) = f_N(Q)^{(p^2-1)/N}$$

where the rational function f_N has $\mathrm{div}(f_N) = N(P) - N(O)$. The pairing has the bilinear property,

$$e(aP, bQ) = e(P, Q)^{ab}$$

for all $P \in E(\mathbb{F}_p)[N], Q \in E(\mathbb{F}_{p^2})$ and $a, b \in \mathbb{Z}$.

2.4   Miller's Algorithm

Here, we describe the Miller Algorithm (Algorithm 1). First, let us define $n = \lfloor \log_2 N \rfloor + 1$ as the bit length of N.

Algorithm 1. Miller's Algorithm

Input: prime : p, composite order : N, $P \in E(\mathbb{F}_p)[N]$, $Q \in E(\mathbb{F}_{p^2})$

Output: $e(P,Q) \in \mathbb{F}_{p^2}^{\times}/(\mathbb{F}_{p^2}^{\times})^N$

1: $f \leftarrow 1, V \leftarrow P$ and $N = \sum_{i=0}^{n-1} N_i 2^i$ ($N_{n-1} = 1$, $N_i \in \{0,1\}$ ($i = 0,1,\ldots,n-2$))
2: **for** $j \leftarrow n-2$ to 0 **do**
3:     $(f,V) \leftarrow [\text{TDBL}(f,V,Q)]$
4:     **if** $N_j = 1$ **then**
5:         $(f,V) \leftarrow [\text{TADD}(f,P,V,Q)]$
6:     **end if**
7: **end for**
8: $e(P,Q) \leftarrow f^{(p^2-1)/N}$

Steps 2 to 7 are called the Miller loop, and step 8 is the final modular exponentiation step. TDBL of step 3 is Algorithm 2, TADD of step 5 is Algorithm 3, and step 8 is Algorithm 4.

Algorithm 2. TDBL

Input: $f \in \mathbb{F}_{p^2}^{\times}$, $V \in E(\mathbb{F}_p)[N]$, $Q \in E(\mathbb{F}_{p^2})$

Output: $f \in \mathbb{F}_{p^2}^{\times}$, $2V \in E(\mathbb{F}_p)[N]$

1: $V \leftarrow 2V$
2: $f \leftarrow f^2 \cdot l_{V,V}(Q)$

Algorithm 3. TADD

Input: $f \in \mathbb{F}_{p^2}^{\times}$, $P, V \in E(\mathbb{F}_p)[N]$, $Q \in E(\mathbb{F}_{p^2})$

Output: $f \in \mathbb{F}_{p^2}^{\times}$, $V + P \in E(\mathbb{F}_p)[N]$

1: $V \leftarrow V + P$
2: $f \leftarrow f \cdot l_{V,P}(Q)$

Algorithm 4. Final modular exponentiation

Input: prime: p, composite order: N, $f \in \mathbb{F}_{p^2}^{\times}$

Output: $f^{(p^2-1)/N} \in \mathbb{F}_{p^2}^{\times}/(\mathbb{F}_{p^2}^{\times})^N$

1: $g \leftarrow f$, $r \leftarrow (p^2-1)/N$ and $r = \sum_{i=0}^{n-1} r_i 2^i$
   ($r_{n-1} = 1$, $r_i \in \{0,1\}$ ($i = 0,1,\ldots,n-2$))
2: **for** $j \leftarrow n-2$ to 0 **do**
3:     $g \leftarrow g^2$
4:     **if** $r_j = 1$ **then**
5:         $g \leftarrow g \cdot f$
6:     **end if**
7: **end for**

2.5   Width-w Radix-r Non-adjacent Form

Now let us explain wrNAF [25]. For a positive integer d, the wrNAF representation of d is

$$d = \sum_{i=0}^{n} d_w[i] r^i \quad (d_w[i] \in D_{w,r})$$

where $D_{w,r} = \{0, \pm 1, \pm 2, \ldots, \pm \lfloor \frac{r^w - 1}{2} \rfloor \} \setminus \{\pm 1 r, \pm 2 r, \ldots, \pm \lfloor \frac{r^{w-1} - 1}{2} \rfloor r \}$ and $d_w[n] > 0$. The wrNAF representation has at most 1 non-zero digit among any of the w adjacent digits. In particular, if $r = 2$, wrNAF is called wNAF. wrNAF has the following properties [25]: (1) Every positive integer d has a unique wrNAF. (2) The wrNAF representation of d has the smallest Hamming weight among all signed representations for d with the digit set $D_{w,r}$. (3) The non-zero density of the wrNAF is asymptotically $(r-1)/\{w(r-1)+1\}$ for $n \to \infty$. We describe the algorithm that generates the wrNAF representation from a positive integer d (Algorithm 5) [25].

Algorithm 5. wrNAF [25]

Input: $d \in \mathbb{Z}_{>0}$, width w, radix r
Output: $d = \sum_{i=0}^{n} d_w[i] r^i$ (wrNAF)
 1: $i \leftarrow 0$
 2: **while** $d > 0$ **do**
 3:     **if** $d \bmod r = 0$ **then**
 4:         $d_w[i] \leftarrow 0$
 5:     **else**
 6:         $d_w[i] \leftarrow d \bmod s\ r^w$ and $d \leftarrow d - d_w[i]$
 7:     **end if**
 8:     $d \leftarrow d/r$ and $i \leftarrow i + 1$
 9: **end while**

2.6   Window Hybrid Binary-Ternary Form

Here, we explain the window hybrid binary-ternary form (w-HBTF)[9], which is a representation mixing bases 2 and 3 and using the window method of a positive integer. It is special case of double-base chains. The w-HBTF is also a representation of the extension of wNAF. The w in w-HBTF is the value of an expression of the form $2^b 3^t$ with $b, t \in \mathbb{N}$. For example, if $b = 1$ and $t = 1$, we get a window of size $2 \cdot 3 = 6$. For a positive integer d, the w-HBTF representation of d is

$$d = \sum_{i=1}^{m} s_i 2^{b_i} 3^{t_i}$$

where $s_i \in D_w = \{a \in \mathbb{Z} \mid 0 < |a| \leq 2^{b-1}3^t$ and $\gcd(a, 2, 3) = 1\}$, $b_1 \geq b_2 \geq \ldots, \geq b_m \geq 0$, $t_1 \geq t_2 \geq, \ldots, \geq t_m \geq 0$. $|D_w|$ is defined as the number of sets D_w, and D_w^+ is defined as the positive integers in D_w. Note that if $t = 0$, w-HBTF is equivalent to b-NAF. Algorithm 6 generates a unique w-HBTF representation of a positive integer d for a width w [1].

Algorithm 6. w-HBTF [1]

Input: $d \in \mathbb{Z}_{>0}$, width w such that $w = 2^b 3^t (b, t \in \mathbb{Z}_{>0})$
Output: $d = \sum_{i=1}^m s_i 2^{b_i} 3^{t_i}$ $(w$-HBTF$)$
1: $i \leftarrow 0$ and $m \leftarrow 0$
2: **while** $d > 0$ **do**
3:    **if** $d \bmod 2 = 0$ **then**
4:       whbt$[i] \leftarrow 0$ and base$[i] \leftarrow 2$
5:    **else if** $d \bmod 3 = 0$ **then**
6:       whbt$[i] \leftarrow 0$ and base$[i] \leftarrow 3$
7:    **else**
8:       whbt$[i] \leftarrow d$ mods w, base$[i] \leftarrow 2$ and $d \leftarrow d-$whbt$[i]$ and $m \leftarrow m + 1$
9:    **end if**
10:   $d \leftarrow d/$base$[i]$ and $i \leftarrow i + 1$
11: **end while**
12: $s_m \leftarrow$ whbt$[0]$
13: **if** $s_m \neq 0$ **then**
14:   $b_m \leftarrow 0$, $t_m \leftarrow 0$ and $m \leftarrow m - 1$
15: **end if**
16: **for** $j \leftarrow 1$ to $i - 1$ **do**
17:   **if** whbt$[j] = 0$ and base$[j - 1] = 2$ **then**
18:      $b_m \leftarrow b_m + 1$
19:   **else if** whbt$[j] = 0$ and base$[j - 1] = 3$ **then**
20:      $t_m \leftarrow t_m + 1$
21:   **else if** whbt$[j] \neq 0$ and base$[j - 1] = 2$ **then**
22:      $b_m \leftarrow b_m + 1$, $s_m \leftarrow$ whbt$[j]$, $b_{m-1} \leftarrow b_m$, $t_{m-1} \leftarrow t_m$ and $m \leftarrow m - 1$
23:   **else if** whbt$[j] \neq 0$ and base$[j - 1] = 3$ **then**
24:      $t_m \leftarrow t_m + 1$, $s_m \leftarrow$ whbt$[j]$, $b_{m-1} \leftarrow b_m$, $t_{m-1} \leftarrow t_m$ and $m \leftarrow m - 1$
25:   **end if**
26: **end for**

Example 1. Let $w = 6 = 2 \cdot 3$ and $d = 12539$. By using Algorithm 6, we can represent d as follows.

$$d = 2^6 \cdot 3^5 - 2^5 \cdot 3^4 - 2^4 \cdot 3^3 + 2^2 \cdot 3^1 - 1$$

3   Miller's Algorithm Using the Window Method

Here, we explain the algorithms for computing pairings using the window method. In order to reduce the computational costs of the Miller loop, Miller's Algorithm using the window method needs a precomputation of points on an elliptic curve and rational functions for the width w.

3.1   Window Miller's Algorithm

Kobayashi et al. proposed an efficient algorithm for computing Miller's algorithm using the window method, called Window Miller's algorithm [17]. Algorithm 7 is Window Miller's Algorithm with wNAF.

Algorithm 7. Window Miller's Algorithm with wNAF [17]

Input: prime: p, order: N, $P = (x_P, y_P) \in E(\mathbb{F}_p)[N]$, $Q = (x_Q, y_Q) \in E(\mathbb{F}_{p^2})$
Output: $e(P,Q) \in \mathbb{F}_{p^2}^\times / (\mathbb{F}_{p^2}^\times)^N$

1: Represent $N = \sum_{i=0}^{n} N_i 2^i$ using wNAF ($N_n > 0$)
2: $P_1 \leftarrow P$ and $f_1 \leftarrow 1$ and $g \leftarrow (y_Q + y_P)/(x_Q - x_P)$
3: $P_2 \leftarrow 2P$ and $f_2 \leftarrow l_{P,P}(Q)$
4: **for** $k \leftarrow 3$ to $2^{w-1} - 1$ step 2 **do**
5:     $P_k \leftarrow P_{k-2} + 2P$ and $f_k \leftarrow f_{k-2} \cdot f_2 \cdot l_{(k-2)P,2P}(Q)$
6: **end for**
7: **for** $k \leftarrow 3$ to $2^{w-1} - 1$ step 2 **do**
8:     $f_{-k} \leftarrow (a_k - b_k i)$ /*$f_k = (a_k + b_k i)$*/
9: **end for**
10: $f \leftarrow f_{N_n}$ and $V \leftarrow P_{N_n}$
11: **for** $j \leftarrow n - 1$ to 0 **do**
12:     $(f, V) \leftarrow [\text{TDBL}(f, V, Q)]$
13:     **if** $N_j = 1$ **then**
14:         $(f, V) \leftarrow [\text{TADD}(f, P, V, Q)]$
15:     **else if** $N_j = -1$ **then**
16:         $(f, V) \leftarrow [\text{TSUB}(f, g, P, V, Q)]$
17:     **else if** $N_j \in (D_{w,2} - \{0, \pm 1\})$ **then**
18:         $(f, V) \leftarrow [\text{TADD with } P_{N_j}(f, f_{N_j}, P_{N_j}, V, Q)]$
19:     **end if**
20: **end for**
21: $e(P,Q) \leftarrow f^{(p^2-1)/N}$

Steps 2 to 9 are called the precomputed part, and steps 11 to 20 are called the Miller loop. The wNAF representation for a width w has a digit set $D_{w,2}$. Hence, we need to precompute the points on an elliptic curve $3P, 5P, \ldots, (2^{w-1} - 1)P$ and rational functions $f_{\pm 3}, f_{\pm 5}, \ldots, f_{\pm(2^{w-1}-1)}$. We also need TSUB and TADD with P_j of Algorithm 7. Note that in the case of $w = 2$, we can skip steps 3 to 9 and 17 to 19. Now let us show how to compute of steps 8 of Algorithm 7 [17]. In fact, we will compute $f_{-k} = 1/(f_k \cdot v_{kP}(Q))$ of step 8 of Algorithm 7. For $f_k = a_k + b_k i$ ($a_k, b_k \in \mathbb{F}_p$), we have

$$f_{-k} = \frac{1}{f_k \cdot v_{kP}(Q)} = \frac{a_k - b_k i}{(a_k + b_k i)(a_k - b_k i) \cdot v_{kP}(Q)} = \frac{a_k - b_k i}{(a_k^2 + b_k^2) \cdot v_{kP}(Q)}$$

The denominator $(a_k^2 + b_k^2) \cdot v_{kP}(Q)$ belongs to $\mathbb{F}_p$, and these product is equal to 1 in the final modular exponentiation because of Fermat small theorem. Thus, ignoring $(a_k^2 + b_k^2) \cdot v_{kP}(Q)$ will not change the results of the pairings. Therefore, for $f_k = a_k + b_k i$ $(a_k, b_k \in \mathbb{F}_p)$, we can regard f_{-k} as $f_{-k} = a_k - b_k i$.

TSUB is Algorithm 8, and TADD with P_k is Algorithm 9.

Algorithm 8. TSUB [17]

Input: $f, g \in \mathbb{F}_{p^2}^{\times}$, $P, V \in E(\mathbb{F}_p)[N]$, $Q \in E(\mathbb{F}_{p^2})$
Output: $f \in \mathbb{F}_{p^2}^{\times}$, $V - P \in E(\mathbb{F}_p)[N]$
  1: $V \leftarrow V - P$
  2: $f \leftarrow f \cdot (g - \alpha)$ (α is the slope of the line through point V and $-P$.)

Note that $f_{-1}(Q) = 1/v_P(Q)$, because of equation (2) in Section 2.3. Thus, we can compute $f_{-1} \cdot l_{V,-P}(Q)$ (equation (4) of Section 2.3) as

$$f_{-1} \cdot l_{V,-P}(Q) = \frac{(y_Q + y_P) - \alpha(x_Q - x_P)}{x_Q - x_P}$$
$$= \frac{y_Q + y_P}{x_Q - x_P} - \alpha$$

where α is the slope of the line through the points V and $-P$. We precompute $(y_Q + y_P)/(x_Q - x_P)$ as g in step 2 of Algorithm 7. After that, we compute $g - \alpha$ in step 2 of Algorithm 8. In this way, we can omit the computation of $l_{V,-P}$ and eliminate one multiplication from $\mathbb{F}_p$ in the TSUB algorithm.

Algorithm 9. TADD with P_k [17]

Input: $f, f_{\pm k} \in \mathbb{F}_{p^2}^{\times}$, $P_k, V \in E(\mathbb{F}_p)[N]$, $Q \in E(\mathbb{F}_{p^2})$
Output: $f \in \mathbb{F}_{p^2}^{\times}$, $V \pm P_k \in E(\mathbb{F}_p)[N]$
  1: $V \leftarrow V \pm P_k$
  2: $f \leftarrow f \cdot f_{\pm k} \cdot l_{V, \pm kP}(Q)$

In order to evaluate the computational costs of TADD, TSUB and TADD with P_k (see Section 4.1), we use Algorithm 9 in the case $k > 1$.

3.2   Proposed Algorithm

We devised an efficient version of Miller's Algorithm (Algorithm 10) using w-HBTF for the pairings.

Steps 2 to 11 are the precomputed part, and steps 13 to 28 are the Miller loop. The w-HBTF representation for a width w has a digit set D_w. Hence, we need to precompute the points on an elliptic curve kP and rational functions $f_{\pm k}$ for $k \in D_w$. Algorithm 10 also needs TTRL, TSUB, and TADD with P_j. Note that in the case of $w = 6$, we can skip steps 3 to 11 and 25 to 27.

TTRL is Algorithm 11. Step 1 of Algorithm 11 computes the triple $3P_1 = (x_3, y_3)$ for all $P_1 = (x_1, y_1)$ $(P_1 \neq -P_1, -2P_1)$, as follows. First, it precomputes

---

Algorithm 10. Proposed Algorithm using w-HBTF

Input: prime : p, order : N, width : w, $P \in E(\mathbb{F}_p)[N]$, $Q \in E(\mathbb{F}_{p^2})$
Output: $e(P,Q) \in \mathbb{F}_{p^2}^{\times}/(\mathbb{F}_{p^2}^{\times})^N$
1: Represent $N = \sum_{i=1}^{m} s_i 2^{b_i} 3^{t_i}$ using w-HBTF
2: $P_1 \leftarrow P$, $f_1 \leftarrow 1$ and $g \leftarrow (y_Q + y_P)/(x_Q - x_P)$
3: $P_2 \leftarrow 2P$ and $f_2 \leftarrow l_{P,P}(Q)$
4: **for** $k \leftarrow 3$ to $\max\{D_w^+\}$ step 2 **do**
5:     $P_k \leftarrow P_{k-2} + 2P$ and $f_k \leftarrow f_{k-2} \cdot f_2 \cdot l_{(k-2)P,2P}(Q)$
6: **end for**
7: **for** $k \leftarrow 5$ to $\max\{D_w^+\}$ step 2 **do**
8:    **if** $k \in D_w$ **then**
9:        $f_{-k} \leftarrow (a_k - b_k i)$ /*$f_k = (a_k + b_k i)$*/
10:    **end if**
11: **end for**
12: $f \leftarrow f_{s_1}$ and $V \leftarrow P_{s_1}$
13: **for** $i \leftarrow 1$ to $m-1$ **do**
14:    $u \leftarrow b_i - b_{i+1}$ and $v \leftarrow t_i - t_{i+1}$
15:    **for** $j \leftarrow 1$ to u **do**
16:        $(f, V) \leftarrow [\text{TDBL}(f, V, Q)]$
17:    **end for**
18:    **for** $j \leftarrow 1$ to v **do**
19:        $(f, V) \leftarrow [\text{TTRL}(f, V, Q)]$
20:    **end for**
21:    **if** $s_{i+1} = 1$ **then**
22:        $(f, V) \leftarrow [\text{TADD}(f, P, V, Q)]$
23:    **else if** $s_{i+1} = -1$ **then**
24:        $(f, V) \leftarrow [\text{TSUB}(f, g, P, V, Q)]$
25:    **else if** $s_{i+1} \in D_w - \{\pm 1\}$ **then**
26:        $(f, V) \leftarrow [\text{TADD with } P_{s_{i+1}}(f, f_{s_{i+1}}, P_{s_{i+1}}, V, Q)]$
27:    **end if**
28: **end for**
29: $e(P,Q) \leftarrow f^{(p^2-1)/N}$

$$X = (2y_1)^2, \ Z = 3x_1^2 + 1, \ Y = Z^2, \ d = X(3x_1) - Y, \ D = d(2y_1), \ I = D^{-1},$$
$$\alpha_1 = dIZ, \ \alpha_2 = X^2 I - \alpha_1.$$

Next, it computes

$$x_3 = (\alpha_2 - \alpha_1)(\alpha_1 + \alpha_2) + x_1, \ y_3 = (x_1 - x_3)\alpha_2 - y_1$$

where α_1 is the slope of the tangent line through P_1 and α_2 is the slope of the line through P_1 and $2P_1$. Step 2 of Algorithm 11 computes l_1 by using the method in [12]. For $V = (x_V, y_V) \in E(\mathbb{F}_p)[N], Q = (x_Q, y_Q) \in E(\mathbb{F}_{p^2})$, we have

$$\frac{l_{V,V}(Q) \cdot l_{V,2V}(Q)}{v_{2V}(Q)} = (x_Q - x_V)(x_Q - x_V + \alpha_1(\alpha_1 + \alpha_2)) - (\alpha_1 + \alpha_2)(y_Q - y_V)$$

Algorithm 11. TTRL

Input: $f \in \mathbb{F}_{p^2}^{\times}$, $V \in E(\mathbb{F}_p)[N]$, $Q \in E(\mathbb{F}_{p^2})$

Output: $f \in \mathbb{F}_{p^2}^{\times}$, $3V \in E(\mathbb{F}_p)[N]$

1: $V \leftarrow 3V$
2: $l_1 \leftarrow (l_{V,V}(Q) \cdot l_{V,2V}(Q))/v_{2V}(Q)$
3: $f \leftarrow f^3 \cdot l_1$

The computation of f^3 for $f = a_0 + a_1 i \in \mathbb{F}_{p^2}$ ($a_0, a_1 \in \mathbb{F}_p$) is as follows:

$$f^3 = a_0(X - 3Y) + a_1(3X - Y)i$$

where $X = a_0^2$, $Y = a_1^2$, which needs two multiplications, two squares, and two additions of $\mathbb{F}_p$.

4 Comparison

Let us start by explaining the theoretical computational costs and memory size of the precomputed table of each algorithm. Next, we compare the proposed algorithm and the Window Miller's Algorithm using the supersingular elliptic curve $y^2 = x^3 + x$ and a composite order of size of 2048 bits. The supersingular elliptic curve is suitable for efficiently calculating pairings of composite order, because $k = 2$ is the smallest embedding degree among pairings satisfying $\rho = 1$.

4.1 Computational Costs

We denote the computational cost of multiplication, squaring, addition, and taking the inverse of $\mathbb{F}_p$ as M, S, A and I. Moreover, we denote the computational costs of multiplication, squaring, and cubing $\mathbb{F}_{p^2}$ as M_2, S_2, and C_2. Table 1 shows the computational costs of TDBL, TTRL, TADD, TSUB, and TADD with P_j. Here, $[l]$=1M+2A represents the computational cost of l in step 2 of Algorithms TDBL, TADD, and TADD with P_j.

Table 2 shows the computational costs of the Miller loop. The values of TDBL, TTRL and so on refer to Table 1 of [1] for $w = 6, 12, 18, 24, 36$. $B_{w,n}$, $T_{w,n}$ and $m_{w,n}$ are b_1, t_1 and m in Section 2.6 for w and n, and they correspond to "#base 2 digits", "#base 3 digits", and "#non-zero digits" of Table 1 in [1]. Table 3 lists the number of precomputations.

Table 4 shows the precomputational costs of each algorithm. Note that we don't need to precompute $-kP(k \geq 3)$ because $-kP = (x, -y)$ for all $kP = (x, y)$. In the precomputed part of Algorithm 7 and 10, the computational costs of g, $f_{-k}(k \geq 3)$ are $[g]$=2M+1A+1I, $[f_{-k}]$=2M respectively.

The computational cost of the final modular exponentiation of Algorithm 4 is

$$\lfloor \log_2((p^2 - 1)/N) \rfloor S_2 + \lfloor \log_2((p^2 - 1)/N)/2 \rfloor M_2. \tag{5}$$

Table 1. Computational costs of TDBL, TTRL, TADD, TSUB, and TADD with P_k

	STEP 1	$2M+2S+4A+1I$
TDBL	STEP 2	$1M_2+1S_2+[l]$
	SUM	$8M+2S+13A+1I$
	STEP 1	$7M+4S+8A+1I$
TTRL	STEP 2	$4M+4A$
	STEP 3	$1C_2+1M_2$
	SUM	$16M+6S+19A+1I$
	STEP 1	$2M+1S+6A+1I$
TADD	STEP 2	$1M_2+[l]$
	SUM	$6M+1S+13A+1I$
	STEP 1	$2M+1S+6A+1I$
TSUB	STEP 2	$1M_2+1A$
	SUM	$5M+1S+12A+1I$
	STEP 1	$2M+1S+6A+1I$
TADD with P_k	STEP 2	$2M_2+[l]$
	SUM	$9M+1S+18A+1I$

Table 2. Computational costs of the Miller loop

Algorithm	Computational cost of Miller loop				
	TDBL	TTRL	TADD	TSUB	TADD with P_k
1 [22]	n	–	$n/2$	–	–
7 [17]	$n+1$	–	$n/(w+1) \cdot 1/2^{w-1}$	$n/(w+1) \cdot 1/2^{w-1}$	$n/(w+1) \cdot (2^{w-1}-2)/2^{w-1}$
10	$B_{w,n}$	$T_{w,n}$	$m_{w,n}/|D_w|$	$m_{w,n}/|D_w|$	$m_{w,n} \cdot (|D_w|-2)/|D_w|$

4.2   Parameter and Implementation Environment

The composite order N is the product of two random prime numbers whose bit lengths are the same. In this study, we used a composite order of size of 2048 bits. Let $p = 4N - 1$ be a prime satisfying $p \equiv 3 \mod 4$ and $N \mid p+1$. Accordingly, the size of p is 2050 bits on average.

In our implementation, we used a PC with the following specifications: OS: Mac OS X Lion 10.7.5, CPU: Intel Core i7 2.7GHz, RAM: 4GB, Language: Java, compiler: JDK (Java Development Kit) 6, and virtual machine: JRE (Java Runtime Environment) 6.

We used BigInteger class in our implementation for each operation of $\mathbb{F}_p$. Table 5 shows the time for each operation of $\mathbb{F}_p$.

The proportion of M, S and I was $1 : 1 : 21$.

4.3   Evaluation

We compared the proposed algorithm with Window Miller's Algorithm. For the parameter described in section 4.2, Table 6 shows the theoretical computational

Table 3. Number of precomputations

Algorithm	kP $(k \geq 3)$	g	$f_{\pm k}$ $(k \geq 3)$
1 [22]	–	–	–
7 [17]	$2^{w-2} - 1$	1	$2^{w-1} - 2$
10	$|D_w^+ - \{1\}|$	1	$|D_w - \{\pm 1\}|$

Table 4. Precomputational costs

Algorithm	[TDBL]	[TADD]	[TADD with P_k]	[g]	[f_{-k}]
1 [22]	–	–	–	–	–
7 [17]	1^*	1^*	$2^{w-2} - 2^*$	1	$2^{w-2} - 1^*$
10	1^*	1^*	$(\max\{D_w^+\} - 3)/2^*$	1	$|D_w^+ - \{1\}|^*$

* In Algorithm 7, the * part is 0 when $w = 2$, and in Algorithm 10, the * part is 0 when $w = 6$.

cost, implementation time, and memory size of the precomputed table for each algorithm.

We explain the computational cost and memory size of the precomputed table in Table 6. A, M, S and I of $\mathbb{F}_p$ for each algorithm were computed using Tables 1, 2, 3, 4, equation (5) of Section 4.1, and the parameter described in Section 4.2. We ignored A and estimated the theoretical computational cost of multiplication M by using S=M and I=21M. Note that if $w = 48, 54, 72, 96, 108$ and 162, we estimated the theoretical computational cost of M by using S=M and I=21M. The memory size of the precomputed table was computed using the parameter of Section 4.2 and the following expression:

([The number of precomputation of elliptic curve points] $\times \overline{2} +$ [The number of precomputation of rational functions] $\times \underline{2} +$ [The number of precomputation of g] $\times \underline{2}) \times \lfloor \log_2 p \rfloor$ (6)

Here, $\overline{2}$ means two coordinates (x and y of an elliptic curve point $(x, y) \in E(\mathbb{F}_p)$), and $\underline{2}$ means the embedding degree.

In Miller's Algorithm using the window method, as the width w increases, the computational cost of the Miller loop decreases but the computational cost of the precomputed part increases.

Therefore, the proposed algorithm with $w = 6 = 2 \cdot 3$ was about 12% faster than Window Miller's Algorithm with $w = 2$ for the smallest precomputed tables having the same memory size, and the proposed algorithm with $w = 162 = 2 \cdot 3^4$ was about 8.5% faster than Window Miller's Algorithm with $w = 7$.

In the case of $N =1024$ bits and $p =1026$ bits, the proposed algorithm with $w = 54 = 2 \cdot 3^3$ (257ms) was about 7.9% faster than Window Miller's Algorithm with $w = 6$ (279ms) and the proportion of M, S and I was $1 : 1 : 23$. Moreover, in the case of $N =3072$ bits and $p =3074$ bits, the proposed algorithm with

Table 5. Time for each operation on $\mathbb{F}_p$ (μs)

Addition	Multiplication	Square	Inverse
0.2	20.8	20.8	437

Table 6. Theoretical computational cost, implementation time, and memory size of the precomputed tables

Algorithm	w	Computational cost (M)	time (s)	memory (bits)
1 [22]	–	99342	2.125	0
7 [17]	2	89497	1.920	4104
	3	85761	1.835	16416
	4	83190	1.787	41040
	5	81415	1.751	90288
	6	80252	1.719	188784
	7	**79683**	**1.715**	385776
	8	79872	1.717	779000
10	6 ($2 \cdot 3$)	79304	1.699	4104
	12 ($2^2 \cdot 3$)	78889	1.692	16416
	18 ($2 \cdot 3^2$)	76289	1.634	28728
	24 ($2^3 \cdot 3$)	78398	1.664	41040
	36 ($2^2 \cdot 3^2$)	75214	1.618	65664
	48 ($2^4 \cdot 3$)	76960	1.650	90288
	54 ($2 \cdot 3^3$)	73586	1.577	102600
	72 ($2^3 \cdot 3^2$)	75215	1.617	139536
	96 ($2^5 \cdot 3$)	76664	1.649	188784
	108 ($2^2 \cdot 3^3$)	73829	1.590	213408
	162 ($2 \cdot 3^4$)	**72727**	**1.569**	324218

$w = 162 = 2 \cdot 3^4$ (4.69s) was about 8.2% faster than Window Miller's Algorithm with $w = 7$ (5.11s) and the proportion of M, S and I was $1 : 1 : 20$.

5   Conclusion

We proposed an efficient version of Miller's Algorithm that uses the window hybrid binary-ternary form (w-HBTF) which is a representation mixing base 2 and base 3 in order to speed up the computation of the Tate pairing. We implemented this algorithm and Window Miller's Algorithm in Java on a PC and compared the proposed algorithm with Window Miller's Algorithm. In this implementation, we used the supersingular elliptic curve $y^2 = x^3 + x$ of embedding degree 2 and a composite order of size of 2048 bits. The proposed algorithm with $w = 6 = 2 \cdot 3$ turned out to be about 12% faster than Window Miller's

Algorithm with $w = 2$ given smallest precomputed tables of the same memory size. Moreover, the proposed algorithm with $w = 162 = 2 \cdot 3^4$ was about 8.5% faster than Window Miller's Algorithm with $w = 7$ on each fastest algorithm.

References

1. Adikari, J., Dimitrov, V.S., Imbert, L.: Hybrid Binary-Ternary Number System for Elliptic Curve Cryptosystems. IEEE Transactions on Computers 60(2), 254–265 (2011)
2. Barreto, P.S.L.M., Kim, H.Y., Lynn, B., Scott, M.: Efficient Algorithms for Pairing-Based Cryptosystems. In: Yung, M. (ed.) CRYPTO 2002. LNCS, vol. 2442, pp. 354–369. Springer, Heidelberg (2002)
3. Barreto, P.S.L.M., Galbraith, S., ÒhÉigeartaigh, C., Scott, M.: Efficient Pairing Computation on Supersingular Abelian Varieties. Designs, Codes and Cryptography 42(3), 239–271 (2007)
4. Boneh, D., Goh, E.-J., Nissim, K.: Evaluating 2-DNF Formulas on Ciphertexts. In: Kilian, J. (ed.) TCC 2005. LNCS, vol. 3378, pp. 325–341. Springer, Heidelberg (2005)
5. Boneh, D., Franklin, M.: Identity-Based Encryption from the Weil Pairing. In: Kilian, J. (ed.) CRYPTO 2001. LNCS, vol. 2139, pp. 213–229. Springer, Heidelberg (2001)
6. Boneh, D., Di Crescenzo, G., Ostrovsky, R., Persiano, G.: Public Key Encryption with Keyword Search. In: Cachin, C., Camenisch, J.L. (eds.) EUROCRYPT 2004. LNCS, vol. 3027, pp. 506–522. Springer, Heidelberg (2004)
7. Ciet, M., Joye, M., Lauter, K., Montgomery, P.L.: Trading Inversions for Multiplications in Elliptic Curve Cryptography. Designs, Codes and Cryptography 39(2), 189–206 (2006)
8. Devegili, A., Eigeartaigh, C., Scott, M., Dahab, R.: Multiplication and Squaring on Pairing-Friendly Fields. Cryptography ePrint Archive, Report 2006/471 (2006)
9. Dimitrov, V.S., Imbert, L., Mishra, P.K.: Efficient and Secure Elliptic Curve Point Multiplication Using Double-Base Chains. In: Roy, B. (ed.) ASIACRYPT 2005. LNCS, vol. 3788, pp. 59–78. Springer, Heidelberg (2005)
10. Dimitrov, V.S., Imbert, L., Mishra, P.K.: The Double-Base Number System and Its Application to Elliptic Curve Cryptography. Mathmatics of Computation 77(262), 1075–1104 (2008)
11. Doche, C., Imbert, L.: Extended Double-Base Number System with Applications to Elliptic Curve Cryptography. In: Barua, R., Lange, T. (eds.) INDOCRYPT 2006. LNCS, vol. 4329, pp. 335–348. Springer, Heidelberg (2006)
12. Eisenträger, K., Lauter, K., Montgomery, P.L.: Fast Elliptic Curve Arithmetic and Improved Weil Pairing Evaluation. In: Joye, M. (ed.) CT-RSA 2003. LNCS, vol. 2612, pp. 343–354. Springer, Heidelberg (2003)
13. Guillevic, A.: Comparing the Pairing Efficiency over Composite-Order and Prime-Order Elliptic Curves. Cryptography ePrint Archive, Report 2013/218 (2013)
14. Hankerson, D., Menezes, A., Vanstone, S.: Guide to Elliptic Curve Cryptography. Springer (2004)
15. Hess, F., Smart, N., Vercauteren, F.: The Eta Pairing Revisited. IEEE Transactions on Information Theory 52(10), 4595–4602 (2006)
16. Joye, M.: RSA Moduli with a Predetermined Portion: Techniques and Applications. In: Chen, L., Mu, Y., Susilo, W. (eds.) ISPEC 2008. LNCS, vol. 4991, pp. 116–130. Springer, Heidelberg (2008)

17. Kobayashi, T., Aoki, K., Imai, H.: Efficient Algorithm for Tate Pairing. IEICE Transaction on Fundamentals of Electronics, Communications and Computer Sciences E89-A(1), 134–143 (2006)
18. Koblitz, N., Menezes, A.: Pairing-Based Cryptography at High Security Levels. In: Smart, N.P. (ed.) Cryptography and Coding 2005. LNCS, vol. 3796, pp. 13–36. Springer, Heidelberg (2005)
19. Le, D., Tan, C.: Speeding Up Ate Pairing Computation in Affine Coordinates. Cryptography and Coding, Cryptography ePrint Archive, Report 2013/119 (2013)
20. Lee, E., Lee, H.S., Park, C.M.: Efficient and Generalized Pairing Computation on Abelian Varieties. IEEE Transactions on Information Theory 55(4), 1793–1803 (2009)
21. Meiklejohn, S., Shacham, H., Freeman, D.M.: Limitations on Transformations from Composite-Order to Prime-Order Groups: The Case of Round-Optimal Blind Signatures. In: Abe, M. (ed.) ASIACRYPT 2010. LNCS, vol. 6477, pp. 519–538. Springer, Heidelberg (2010)
22. Miller, V.: Short Programs for Functions on Curves (1986) (unpublished manuscript)
23. Ostrovsky, R., Skeith III, W.E.: Private Searching on Streaming Data. In: Shoup, V. (ed.) CRYPTO 2005. LNCS, vol. 3621, pp. 223–240. Springer, Heidelberg (2005)
24. Shacham, H., Waters, B.: Efficient ring signatures without random oracles. In: Okamoto, T., Wang, X. (eds.) PKC 2007. LNCS, vol. 4450, pp. 166–180. Springer, Heidelberg (2007)
25. Takagi, T., Yen, S.-M., Wu, B.-C.: Radix-r non-adjacent form. In: Zhang, K., Zheng, Y. (eds.) ISC 2004. LNCS, vol. 3225, pp. 99–110. Springer, Heidelberg (2004)
26. Vercauteren, F.: Optimal Pairings. IEEE Transactions on Information Theory 56(1), 455–461 (2010)

How to Factor N_1 and N_2 When $p_1 = p_2$ mod 2^t

Kaoru Kurosawa and Takuma Ueda

Ibaraki University, Japan
kurosawa@mx.ibaraki.ac.jp

Abstract. Let $N_1 = p_1q_1$ and $N_2 = p_2q_2$ be two different RSA moduli. Suppose that

$$p_1 = p_2 \bmod 2^t$$

for some t, and q_1 and q_2 are α bit primes. Then May and Ritzenhofen showed that N_1 and N_2 can be factored in quadratic time if

$$t \geq 2\alpha + 3.$$

In this paper, we improve this lower bound on t. Namely we prove that N_1 and N_2 can be factored in quadratic time if

$$t \geq 2\alpha + 1.$$

Further our simulation result shows that our bound is tight as far as the factoring method of May and Ritzenhofen is used.

Keywords: factoring, Gaussian reduction algorithm, lattice.

1 Introduction

Factoring $N = pq$ is a fundamental problem in modern cryptography, where p and q are large primes. Since RSA was invented, some factoring algorithms which run in subexponential time have been developed, namely the quadratic sieve [10], the elliptic curve [4] and number field sieve [5]. However, no polynomial time algorithm is known.

On the other hand, the so called oracle complexity of the factorization problem were studied by Rivest and Shamir [11], Maurer [6] and Coppersmith [1]. In particular, Coppersmith [1] showed that one can factor N if a half of the most significant bits of p are given.

Recently, May and Ritzenhofen [7] considered another approach (which received the "Best Paper Award" of PKC 2009). Suppose that we are given $N_1 = p_1q_1$ and $N_2 = p_2q_2$. If

$$p_1 = p_2,$$

then it is easy to factor N_1, N_2 by using Euclidean algorithm. May and Ritzenhofen showed that it is easy to factor N_1, N_2 even if

$$p_1 = p_2 \bmod 2^t$$

K. Sakiyama and M. Terada (Eds.): IWSEC 2013, LNCS 8231, pp. 217–225, 2013.

for sufficiently large t. More precisely suppose that q_1 and q_2 are α bit primes. Then they showed that N_1 and N_2 can be factored in quadratic time if

$$t \geq 2\alpha + 3.$$

In this paper, we improve the above lower bound on t. We prove that N_1 and N_2 can be factored in quadratic time if

$$t \geq 2\alpha + 1.$$

Further our simulation result shows that our bound is tight as far as the factoring method of May and Ritzenhofen [7] is used.

Also our proof is conceptually simpler than that of May and Ritzenhofen [7]. In particular, we do not use the Minkowski bound whereas it is required in their proof.

As written in [7], one application of our result is malicious key generation of RSA moduli, i.e. the construction of backdoored RSA moduli [2,13]. In [7], the authors also suggest the following constructive cryptographic applications. Consider the one more RSA modulus problem such that on input $N_1 = p_1 q_1$, one has to produce $N_2 = p_1 q_2$ with $p_1 = p_2$ mod 2^t. Our result shows that this problem is equivalent to the factorization problem as long as $t \geq 2\alpha + 1$. So the one more RSA modulus problem might serve as a basis for various cryptographic primitives, whose security is then in turn directly based on factoring (imbalanced) integers.

(Related work) Sarkar and Maitra [12] extended the result of May and Ritzenhofen [7] under a *heuristic* assumption (see Assumption 1 of [12, page 4003]). However, this assumption is *heuristic* only as they wrote in [12].

2   Preliminaries

2.1   Lattice

An integer lattice L is a discrete additive subgroup of Z^n. An alternative equivalent definition of an integer lattice can be given via a basis. Let d, n be integers such that $0 < d \leq n$. Let $\mathbf{b}_1, \cdots, \mathbf{b}_d \in Z^n$ be linearly independent vectors. Then the set of all integer linear combinations of the $\mathbf{b}_i$ spans an integer lattice L, i.e.

$$L = \left\{ \sum_{i=1}^{d} a_i \mathbf{b}_i \mid a_i \in Z \right\}.$$

We call $B = \begin{pmatrix} \mathbf{b}_1 \\ \vdots \\ \mathbf{b}_d \end{pmatrix}$ a basis of the lattice, the value d denotes the dimension or rank of the basis. The lattice is said to have full rank if $d = n$. The determinant $\det(L)$ of a lattice is the volume of the parallelepiped spanned by the basis

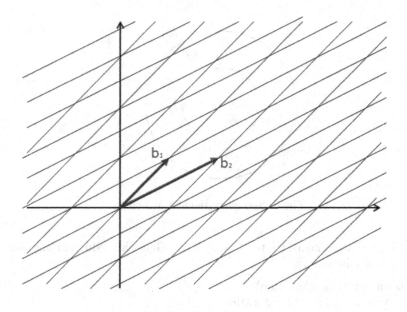

Fig. 1. Lattice

vectors. The determinant $\det(L)$ is invariant under unimodular basis transformations of B. In case of a full rank lattice $\det(L)$ is equal to the absolute value of the Gramian determinant of the basis B. Let us denote by $||\mathbf{v}||$ the Euclidean ℓ_2-norm of a vector $\mathbf{v}$. Hadamardfs inequality [8] relates the length of the basis vectors to the determinant.

Proposition 1. *Let* $B = \begin{pmatrix} \mathbf{b_1} \\ \vdots \\ \mathbf{b_d} \end{pmatrix} \in Z^{n \times n}$ *be an arbitrary non-singular matrix.*

Then

$$\det(B) \leq \prod_{i=1}^{n} ||\mathbf{b}_i||.$$

The successive minima λ_i of the lattice L are defined as the minimal radius of a ball containing i linearly independent lattice vectors of L (see Fig.2).

Proposition 2. *(Minkowski [9]). Let* $L \subseteq Z^n$ *be an integer lattice. Then* L *contains a non-zero vector* $\mathbf{v}$ *with*

$$||\mathbf{v}|| = \lambda_1 \leq \sqrt{n} \det(L)^{1/n}$$

2.2 Gaussian Reduction Algorithm

In a two-dimensional lattice L, basis vectors $\mathbf{v}_1, \mathbf{v}_2$ with lengths $||\mathbf{v}_1|| = \lambda_1$ and $||\mathbf{v}_2|| = \lambda_2$ are efficiently computable by using Gaussian reduction algorithm.

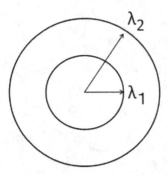

Fig. 2. Successive minima λ_1 and λ_2

Let $\lfloor x \rceil$ denote the nearest integer to x. Then Gaussian reduction algorithm is described as follows.

(Gaussian reduction algorithm)
Input: Basis $\mathbf{b}_1, \mathbf{b}_2 \in Z^2$ for a lattice L.
Output: Basis $(\mathbf{v}_1, \mathbf{v}_2)$ for L such that $||\mathbf{v}_1|| = \lambda_1$ and $||\mathbf{v}_2|| = \lambda_2$.

1. Let $\mathbf{v}_1 := \mathbf{b}_1$ and $\mathbf{v}_2 := \mathbf{b}_2$.
2. Compute $\mu := (\mathbf{v}_1, \mathbf{v}_2)/||\mathbf{v}_1||^2$,
   $$\mathbf{v}_2 := \mathbf{v}_2 - \lfloor \mu \rceil \cdot \mathbf{v}_1.$$
3. while $||\mathbf{v}_2|| < ||\mathbf{v}_1||$ do:
4.   Swap $\mathbf{v}_1$ and $\mathbf{v}_2$.
5.   Compute $\mu := (\mathbf{v}_1, \mathbf{v}_2)/||\mathbf{v}_1||^2$,
      $$\mathbf{v}_2 := \mathbf{v}_2 - \lfloor \mu \rceil \cdot \mathbf{v}_1.$$
6. end while
7. return $(\mathbf{v}_1, \mathbf{v}_2)$.

Proposition 3. *The above algorithm outputs a basis $(\mathbf{v}_1, \mathbf{v}_2)$ for L such that $||\mathbf{v}_1|| = \lambda_1$ and $||\mathbf{v}_2|| = \lambda_2$. Further they can be determined in time $O(\log^2(\max\{||\mathbf{b}_1||, ||\mathbf{b}_2||\})$.*

Information on Gaussian reduction algorithm and its running time can be found in [8,3].

3   Previous Implicit Factoring of Two RSA Moduli

Let $N_1 = p_1 q_1$ and $N_2 = p_2 q_2$ be two different RSA moduli. Suppose that

$$p_1 = p_2(= p) \bmod 2^t \tag{1}$$

for some t, and q_1 and q_2 are α bit primes. This means that p_1, p_2 coincide on the t least significant bits. I.e.,

$$p_1 = p + 2^t \tilde{p}_1 \text{ and } p_2 = p + 2^t \tilde{p}_2$$

for some common p that is unknown to us. Then May and Ritzenhofen [7] showed that N_1 and N_2 can be factored in quadratic time if $t \geq 2\alpha + 3$. In this section, we present their idea.

From eq.(1), we have

$$N_1 = pq_1 \bmod 2^t$$
$$N_2 = pq_2 \bmod 2^t$$

Since q_1, q_2 are odd, we can solve both equations for p. This leaves us with

$$N_1/q_1 = N_2/q_2 \bmod 2^t$$

which we write in form of the linear equation

$$(N_2/N_1)q_1 - q_2 = 0 \bmod 2^t \qquad (2)$$

The set of solutions

$$L = \{(x_1, x_2) \in Z^2 \mid (N_2/N_1)x_1 - x_2 = 0 \bmod 2^t\}$$

forms an additive, discrete subgroup of Z^2. Thus, L is a 2-dimensional integer lattice. L is spanned by the row vectors of the basis matrix

$$B_L = \begin{pmatrix} 1, & (N_2/N_1 \bmod 2^t) \\ 0, & 2^t \end{pmatrix} \qquad (3)$$

The integer span of B_L, denoted by $span(B_L)$, is equal to L. To see why, let

$$\mathbf{b_1} = (1, (N_2/N_1 \bmod 2^t))$$
$$\mathbf{b_2} = (0, 2^t)$$

Then they are solutions of

$$(N_2/N_1)x_1 - x_2 = 0 \bmod 2^t$$

Thus, every integer linear combination of $\mathbf{b_1}$ and $\mathbf{b_2}$ is a solution which implies that $span(B_L) \subseteq L$.

Conversely, let $(x_1, x_2) \in L$, i.e.

$$(N_2/N_1)x_1 - x_2 = k \cdot 2^t$$

for some $k \in Z$. Then

$$(x_1, -k)B_L = (x_1, x_2) \in span(B_L)$$

and thus $L \subseteq span(B_L)$.

Notice that by Eq. (2), we have

$$\mathbf{q} = (q_1, q_2) \in L. \qquad (4)$$

If we were able to find this vector in L, then we could factor N_1, N_2 easily. We know that the length of the shortest vector is upper bounded by the Minkowski bound

$$\sqrt{2} \cdot \det(L)^{1/2} = \sqrt{2} \cdot 2^{t/2}.$$

Since we assume that q_1, q_2 are α-bit primes, we have $q_1, q_2 \leq 2^\alpha$. If α is sufficiently small, then $\|\mathbf{q}\|$ is smaller than the Minkowski bound and, therefore, we can expect that q is among the shortest vectors in L. This happens if

$$\|\mathbf{q}\| \leq \sqrt{2} \cdot 2^\alpha \leq \sqrt{2} \cdot 2^{t/2}$$

So if $t \geq 2\alpha$, we expect that $\mathbf{q}$ is a short vector in L. We can find a shortest vector in L using Gaussian reduction algorithm on the lattice basis B in time

$$O(\log^2(2^t)) = O(\log^2(\min\{N_1, N_2\})).$$

By elaborating the above argument, May and Ritzenhofen [7] proved the following.

Proposition 4. *Let $N_1 = p_1 q_1$ and $N_2 = p_2 q_2$ be two different RSA moduli such that $p_1 = p_2 \bmod 2^t$ for some t, and q_1 and q_2 are α bit primes. If*

$$t \geq 2\alpha + 3, \tag{5}$$

then N_1, N_2 can be factored in time $O(\log^2(\min\{N_1, N_2\}))$.

4   Improvement

In this section, we improve the lower bound on t of Proposition 4.

Lemma 1. *If $\|\mathbf{q}\| < \lambda_2$, then $\mathbf{q} = c \cdot \mathbf{v}_1$ for some integer c, where $\mathbf{v}_1$ is the shortest vector in L.*

(Proof) Suppose that $\mathbf{q} \neq c \cdot \mathbf{v}_1$ for any integer c. This means that $\mathbf{v}_1$ and $\mathbf{q}$ are linearly independent vectors. Therefore it must be that $\|\mathbf{q}\| \geq \lambda_2$ from the definition of λ_2. However, this is against our assumption that $\|\mathbf{q}\| < \lambda_2$. Therefore we have $\mathbf{q} = c \cdot \mathbf{v}_1$ for some integer c.

Q.E.D.

Lemma 2. *If q_1 and q_2 are α bits long, then*

$$\|\mathbf{q}\| < 2^{\alpha+0.5}$$

(Proof) Since q_1 and q_2 are α-bits long, we have

$$q_i \leq 2^\alpha - 1$$

for $i = 1, 2$. Therefore

$$\|\mathbf{q}\| \leq \sqrt{2}(2^\alpha - 1) < \sqrt{2} \cdot 2^\alpha = 2^{\alpha+0.5}$$

Q.E.D.

Theorem 1. *Let $N_1 = p_1 q_1$ and $N_2 = p_2 q_2$ be two different RSA moduli such that*

$$p_1 = p_2 \bmod 2^t$$

for some t, and q_1 and q_2 are α-bit primes. If

$$t \geq 2\alpha + 1, \tag{6}$$

then N_1, N_2 can be factored in time $O(\log^2(\min\{N_1, N_2\}))$.

(Proof) If $q_1 = q_2$, the we can factor N_1, N_2 by using Euclidean algorithm easily. Therefore we assume that $q_1 \neq q_2$.

Apply Gaussian reduction algorithm to B_L. Then we obtain

$$B_0 = \begin{pmatrix} \mathbf{v}_1 \\ \mathbf{v}_2 \end{pmatrix}$$

such that

$$||\mathbf{v}_1|| = \lambda_1 \text{ and } ||\mathbf{v}_2|| = \lambda_2.$$

We will show that $\mathbf{q} = \mathbf{v}_1$ or $\mathbf{q} = -\mathbf{v}_1$, where $\mathbf{q} = (q_1, q_2)$.

From Hadamard's inequality, we have

$$||\mathbf{v}_2||^2 \geq ||\mathbf{v}_1|| ||\mathbf{v}_2|| \geq \det(B_0) = \det(B_L) = 2^t,$$

where $\det(B_0) = \det(B_L)$ because B_0 and B_L span the same lattice L. The last equality comes from eq.(3). Therefore we obtain that

$$\lambda_2 = ||\mathbf{v}_2|| \geq 2^{t/2}.$$

Now suppose that

$$t \geq 2\alpha + 1$$

Then

$$t/2 \geq \alpha + 0.5.$$

Therefore

$$\lambda_2 = ||\mathbf{v}_2|| \geq 2^{t/2} \geq 2^{\alpha+0.5} > ||\mathbf{q}||$$

from Lemma 2. This means that

$$(q_1, q_2) = \mathbf{q} = c \cdot \mathbf{v}_1$$

for some integer c from Lemma 1. Further since $\gcd(q_1, q_2) = 1$, it must be that $c = 1$ or -1. Therefore $\mathbf{q} = \mathbf{v}_1$ or $\mathbf{q} = -\mathbf{v}_1$ (see Fig.3).

Finally from Proposition 3, Gaussian reduction algorithm runs in time

$$O(\log^2(2^t)) = O(\log^2(\min\{N_1, N_2\})).$$

Q.E.D.

Compare eq.(6) and eq.(5), and notice that we have improved the previous lower bound on t.

Also our proof is conceptually simpler than that of May and Ritzenhofen [7]. In particular, we do not use the Minkowski bound whereas it is required in their proof.

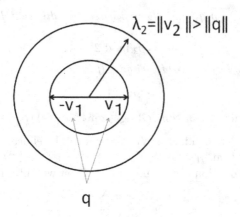

Fig. 3. Proof of Theorem 1

5   Generalization

Theorem 1 can be generalized as follows.

Corollary 1. *Let $N_1 = p_1 q_1$ and $N_2 = p_2 q_2$ be two different RSA moduli such that*

$$p_1 = p_2 \bmod T$$

for some T. Let q_1 and q_2 be α-bits long primes. Then if

$$T \geq 2^{2\alpha+1} \tag{7}$$

then N_1, N_2 can be factored in time $O(\log^2(\min\{N_1, N_2\}))$.

Corollary 2. *Let $N_1 = p_1 q_1$ and $N_2 = p_2 q_2$ be two different RSA moduli such that*

$$p_1 = p_2 \bmod T$$

for some T. If

$$T > q_1^2 + q_2^2 \tag{8}$$

then N_1, N_2 can be factored in time $O(\log^2(\min\{N_1, N_2\}))$.

The proofs are almost the same as that of Theorem 1.

6   Simulation

We verified Theorem 1 by computer simulation. We considered the case such that q_1 and q_2 are $\alpha = 250$ bits long. Theorem 1 states that if

$$t \geq 2\alpha + 1 = 501,$$

then we can factor N_1 and N_2 by using Gaussian reduction algorithm. The simulation results are shown in Table 6, where p_1 and p_2 are 750 bits long. For each value of t, the success rate is computed over 100 samples.

From this table, we can see that we can indeed factor N_1 and N_2 if $t \geq 501$. We can also see that we fail to factor N_1 and N_2 if $t \leq 500$. This shows that our bound is tight as far as the factoring method of May and Ritzenhofen [7] is used.

Table 1. Computer Simulation

number of shared bits t	success rate
503	100%
502	100%
501	100%
500	40%
499	0%
498	0%

References

1. Coppersmith, D.: Finding a small root of a bivariate integer equation; factoring with high bits known. In: Maurer, U.M. (ed.) EUROCRYPT 1996. LNCS, vol. 1070, pp. 178–189. Springer, Heidelberg (1996)
2. Crépeau, C., Slakmon, A.: Simple backdoors for RSA key generation. In: Joye, M. (ed.) CT-RSA 2003. LNCS, vol. 2612, pp. 403–416. Springer, Heidelberg (2003)
3. Galbraith, S.D.: Mathematics of Public Key Cryptography. Cambridge University Press (2012)
4. Lenstra Jr., H.W.: Factoring Integers with Elliptic Curves. Ann. Math. 126, 649–673 (1987)
5. Lenstra, A.K., Lenstra Jr., H.W.: The Development of the Number Field Sieve. Springer, Heidelberg (1993)
6. Maurer, U.M.: Factoring with an oracle. In: Rueppel, R.A. (ed.) EUROCRYPT 1992. LNCS, vol. 658, pp. 429–436. Springer, Heidelberg (1993)
7. May, A., Ritzenhofen, M.: Implicit factoring: On polynomial time factoring given only an implicit hint. In: Jarecki, S., Tsudik, G. (eds.) PKC 2009. LNCS, vol. 5443, pp. 1–14. Springer, Heidelberg (2009)
8. Meyer, C.D.: Matrix Analysis and Applied Linear Algebra. Cambridge University Press, Cambridge (2000)
9. Minkowski, H.: Geometrie der Zahlen. Teubner-Verlag (1896)
10. Pomerance, C.: The quadratic sieve factoring algorithm. In: Beth, T., Cot, N., Ingemarsson, I. (eds.) EUROCRYPT 1984. LNCS, vol. 209, pp. 169–182. Springer, Heidelberg (1985)
11. Rivest, R.L., Shamir, A.: Efficient factoring based on partial information. In: Pichler, F. (ed.) EUROCRYPT 1985. LNCS, vol. 219, pp. 31–34. Springer, Heidelberg (1986)
12. Sarkar, S., Maitra, S.: Approximate Integer Common Divisor Problem Relates to Implicit Factorization. IEEE Transactions on Information Theory 57(6), 4002–4013 (2011)
13. Young, A., Yung, M.: The prevalence of kleptographic attacks on discrete-log based cryptosystems. In: Kaliski Jr., B.S. (ed.) CRYPTO 1997. LNCS, vol. 1294, pp. 264–276. Springer, Heidelberg (1997)

Achieving Chosen Ciphertext Security from Detectable Public Key Encryption Efficiently via Hybrid Encryption

Takahiro Matsuda and Goichiro Hanaoka

Research Institute for Secure Systems,
National Institute of Advanced Industrial Science and Technology (AIST), Japan
{t-matsuda,hanaoka-goichiro}@aist.go.jp

Abstract. In EUROCRYPT'12, Hohenberger, Lewko, and Waters proposed a new paradigm for constructing chosen ciphertext secure public key encryption (PKE) schemes from a new concept of detectable PKE. In this paper, we propose an efficient variant of the Hohenberger-Lewko-Waters (HLW) construction, based on the techniques and results from hybrid encryption. On the technical side, our security proof avoids using the notion of nested-indistinguishability that was used in the original proof by Hohenberger et al., and we believe that what role each building block plays is clearer, leading to better understanding of the HLW paradigm.

Keywords: public key encryption, key encapsulation mechanism, chosen ciphertext security, detectable public key encryption.

1 Introduction

Background and Motivation. For public key encryption (PKE), security (indistinguishability) against chosen ciphertext attacks (CCA) [14,17,6] is nowadays considered as a de-facto standard security notion required in most practical situations/applications in which PKE schemes are used, due to its resilience against practical attacks (e.g. Bleichenbacher's attack [1]) and its implication to strong security notions (e.g. non-malleability [6] and universal composability [2]). Thus, constructing and understanding CCA secure PKE schemes is one of the central research themes in the area of cryptography.

In EUROCRYPT'12, Hohenberger, Lewko, and Waters [7] proposed a new paradigm for constructing CCA secure PKE schemes from a new concept of *detectable PKE* (its formal definition is given in Section 2.1). Their construction (which we call "HLW" construction) is based on a mixture of the double-layered encryption of Myers and Shelat [13] and the double (parallel) encryption of Naor and Yung [14], and it uses three PKE schemes as building blocks; one for "inner" encryption that encrypts a plaintext together with a randomness for "outer" encryption, and the remaining two for "outer" encryption that encrypt the "inner" ciphertext. (We review the HLW construction in Section 3.)

K. Sakiyama and M. Terada (Eds.): IWSEC 2013, LNCS 8231, pp. 226–243, 2013.

The HLW construction is elegant, and clarifies that detectable PKE serves as a new target for constructing a CCA secure PKE scheme from general cryptographic assumptions. Like other generic constructions of CCA secure PKE, however, a PKE scheme obtained from the HLW construction will not be so practical. In particular, the ciphertext size of the resulting scheme will be large, even if we use building block schemes with small ciphertext size. Therefore, it is rather a feasibility result, than a method for obtaining practical schemes.

The main purpose of this paper is to show that the HLW construction can be made more (space-)efficient, so that it is also useful for obtaining (space-)efficient CCA secure schemes.

Our Contribution. In this paper, we propose a variant of the HLW construction, based on the techniques and results from hybrid encryption [5]. Specifically, we show that we can make the HLW construction more efficient by implementing two building blocks with an appropriate combination of a key encapsulation mechanism (KEM) and a symmetric key encryption (SKE) scheme. Our construction is based on a simple observation that the double (parallel) encryption mechanism for proving the CCA security of the HLW construction can be efficiently realized by employing the hybrid encryption methodology. On the technical side, we believe that our security proof is more direct and modular than that of the original HLW construction. Specifically, our security proof avoids using the notion of "nested-indistinguishability" which is a construction-specific security notion (i.e. a security notion that can be considered only for the HLW construction) used by Hohenberger et al. in their original security proof, and thus we believe that what role each building block plays is clearer, leading to better understanding of the HLW paradigm.

Our proposed construction is given in Section 4 where we also explain the ideas towards our proposed construction in more details. In Section 5, we compare the ciphertext size between our scheme and the original HLW construction.

Related Work. Our construction and the HLW construction construct a CCA secure PKE scheme generically from other cryptographic primitives. Here, we briefly review other generic constructions of CCA secure PKE.

Dolev, Dwork, and Naor [6] showed the first construction of a CCA secure PKE scheme, from a chosen plaintext secure (CPA secure) PKE scheme and a non-interactive zero-knowledge (NIZK) proof system, based on the construction by Naor and Yung [14] that achieves weaker non-adaptive CCA security. These NIZK-based constructions were further improved in [19,11].

Canetti, Halevi, and Katz [3] showed how to transform an identity-based encryption [20] scheme into a CCA secure PKE scheme. Kiltz [8] showed that the transform of [3] is applicable to a weaker primitive called tag-based encryption.

Peikert and Waters [15] showed how to construct a CCA secure PKE scheme from a *lossy* trapdoor function (TDF). Subsequent works showed that TDFs with weaker security/functionality properties are sufficient for obtaining CCA secure PKE schemes [18,12,9,21].

Myers and Shelat [13] showed that a CCA secure PKE scheme for 1-bit message can be turned into one with an arbitrarily large plaintext space. Recently, Lin and Tessaro [10] showed how to amplify weak CCA security into strong (ordinary) CCA secure one.

2   Preliminaries

In this section, we review the basic notation and the definitions of primitives.

Basic Notation. $\mathbb{N}$ denotes the set of all natural numbers, and for $n \in \mathbb{N}$ we let $[n] := \{1, \ldots, n\}$. "$x \leftarrow y$" denotes that x is chosen uniformly at random from y if y is a finite set, or y is assigned to x otherwise. If x and y are bit-strings, then "$|x|$" denotes the bit-length of x, "$(x\|y)$" denotes the concatenation of x and y, and the operation "$(x \overset{?}{=} y)$" returns 1 if $x = y$ or returns 0 otherwise. "PPTA" denotes a *probabilistic polynomial time algorithm*. If $\mathcal{A}$ is a probabilistic algorithm, then "$y \leftarrow \mathcal{A}(x; r)$" denotes that $\mathcal{A}$ computes y as output by taking x as input and using r as randomness, and "$\mathcal{A}^{\mathcal{O}}$" denotes an algorithm $\mathcal{A}$ with oracle access to $\mathcal{O}$. The character "k" denotes the security parameter. The symbol "$\perp$" (which will be output by a decryption algorithm) denotes that a ciphertext is invalid. A function $f : \mathbb{N} \rightarrow [0, 1]$ is said to be *negligible* if for all positive polynomials $p(\cdot)$ and all sufficiently large k, we have $f(k) < 1/p(k)$.

2.1   (Detectable) Public Key Encryption

A public key encryption (PKE) scheme Π consists of the three PPTAs (PKG, PEnc, PDec) with the following interface:

Key Generation:	**Encryption:**	**Decryption:**
$(pk, sk) \leftarrow \mathsf{PKG}(1^k)$	$c \leftarrow \mathsf{PEnc}(pk, m)$	m (or $\perp$) $\leftarrow \mathsf{PDec}(sk, c)$

where PDec is a deterministic algorithm, (pk, sk) is a public/secret key pair, and c is a ciphertext of a plaintext m under pk.

A PKE scheme is required to satisfy *correctness*: for all $k \in \mathbb{N}$, all keys (pk, sk) output from $\mathsf{PKG}(1^k)$ and all plaintexts m, it holds that $\mathsf{PDec}(sk, \mathsf{PEnc}(pk, m)) = m$.

A tuple of PPTAs $\Pi = (\mathsf{PKG}, \mathsf{PEnc}, \mathsf{PDec}, \mathsf{F})$ is said to be a *detectable* PKE scheme if $(\mathsf{PKG}, \mathsf{PEnc}, \mathsf{PDec})$ constitutes PKE, and F is a predicate that takes a public key pk and two ciphertexts c, c' as input and outputs either 0 or 1. This predicate is used to define the security notions (*detectable CCA security* and *unpredictability*) for detectable PKE that we explain below.

Security Notions. For the security notions of a PKE scheme, we recall chosen plaintext security (CPA), chosen ciphertext security (CCA), and its 1-bounded CCA-analogue [4] (1-CCA). We also recall *detectable CCA* (DCCA) *security* and *unpredictability* for detectable PKE [7].

$\mathsf{Expt}_{\Pi,\mathcal{A}}^{\mathrm{ATK}}(k):$
  $b \leftarrow \{0,1\}$
  $(pk, sk) \leftarrow \mathsf{PKG}(1^k)$
  $(m_0, m_1, \mathsf{st}) \leftarrow \mathcal{A}_1^{\mathcal{O}}(pk)$
  $c^* \leftarrow \mathsf{PEnc}(pk, m_b)$
  $b' \leftarrow \mathcal{A}_2^{\mathcal{O}}(c^*, \mathsf{st})$
  Return $(b' \stackrel{?}{=} b)$.

$\mathsf{Expt}_{\Gamma,\mathcal{A}}^{\mathrm{ATK}}(k):$
  $b \leftarrow \{0,1\}$
  $(pk, sk) \leftarrow \mathsf{KKG}(1^k)$
  $\mathsf{st} \leftarrow \mathcal{A}_1^{\mathcal{O}}(pk)$
  $K_0^* \leftarrow \mathcal{K}$
  $(c^*, K_1^*) \leftarrow \mathsf{KEnc}(pk)$
  $b' \leftarrow \mathcal{A}_2^{\mathcal{O}}(c^*, K_b^*, \mathsf{st})$
  Return $(b' \stackrel{?}{=} b)$.

$\mathsf{Expt}_{E,\mathcal{A}}^{\mathrm{ATK}}(k):$
  $b \leftarrow \{0,1\}$
  $K \leftarrow \{0,1\}^k$
  $(m_0, m_1, \mathsf{st}) \leftarrow \mathcal{A}_1(1^k)$
  $c^* \leftarrow \mathsf{SEnc}(K, m_b)$
  $b' \leftarrow \mathcal{A}_2^{\mathcal{O}}(c^*, \mathsf{st})$
  Return $(b' \stackrel{?}{=} b)$.

$\mathsf{Expt}_{\Pi,\mathcal{A}}^{\mathrm{UNP}}(k):$
  $(pk, sk) \leftarrow \mathsf{PKG}(1^k)$
  $(m, c) \leftarrow \mathcal{A}^{\mathsf{PDec}(sk,\cdot)}(pk)$
  $c^* \leftarrow \mathsf{PEnc}(pk, m)$
  Return $\mathsf{F}(pk, c^*, c)$.

$\mathsf{Expt}_{\Gamma,\mathcal{A}}^{\mathrm{UNP}}(k):$
  $(pk, sk) \leftarrow \mathsf{KKG}(1^k)$
  $c \leftarrow \mathcal{A}^{\mathsf{KDec}(sk,\cdot)}(pk)$
  $(c^*, K^*) \leftarrow \mathsf{KEnc}(pk)$
  Return $\mathsf{F}(pk, c^*, c)$.

$\mathsf{Adv}_{\mathcal{P},\mathcal{A}}^{\mathrm{ATK}}(k) :=$
  $2 \cdot | \Pr[\mathsf{Expt}_{\mathcal{P},\mathcal{A}}^{\mathrm{ATK}}(k) = 1] - 1/2|$
$\mathsf{Adv}_{\mathcal{P},\mathcal{A}}^{\mathrm{UNP}}(k) :=$
    $\Pr[\mathsf{Expt}_{\mathcal{P},\mathcal{A}}^{\mathrm{UNP}}(k) = 1]$

Fig. 1. Security experiments for a (detectable) PKE scheme Π (left), those for a (detectable) KEM Γ (center), and those for a SKE scheme E (top-right), and the definitions of the advantage of an adversary $\mathcal{A}$ against the security of primitive $\mathcal{P} \in \{\Pi, \Gamma, E\}$ (bottom-right)

Let $\mathsf{ATK} \in \{\mathsf{CPA}, 1\text{-}\mathsf{CCA}, \mathsf{CCA}, \mathsf{DCCA}\}$. For a (detectable) PKE scheme Π and an adversary $\mathcal{A} = (\mathcal{A}_1, \mathcal{A}_2)$, consider the ATK experiment $\mathsf{Expt}_{\Pi,\mathcal{A}}^{\mathrm{ATK}}(k)$ as described in Fig. 1 (top-left). In the experiment, it is required that $|m_0| = |m_1|$ holds, and the oracle $\mathcal{O}$ is defined depending on ATK as follows: If $\mathsf{ATK} = \mathsf{CPA}$, then $\mathcal{O}$ is unavailable (it returns $\bot$ for any input); If $\mathsf{ATK} \in \{1\text{-}\mathsf{CCA}, \mathsf{CCA}, \mathsf{DCCA}\}$, then $\mathcal{O}$ is the decryption oracle $\mathcal{O}(\cdot) = \mathsf{PDec}(sk, \cdot)$, but $\mathcal{A}_2$ is not allowed to submit "prohibited" queries (see below) to $\mathcal{O}$, and in addition, in the $1\text{-}\mathsf{CCA}$ case $\mathcal{A}_2$ can use $\mathcal{O}$ only once. The prohibited query for $\mathsf{ATK} \in \{1\text{-}\mathsf{CCA}, \mathsf{CCA}\}$ is c^*, while that for DCCA is c satisfying $\mathsf{F}(pk, c^*, c) = 1$. We say that a (detectable) PKE scheme Π is ATK secure if $\mathsf{Adv}_{\Pi,\mathcal{A}}^{\mathrm{ATK}}(k)$ is negligible for any PPTA $\mathcal{A}$.

For a detectable PKE scheme Π (with predicate F) and an adversary $\mathcal{A}$, consider the unpredictability experiment $\mathsf{Expt}_{\Pi,\mathcal{A}}^{\mathrm{UNP}}(k)$ as described in Fig. 1 (bottom-left). We say that a detectable PKE scheme Π is *unpredictable* if $\mathsf{Adv}_{\Pi,\mathcal{A}}^{\mathrm{UNP}}(k)$ is negligible for any PPTA $\mathcal{A}$.

2.2 (Detectable) Key Encapsulation Mechanisim

A key encapsulation mechanism (KEM) Γ consists of the three PPTAs (KKG, KEnc, KDec) with the following interface:

Key Generation:	**Encapsulation:**	**Decapsulation:**
$(pk, sk) \leftarrow \mathsf{KKG}(1^k)$	$(c, K) \leftarrow \mathsf{KEnc}(pk)$	$K \ (\text{or } \bot) \leftarrow \mathsf{KDec}(sk, c)$

where KDec is a deterministic algorithm, (pk, sk) is a public/secret key pair that defines a session-key space $\mathcal{K}$, and c is a ciphertext of a session-key $K \in \mathcal{K}$ under pk.

A KEM is required to satisfy *correctness*: for all $k \in \mathbb{N}$, all keys (pk, sk) output from $\mathsf{KKG}(1^k)$ and all ciphertext/session-key pairs (c, K) output from $\mathsf{KEnc}(pk)$, it holds that $\mathsf{KDec}(sk, c) = K$.

We also define a KEM-analogue of detectable PKE, which we call *detectable KEM*, as a KEM that has an efficiently computable predicate F whose interface is the same as that of detectable PKE.

Security Notions. As in the PKE case, we consider CCA, 1-CCA, and CPA security for a KEM, and DCCA security and unpredictability for a detectable KEM.

Let $\mathsf{ATK} \in \{\mathsf{CPA}, \mathsf{1\text{-}CCA}, \mathsf{CCA}, \mathsf{DCCA}\}$. For a (detectable) KEM Γ and an adversary $\mathcal{A} = (\mathcal{A}_1, \mathcal{A}_2)$, consider the ATK experiment $\mathsf{Expt}_{\Gamma,\mathcal{A}}^{\mathsf{ATK}}(k)$ as described in Fig. 1 (top-center). In the experiment, the oracle $\mathcal{O}$ is defined depending on ATK as follows: If $\mathsf{ATK} = \mathsf{CPA}$, then $\mathcal{O}$ is unavailable (it returns $\perp$ for any input); If $\mathsf{ATK} \in \{\mathsf{1\text{-}CCA}, \mathsf{CCA}, \mathsf{DCCA}\}$, then $\mathcal{O}$ is the decapsulation oracle $\mathcal{O}(\cdot) = \mathsf{KDec}(sk, \cdot)$, and how it is available (e.g. prohibited queries and how many times it can be used) is defined in exactly the same way as the corresponding ATK experiments for (detectable) PKE. We say that a (detectable) KEM Γ is ATK secure if $\mathsf{Adv}_{\Gamma,\mathcal{A}}^{\mathsf{ATK}}(k)$ is negligible for any PPTA $\mathcal{A}$.

For a detectable KEM Γ (with predicate F) and an adversary $\mathcal{A}$, consider the unpredictability experiment $\mathsf{Expt}_{\Gamma,\mathcal{A}}^{\mathsf{UNP}}(k)$ as described in Fig. 1 (bottom-center). We say that a detectable KEM Γ is *unpredictable* if $\mathsf{Adv}_{\Gamma,\mathcal{A}}^{\mathsf{UNP}}(k)$ is negligible for any PPTA $\mathcal{A}$.

2.3   Symmetric Key Encryption

A symmetric key encryption (SKE) scheme[1] E consists of the two PPTAs (SEnc, SDec) with the following interface:

Encryption:	Decryption:
$c \leftarrow \mathsf{SEnc}(K, m)$	m (or $\perp$) $\leftarrow \mathsf{SDec}(K, c)$

where SDec is a deterministic algorithm, c is a ciphertext of a plaintext m under a key $K \in \{0, 1\}^k$, and $k \in \mathbb{N}$ is a security parameter.

A SKE scheme is required to satisfy *correctness*: for all $k \in \mathbb{N}$, all $K \in \{0, 1\}^k$, and all plaintexts m, it holds that $\mathsf{SDec}(K, \mathsf{SEnc}(K, m)) = m$.

Security Notions. For a SKE scheme, we consider indistinguishability under one-time encryption (OT security), that under chosen ciphertext attacks (CCA security) and its 1-bounded CCA analogue (1-CCA security).

Let $\mathsf{ATK} \in \{\mathsf{OT}, \mathsf{1\text{-}CCA}, \mathsf{CCA}\}$. For a SKE scheme E and an adversary $\mathcal{A} = (\mathcal{A}_1, \mathcal{A}_2)$, consider the ATK experiment $\mathsf{Expt}_{E,\mathcal{A}}^{\mathsf{ATK}}(k)$ as described in Fig. 1 (top-right). In the experiment, it is required that $|m_0| = |m_1|$ holds. Furthermore, the oracle $\mathcal{O}$ is defined as follows: If $\mathsf{ATK} = \mathsf{OT}$ then the oracle $\mathcal{O}$ is unavailable (it returns $\perp$ for any input); If $\mathsf{ATK} \in \{\mathsf{1\text{-}CCA}, \mathsf{CCA}\}$, then $\mathcal{O}$ is the decryption

---

[1] A SKE scheme is also called a "data encapsulation mechanism" (DEM) in the context of hybrid encryption [5].

oracle $\mathcal{O}(\cdot) = \mathsf{SDec}(K, \cdot)$, and the restriction is similar to the corresponding ATK experiment for the PKE case. We say that a SKE scheme E is ATK secure if $\mathsf{Adv}_{E,\mathcal{A}}^{\mathrm{ATK}}(k)$ is negligible for any PPTA $\mathcal{A}$.

2.4   Hybrid Encryption and Its Security

We will use the result regarding hybrid encryption, and thus we recall it here.

Let $\Gamma = (\mathsf{KKG}, \mathsf{KEnc}, \mathsf{KDec})$ and $E = (\mathsf{SEnc}, \mathsf{SDec})$ be a KEM and a SKE scheme, respectively, and let $\Pi[\Gamma, E] = (\mathsf{PKG}, \mathsf{PEnc}, \mathsf{PDec})$ denote the PKE scheme obtained by combining Γ and E in a straightforward manner. Namely, $\mathsf{PKG}(1^k)$ generates keys by $(pk, sk) \leftarrow \mathsf{KKG}(1^k)$; $\mathsf{PEnc}(pk, m)$ runs $(c, K) \leftarrow \mathsf{KEnc}(pk)$ and $\tilde{c} \leftarrow \mathsf{SEnc}(K, m)$, and outputs a ciphertext $C = (c, \tilde{c})$; $\mathsf{PDec}(sk, (c, \tilde{c}))$ runs $K \leftarrow \mathsf{KDec}(sk, c)$, and outputs $m \leftarrow \mathsf{SDec}(K, \tilde{c})$ if $K \neq \bot$ or $\bot$ otherwise. Furthermore, if Γ is a detectable KEM with predicate F, then we view the resulting hybrid scheme $\Pi[\Gamma, E]$ as a detectable PKE scheme that has the predicate $\mathsf{F}'(pk, C = (c, \tilde{c}), C' = (c', \tilde{c}')) := \mathsf{F}(pk, c, c')$. (That is, the predicate F' for $\Pi[\Gamma, E]$ tests the predicate F of the KEM-ciphertexts.)

The following composition results can be shown. The first bullet is a well-known result shown by Cramer and Shoup [5], and we will use the result for the 1–CCA case in the security proof of our proposed scheme. The second and third bullets are not directly used in our proposed scheme, but may be useful for future studies of detectable PKE/KEM. (The proof is straightforward and thus omitted.)

Lemma 1. *Let Γ be a (detectable) KEM and E be a SKE scheme, and let $\Pi[\Gamma, E]$ be the (detectable) PKE scheme obtained by combining Γ and E straightforwardly as above. Then:*

- *([5]) For $\mathsf{ATK} \in \{1\text{–}\mathsf{CCA}, \mathsf{CCA}\}$, if Γ and E are both ATK secure, then so is $\Pi[\Gamma, E]$.*
- *If Γ is DCCA secure and E is OT secure, then $\Pi[\Gamma, E]$ is DCCA secure.*
- *If Γ is unpredictable, then so is $\Pi[\Gamma, E]$ (regardless of the security of E).*

3   The Hohenberger-Lewko-Waters Construction

In this section, we recall the Hohenberger-Lewko-Waters (HLW) construction.

Let $\Pi_{\mathrm{in}} = (\mathsf{PKG}_{\mathrm{in}}, \mathsf{PEnc}_{\mathrm{in}}, \mathsf{PDec}_{\mathrm{in}})$, $\Pi_{\mathsf{A}} = (\mathsf{PKG}_{\mathsf{A}}, \mathsf{PEnc}_{\mathsf{A}}, \mathsf{PDec}_{\mathsf{A}})$, and $\Pi_{\mathsf{B}} = (\mathsf{PKG}_{\mathsf{B}}, \mathsf{PEnc}_{\mathsf{B}}, \mathsf{PDec}_{\mathsf{B}})$ be PKE schemes, where it is assumed that the randomness space of $\mathsf{PEnc}_{\mathsf{A}}$ and that of $\mathsf{PEnc}_{\mathsf{B}}$ are $\{0, 1\}^k$ when they are used with security parameter k. Then the HLW PKE scheme $\Pi_{\mathrm{HLW}} = (\mathsf{PKG}_{\mathrm{HLW}}, \mathsf{PEnc}_{\mathrm{HLW}}, \mathsf{PDec}_{\mathrm{HLW}})$ is constructed as in Fig. 2.

For convenience, we call Π_{in} the *"inner"* scheme, and Π_{A} and Π_{B} the *"outer"* schemes.

It was shown in [7] that if Π_{in} is DCCA secure and unpredictable (with respect to some predicate F), Π_{A} is 1–CCA secure, and Π_{B} is CPA secure, then Π_{HLW} is CCA secure.

$\mathsf{PKG}_{\mathsf{HLW}}(1^k)$:	$\mathsf{PEnc}_{\mathsf{HLW}}(PK, m)$:	$\mathsf{PDec}_{\mathsf{HLW}}(SK, C)$:
$(pk_{\mathsf{in}}, sk_{\mathsf{in}}) \leftarrow \mathsf{PKG}_{\mathsf{in}}(1^k)$	$(pk_{\mathsf{in}}, pk_{\mathsf{A}}, pk_{\mathsf{B}}) \leftarrow PK$	$(sk_{\mathsf{in}}, sk_{\mathsf{A}}, sk_{\mathsf{B}}) \leftarrow SK$
$(pk_{\mathsf{A}}, sk_{\mathsf{A}}) \leftarrow \mathsf{PKG}_{\mathsf{A}}(1^k)$	$r_{\mathsf{A}}, r_{\mathsf{B}} \leftarrow \{0,1\}^k$	$(c_{\mathsf{A}}, c_{\mathsf{B}}) \leftarrow C$
$(pk_{\mathsf{B}}, sk_{\mathsf{B}}) \leftarrow \mathsf{PKG}_{\mathsf{B}}(1^k)$	$\beta \leftarrow (r_{\mathsf{A}} \| r_{\mathsf{B}} \| m)$	$c_{\mathsf{in}} \leftarrow \mathsf{PDec}_{\mathsf{A}}(sk_{\mathsf{A}}, c_{\mathsf{A}})$
$PK \leftarrow (pk_{\mathsf{in}}, pk_{\mathsf{A}}, pk_{\mathsf{B}})$	$c_{\mathsf{in}} \leftarrow \mathsf{PEnc}_{\mathsf{in}}(pk_{\mathsf{in}}, \beta)$	If $c_{\mathsf{in}} = \bot$ then return $\bot$.
$SK \leftarrow (sk_{\mathsf{in}}, sk_{\mathsf{A}}, sk_{\mathsf{B}})$	$c_{\mathsf{A}} \leftarrow \mathsf{PEnc}_{\mathsf{A}}(pk_{\mathsf{A}}, c_{\mathsf{in}}; r_{\mathsf{A}})$	$\beta \leftarrow \mathsf{PDec}_{\mathsf{in}}(sk_{\mathsf{in}}, c_{\mathsf{in}})$
Return (PK, SK).	$c_{\mathsf{B}} \leftarrow \mathsf{PEnc}_{\mathsf{B}}(pk_{\mathsf{B}}, c_{\mathsf{in}}; r_{\mathsf{B}})$	If $\beta = \bot$ then return $\bot$.
	Return $C \leftarrow (c_{\mathsf{A}}, c_{\mathsf{B}})$.	Parse β as $(r_{\mathsf{A}}, r_{\mathsf{B}}, m)$.
		If $\mathsf{PEnc}_{\mathsf{A}}(pk_{\mathsf{A}}, c_{\mathsf{in}}; r_{\mathsf{A}}) = c_{\mathsf{A}}$
		and $\mathsf{PEnc}_{\mathsf{B}}(pk_{\mathsf{B}}, c_{\mathsf{in}}; r_{\mathsf{B}}) = c_{\mathsf{B}}$
		then return m.
		Return $\bot$.

Fig. 2. The Hohenberger-Lewko-Waters Construction Π_{HLW}

4 An Efficient Variant of the HLW Construction

In this section, we show our main result: a more efficient variant of the HLW construction. The ideas towards our scheme are fairly simple:

– We implement the "inner" scheme Π_{in} and one of the "outer" schemes, Π_{A}, by hybrid encryption, i.e. they consist of a KEM and a SKE scheme.
– What is essential in the decryption algorithm of the original HLW construction is that we can decrypt (and check the validity of) a ciphertext in two different ways. Thus, instead of encrypting the entire "inner" ciphertext c_{in} doubly (parallelly) by the "outer" schemes, we realize the property by encrypting only the "session-key" (the value "α" in our construction) used to symmetric-key-encrypt c_{in}.
– Instead of encrypting the randomness r_{A} and r_{B} for the "outer" schemes (together with a plaintext m) by the "inner" scheme, we derive r_{A} and r_{B} from a session-key for "inner" hybrid encryption. As a consequence, the inner SKE scheme now only needs to encrypt m.

Our proposed construction is as follows. Let $\Gamma_{\mathsf{in}} = (\mathsf{KKG}_{\mathsf{in}}, \mathsf{KEnc}_{\mathsf{in}}, \mathsf{KDec}_{\mathsf{in}}, \mathsf{F})$ be a detectable KEM with the session-key space[2] $\{0,1\}^{3k}$. Furthermore, let $\Gamma_{\mathsf{A}} = (\mathsf{KKG}_{\mathsf{A}}, \mathsf{KEnc}_{\mathsf{A}}, \mathsf{KDec}_{\mathsf{A}})$ be a KEM with the session-key space $\{0,1\}^k$, $\Pi_{\mathsf{B}} = (\mathsf{PKG}_{\mathsf{B}}, \mathsf{PEnc}_{\mathsf{B}}, \mathsf{PDec}_{\mathsf{B}})$ be a PKE scheme with the plaintext space $\{0,1\}^k$, and $E_{\mathsf{in}} = (\mathsf{SEnc}_{\mathsf{in}}, \mathsf{SDec}_{\mathsf{in}})$ and $E_{\mathsf{A}} = (\mathsf{SEnc}_{\mathsf{A}}, \mathsf{SDec}_{\mathsf{A}})$ be SKE schemes. We require the randomness space of $\mathsf{KEnc}_{\mathsf{A}}$ and that of $\mathsf{PEnc}_{\mathsf{B}}$ be $\{0,1\}^k$ when they are used with security parameter k.[3] Furthermore, we also require that the size of the ciphertext of PKE and SKE (resp. KEM) can be computed only from the security

---

[2] In our construction, this is without loss of generality because the session-key space of a detectable KEM (with DCCA security and unpredictability) can be freely adjusted by using a pseudorandom generator.

[3] This is again without loss of generality because we can use a pseudorandom generator to obtain longer randomness if necessary.

$\mathsf{PKG}(1^k):$	$\mathsf{PDec}(SK, C):$
$\quad (pk_{\mathsf{in}}, sk_{\mathsf{in}}) \leftarrow \mathsf{KKG}_{\mathsf{in}}(1^k)$	$\quad (sk_{\mathsf{in}}, sk_{\mathsf{A}}, sk_{\mathsf{B}}) \leftarrow SK$
$\quad (pk_{\mathsf{A}}, sk_{\mathsf{A}}) \leftarrow \mathsf{KKG}_{\mathsf{A}}(1^k)$	$\quad (c_{\mathsf{A}}, \widetilde{c}_{\mathsf{A}}, c_{\mathsf{B}}) \leftarrow C$
$\quad (pk_{\mathsf{B}}, sk_{\mathsf{B}}) \leftarrow \mathsf{PKG}_{\mathsf{B}}(1^k)$	$\quad \alpha \leftarrow \mathsf{KDec}_{\mathsf{A}}(sk_{\mathsf{A}}, c_{\mathsf{A}})$
$\quad PK \leftarrow (pk_{\mathsf{in}}, pk_{\mathsf{A}}, pk_{\mathsf{B}})$	$\quad$ If $\alpha = \perp$ then return $\perp$.
$\quad SK \leftarrow (sk_{\mathsf{in}}, sk_{\mathsf{A}}, sk_{\mathsf{B}})$	$\quad (c_{\mathsf{in}} \| \widetilde{c}_{\mathsf{in}}) \leftarrow \mathsf{SDec}_{\mathsf{A}}(\alpha, \widetilde{c}_{\mathsf{A}})$
$\quad$ Return (PK, SK).	$\quad$ If $\mathsf{SDec}_{\mathsf{A}}$ has returned $\perp$ then return $\perp$.
$\mathsf{PEnc}(PK, m):$	$\quad \beta \leftarrow \mathsf{KDec}_{\mathsf{in}}(sk_{\mathsf{in}}, c_{\mathsf{in}})$
$\quad (pk_{\mathsf{in}}, pk_{\mathsf{A}}, pk_{\mathsf{B}}) \leftarrow PK$	$\quad$ If $\beta = \perp$ then return $\perp$.
$\quad (c_{\mathsf{in}}, \beta) \leftarrow \mathsf{KEnc}_{\mathsf{in}}(pk_{\mathsf{in}})$	$\quad$ Parse β as $(r_{\mathsf{A}}, r_{\mathsf{B}}, K) \in \{0, 1\}^{3k}$.
$\quad$ Parse β as $(r_{\mathsf{A}}, r_{\mathsf{B}}, K) \in \{0, 1\}^{3k}$.	$\quad$ If $\mathsf{KEnc}_{\mathsf{A}}(pk_{\mathsf{A}}; r_{\mathsf{A}}) = (c_{\mathsf{A}}, \alpha)$
$\quad \widetilde{c}_{\mathsf{in}} \leftarrow \mathsf{SEnc}_{\mathsf{in}}(K, m)$	$\quad\quad$ and $\mathsf{PEnc}_{\mathsf{B}}(pk_{\mathsf{B}}, \alpha; r_{\mathsf{B}}) = c_{\mathsf{B}}$
$\quad (c_{\mathsf{A}}, \alpha) \leftarrow \mathsf{KEnc}_{\mathsf{A}}(pk_{\mathsf{A}}; r_{\mathsf{A}})$	$\quad\quad\quad$ then return $m \leftarrow \mathsf{SDec}_{\mathsf{in}}(K, \widetilde{c}_{\mathsf{in}})$.
$\quad \widetilde{c}_{\mathsf{A}} \leftarrow \mathsf{SEnc}_{\mathsf{A}}(\alpha, (c_{\mathsf{in}} \| \widetilde{c}_{\mathsf{in}}))$	$\quad$ Return $\perp$.
$\quad c_{\mathsf{B}} \leftarrow \mathsf{PEnc}_{\mathsf{B}}(pk_{\mathsf{B}}, \alpha; r_{\mathsf{B}})$	
$\quad$ Return $C \leftarrow (c_{\mathsf{A}}, \widetilde{c}_{\mathsf{A}}, c_{\mathsf{B}})$.	

Fig. 3. The proposed PKE scheme $\overline{\Pi}$

parameter k and the length of plaintext being encrypted (resp. from k). Then our proposed PKE scheme $\overline{\Pi} = (\overline{\mathsf{PKG}}, \overline{\mathsf{PEnc}}, \overline{\mathsf{PDec}})$ is constructed as in Fig. 3.

Alternative Decryption Algorithm. Before going into the security proof of $\overline{\Pi}$, let us consider the following "alternative" decryption algorithm AltPDec that is useful in our security proof:

AltPDec: This algorithm takes $SK = (sk_{\mathsf{in}}, sk_{\mathsf{A}}, sk_{\mathsf{B}})$ (output from $\overline{\mathsf{PKG}}(1^k)$) and a ciphertext $C = (c_{\mathsf{A}}, \widetilde{c}_{\mathsf{A}}, c_{\mathsf{B}})$ as input, and runs in exactly the same way as $\overline{\mathsf{PDec}}(SK, C)$, except that it executes "$\alpha \leftarrow \mathsf{PDec}_{\mathsf{B}}(sk_{\mathsf{B}}, c_{\mathsf{B}})$" in the third step, instead of "$\alpha \leftarrow \mathsf{KDec}_{\mathsf{A}}(sk_{\mathsf{A}}, c_{\mathsf{A}})$."

Due to the symmetric roles of Γ_{A} and Π_{B} and the validity check of c_{A} and c_{B} at the last step, the following is easy to see.

Lemma 2. *Let SK be a secret key output from $\overline{\mathsf{PKG}}(1^k)$. Then, for any ciphertext C (which could be outside the range of $\overline{\mathsf{PEnc}}$), it holds that $\overline{\mathsf{PDec}}(SK, C) = \mathsf{AltPDec}(SK, C)$.*

Proof of Lemma 2. Let $SK = (sk_{\mathsf{in}}, sk_{\mathsf{A}}, sk_{\mathsf{B}})$ be a secret key generated from $\overline{\mathsf{PKG}}(1^k)$, and let $C = (c_{\mathsf{A}}, \widetilde{c}_{\mathsf{A}}, c_{\mathsf{B}})$ be an arbitrary ciphertext. For notational convenience, let $\alpha_{\mathsf{A}} = \mathsf{KDec}_{\mathsf{A}}(sk_{\mathsf{A}}, c_{\mathsf{A}})$ and $\alpha_{\mathsf{B}} = \mathsf{PDec}_{\mathsf{B}}(sk_{\mathsf{B}}, c_{\mathsf{B}})$. We consider the following two cases, and show that the result of $\overline{\mathsf{PDec}}(SK, C)$ and that of $\mathsf{AltPDec}(SK, C)$ always agree in both cases:

Case $\alpha_{\mathsf{A}} = \alpha_{\mathsf{B}}$: $\overline{\mathsf{PDec}}$ and AltPDec proceed identically after their third step, and thus the outputs of these algorithms agree, regardless of the validity of C.

Case $\alpha_{\mathsf{A}} \neq \alpha_{\mathsf{B}}$: The correctness of Π_{B} implies that there exists no r such that $\mathsf{PEnc}_{\mathsf{B}}(pk_{\mathsf{B}}, \alpha_{\mathsf{A}}; r) = c_{\mathsf{B}}$, and thus $\overline{\mathsf{PDec}}$ returns $\perp$ in its last step at the latest (it returns $\perp$ earlier if $\mathsf{SDec_A}$ or $\mathsf{KDec_{in}}$ outputs $\perp$). Similarly, the correctness of Γ_A implies that there exists no r such that $\mathsf{KEnc_A}(pk_A; r) = (c_A, \alpha_B)$, and thus $\mathsf{AltPDec}$ also returns $\perp$ in its last step at the latest (it may return $\perp$ earlier as above).

This completes the proof of Lemma 2. $\qquad\qquad\qquad\qquad\qquad\qquad\qquad\square$

Security of the proposed PKE scheme. The security of $\overline{\Pi}$ is shown as follows.

Theorem 1. *Assume that the detectable KEM Γ_{in} is DCCA secure and unpredictable, the KEM Γ_A and the SKE scheme E_A are 1–CCA secure, the PKE scheme Π_B is CPA secure, and the SKE scheme E_{in} is OT secure. Then the proposed PKE scheme $\overline{\Pi}$ constructed as in Fig. 3 is CCA secure.*

Proof of Theorem 1. Let $\mathcal{A} = (\mathcal{A}_1, \mathcal{A}_2)$ be any PPTA adversary against the CCA security of $\overline{\Pi}$. Let $Q > 0$ be the number of decryption queries that $\mathcal{A}_2$ submits in the CCA experiment. (Since $\mathcal{A}$ is a PPTA, Q is a polynomial.)

Consider the following sequence of games, where the values with asterisk (*) are those related to the challenge ciphertext $C^* = (c_A^*, \widetilde{c}_A^*, c_B^*)$.

Game 1: This is the ordinary CCA experiment in which $\mathcal{A}$ runs, i.e. $\mathsf{Expt}_{\overline{\Pi}, \mathcal{A}}^{\mathsf{CCA}}(k)$.

Game 2: Same as Game 1, except that if $\mathcal{A}_2$ submits a decryption query $C = (c_A, \widetilde{c}_A, c_B)$ satisfying $(c_A, \widetilde{c}_A) = (c_A^*, \widetilde{c}_A^*)$, then it is answered with $\perp$.

Game 3: Same as Game 2, except that we pick $r_A^*, r_B^*, K^* \in \{0, 1\}^k$ uniformly at random, instead of using those produced from $\mathsf{KEnc_{in}}(pk_{in})$. That is, the step "$(c_{in}^*, \beta^*) \leftarrow \mathsf{KEnc_{in}}(pk_{in})$" in Game 2 is replaced with the steps "$(c_{in}^*, \beta') \leftarrow \mathsf{KEnc_{in}}(pk_{in})$; $r_A^*, r_B^*, K^* \leftarrow \{0, 1\}^k$."
Since the randomness r_A^* and r_B^* used for generating c_A^* and c_B^* are now (and in all subsequent games) chosen uniformly at random and made independent of c_{in}^*, from here on we do not write them explicitly. (Instead, they are always chosen randomly when c_A^* and c_B^* are generated.)

Game 4: Same as Game 3, except that the information on α^* is erased from c_B^*. That is, the step "$c_B^* \leftarrow \mathsf{PEnc_B}(pk_B, \alpha^*)$" in Game 3 is replaced with the step "$c_B^* \leftarrow \mathsf{PEnc_B}(pk_B, 0^k)$."

Game 5: Same as Game 4, except that we use the alternative decryption algorithm $\mathsf{AltPDec}$ for answering to $\mathcal{A}$'s decryption queries. More precisely, in Game 5, $\mathcal{A}_1$'s queries C are answered with $\mathsf{AltPDec}(SK, C)$.; $\mathcal{A}_2$'s queries $C = (c_A, \widetilde{c}_A, c_B)$ are also answered with $\mathsf{AltPDec}(SK, C)$, except that if $(c_A, \widetilde{c}_A) = (c_A^*, \widetilde{c}_A^*)$, then the query is answered with $\perp$.

Game 6: Same as Game 5, except that the information on $(c_{in}^* \| \widetilde{c}_{in}^*)$ is erased from $\widetilde{c}_A^*$. That is, the step "$\widetilde{c}_A^* \leftarrow \mathsf{SEnc_A}(\alpha^*, (c_{in}^* \| \widetilde{c}_{in}^*))$" in Game 5 is replaced with the step "$\widetilde{c}_A^* \leftarrow \mathsf{SEnc_A}(\alpha^*, 0^{|c_{in}^*| + |\widetilde{c}_{in}^*|})$."

We call a decryption query $C = (c_A, \widetilde{c}_A, c_B)$ submitted by $\mathcal{A}_2$ in the above games *dangerous* if the following four conditions are satisfied:

1. $(c_A, \widetilde{c}_A) \neq (c_A^*, \widetilde{c}_A^*)$
2. $\mathsf{KDec_A}(pk_A, c_A) = \alpha \neq \perp$

3. $\mathsf{SDec_A}(\alpha, \widetilde{c}_A) = (c_{\mathrm{in}} \| \widetilde{c}_{\mathrm{in}}) \neq \bot$
4. $\mathsf{F}(pk_{\mathrm{in}}, c^*_{\mathrm{in}}, c_{\mathrm{in}}) = 1$

For $i \in [6]$, we define the following events in Game i:

S_i: $\mathcal{A}$ succeeds in guessing the challenge bit (i.e. $b' = b$ occurs).
D_i: $\mathcal{A}_2$ submits at least one dangerous decryption query.
$\mathsf{D}_i^{(j)}$: (where $j \in [Q]$) $\mathcal{A}_2$'s j-th decryption query is dangerous.

By the definitions of the games and events, a simple manipulation of probabilities gives us the following. (The proof is given in Appendix A.)

Claim 1. *$\mathcal{A}$'s CCA advantage $\mathsf{Adv}^{CCA}_{\overline{\Pi}, \mathcal{A}}(k)$ can be upperbounded as follows:*

$$\mathsf{Adv}^{CCA}_{\overline{\Pi}, \mathcal{A}}(k) \leq$$

$$2 \cdot |\Pr[\mathsf{S}_1] - \Pr[\mathsf{S}_2]| + 2 \cdot |\Pr[\mathsf{S}_2 \wedge \overline{\mathsf{D}_2}] - \Pr[\mathsf{S}_3 \wedge \overline{\mathsf{D}_3}] + \frac{1}{2}(\Pr[\mathsf{D}_2] - \Pr[\mathsf{D}_3])|$$

$$+ 2 \cdot |\Pr[\mathsf{S}_3 \wedge \overline{\mathsf{D}_3}] + \frac{1}{2}\Pr[\mathsf{D}_3] - \frac{1}{2}| + \sum_{i \in \{2,3,4\}} |\Pr[\mathsf{D}_i] - \Pr[\mathsf{D}_{i+1}]|$$

$$+ |\sum_{j \in [Q]} (\Pr[\mathsf{D}_5^{(j)}] - \Pr[\mathsf{D}_6^{(j)}])| + \sum_{j \in [Q]} \Pr[\mathsf{D}_6^{(j)}]. \quad (1)$$

In the following, we will upperbound each term in the inequality (1).

Claim 2. $\Pr[\mathsf{S}_1] = \Pr[\mathsf{S}_2]$.

Proof of Claim 2. Note that the difference between Game 1 and Game 2 is in how $\mathcal{A}_2$'s decryption query $C = (c_A, \widetilde{c}_A, c_B)$ satisfying $(c_A, \widetilde{c}_A) = (c^*_A, \widetilde{c}^*_A)$ is answered. By definition, it is answered with $\bot$ in Game 2. We show that such query is answered with $\bot$ in Game 1 as well and thus the decryption oracles behave identically in both games, which in particular implies $\Pr[\mathsf{S}_1] = \Pr[\mathsf{S}_2]$.

Firstly, $\mathcal{A}_2$'s query must satisfy $C \neq C^*$, and thus $(c_A, \widetilde{c}_A) = (c^*_A, \widetilde{c}^*_A)$ implies $c_B \neq c^*_B$. Secondly, $(c_A, \widetilde{c}_A) = (c^*_A, \widetilde{c}^*_A)$ implies that α^* and $\beta^* = (r^*_A, r^*_B, K^*)$ are recovered in the third and the seventh steps of $\overline{\mathsf{PDec}}(SK, C)$, respectively. Therefore, in order for the query to pass the validity check and be answered with non-$\bot$, we must have $\mathsf{PEnc_B}(pk_B, \alpha^*; r^*_B) = c_B$. However, this is impossible because $c_B \neq c^*_B = \mathsf{PEnc_B}(pk_B, \alpha^*; r^*_B)$. Therefore, the query C with $(c_A, \widetilde{c}_A) = (c^*_A, \widetilde{c}^*_A)$ is answered with $\bot$ in Game 1 as well. This completes the proof of Claim 2. □

Claim 3. *For any constants $p, q \in [0, 1]$, there exists a PPTA $\mathcal{B}$ such that $\mathsf{Adv}^{DCCA}_{\Gamma_{\mathrm{in}}, \mathcal{B}}(k) = |p \cdot (\Pr[\mathsf{S}_2 \wedge \overline{\mathsf{D}_2}] - \Pr[\mathsf{S}_3 \wedge \overline{\mathsf{D}_3}]) + q \cdot (\Pr[\mathsf{D}_2] - \Pr[\mathsf{D}_3])|$.*

Proof of Claim 3. Fix arbitrarily $p, q \in [0, 1]$. Using $\mathcal{A}$ as a building block, we show how to construct a PPTA adversary $\mathcal{B}$ that attacks the DCCA security of the detectable KEM Γ_{in} with the claimed advantage. The description of $\mathcal{B} = (\mathcal{B}_1, \mathcal{B}_2)$ is as follows:

$\mathcal{B}_1^{\mathcal{O}}(pk_{in})$: $\mathcal{B}_1$ first runs $(pk_A, sk_A) \leftarrow \mathsf{KKG}_A(1^k)$ and $(pk_B, sk_B) \leftarrow \mathsf{PKG}_B(1^k)$. Then $\mathcal{B}_1$ sets $PK \leftarrow (pk_{in}, pk_A, pk_B)$ and $SK \leftarrow (\perp, sk_A, sk_B)$, and runs $(m_0, m_1, \mathsf{st}) \leftarrow \mathcal{A}_1^{\overline{\mathsf{PDec}}(SK, \cdot)}(PK)$, where $\mathcal{B}_1$ uses its own decapsulation oracle $\mathcal{O}$ as a substitute for $\mathsf{KDec}_{in}(sk_{in}, \cdot)$. Next, $\mathcal{B}_1$ sets state information $\mathsf{st}_\mathcal{B}$ as all the values known to $\mathcal{B}_1$, and terminates with output $\mathsf{st}_\mathcal{B}$.

$\mathcal{B}_2^{\mathcal{O}}(c_{in}^*, \beta_b^*, \mathsf{st}_\mathcal{B})$: (where b is $\mathcal{B}$'s challenge bit) $\mathcal{B}_2$ first picks two coins $b_p, b_q \in \{0,1\}$ such that $b_p = 1$ (resp. $b_q = 1$) with probability p (resp. q). Then $\mathcal{B}_2$ parses β_b^* as $(r_A^*, r_B^*, K^*) \in \{0,1\}^{3k}$, flips a fair coin $\gamma \in \{0,1\}$, and then runs $\widetilde{c}_{in}^* \leftarrow \mathsf{SEnc}_{in}(K^*, m_\gamma)$, $(c_A^*, \alpha^*) \leftarrow \mathsf{KEnc}_A(pk_A; r_A^*)$, $\widetilde{c}_A^* \leftarrow \mathsf{SEnc}_A(\alpha^*, (c_{in}^* \| \widetilde{c}_{in}^*))$, and $c_B^* \leftarrow \mathsf{PEnc}_B(pk_B, \alpha^*; r_B^*)$. Then $\mathcal{B}_2$ sets $C^* \leftarrow (c_A^*, \widetilde{c}_A^*, c_B^*)$, and runs $\mathcal{A}_2(C^*, \mathsf{st})$. $\mathcal{B}_2$ answers $\mathcal{A}_2$'s decryption queries as specified in Game 2 where again $\mathcal{B}_2$ uses its own oracle $\mathcal{O}$ as a substitute for $\mathsf{KDec}_{in}(sk_{in}, \cdot)$, except that if the query is dangerous (which can be detected by using SK and F), then $\mathcal{B}_2$ terminates with output $b' \leftarrow b_q$. When $\mathcal{A}_2$ terminates with output its guess bit γ', $\mathcal{B}_2$ sets $b' \leftarrow 1$ if $\gamma' = \gamma$ and $b_p = 1$, otherwise (i.e. $\gamma' \neq \gamma$ or $b_p = 0$) $\mathcal{B}_2$ sets $b' \leftarrow 0$, and terminates with output b'.

The above completes the description of $\mathcal{B}$. Note that $\mathcal{B}_2$ never submits a prohibited decapsulation query c_{in} satisfying $\mathsf{F}(pk_{in}, c_{in}^*, c_{in}) = 1$.

Let $\mathsf{D}_\mathcal{B}$ be the event that $\mathcal{A}$ submits a dangerous decryption query in the experiment simulated by $\mathcal{B}$. Consider the case when $b = 1$. It is easy to see that in this case, $\mathcal{B}$ simulates Game 2 perfectly for $\mathcal{A}$ in which $\mathcal{A}$'s challenge bit is γ, until the point $\mathcal{A}_2$ submits a dangerous query. In particular, the value β_1^* associated with c_{in}^* is used as (r_A^*, r_B^*, K^*) as is done in Game 2. All other values $(pk, C^*,$ and the answers to decryption queries) are distributed identically to those of Game 2. Furthermore, $\mathcal{B}_2$ can detect whether $\mathcal{A}_2$'s query is dangerous by using F. These imply that $\Pr[\gamma' = \gamma \wedge \overline{\mathsf{D}_\mathcal{B}}|b = 1] = \Pr[\mathsf{S}_2 \wedge \overline{\mathsf{D}_2}]$ and $\Pr[\mathsf{D}_\mathcal{B}|b = 1] = \Pr[\mathsf{D}_2]$. Recall that $\mathcal{B}$ outputs $b' = 1$ only if either (1) $\mathcal{A}_2$ succeeds in guessing γ without making any dangerous queries (i.e. $\gamma' = \gamma \wedge \overline{\mathsf{D}_\mathcal{B}}$ occurs) and $b_p = 1$, or (2) $\mathcal{A}_2$ makes a dangerous query ($\mathsf{D}_\mathcal{B}$ occurs) and $b_q = 1$. Furthermore, the choice of b_p and b_q is independent of the behavior of $\mathcal{A}$ and $\mathcal{B}$'s challenge bit. These imply

$$\Pr[b' = 1|b = 1] = \Pr[b_p = 1 \wedge \gamma' = \gamma \wedge \overline{\mathsf{D}_\mathcal{B}}|b = 1] + \Pr[b_q = 1 \wedge \mathsf{D}_\mathcal{B}|b = 1]$$
$$= p \cdot \Pr[\gamma' = \gamma \wedge \overline{\mathsf{D}_\mathcal{B}}|b = 1] + q \cdot \Pr[\mathsf{D}_\mathcal{B}|b = 1] = p \cdot \Pr[\mathsf{S}_2 \wedge \overline{\mathsf{D}_2}] + q \cdot \Pr[\mathsf{D}_2].$$

On the other hand, when $b = 0$, $\mathcal{B}$ simulates Game 3 perfectly for $\mathcal{A}$ in which $\mathcal{A}$'s challenge bit is γ, until the point $\mathcal{A}_2$ submits a dangerous query. In particular, the uniformly chosen random value β_0^* (independent of c_{in}^*) is used as (r_A^*, r_B^*, K^*), which is exactly how they are distributed in Game 3. The rest is unchanged from the case of $b = 1$, and thus, with a similar argument, we have $\Pr[b' = 1|b = 0] = p \cdot \Pr[\mathsf{S}_3 \wedge \overline{\mathsf{D}_3}] + q \cdot \Pr[\mathsf{D}_3]$.

In summary, we can calculate $\mathcal{B}$'s DCCA advantage as follows:

$$\mathsf{Adv}_{\Gamma_{in}, \mathcal{B}}^{\mathsf{DCCA}}(k) = |\Pr[b' = 1|b = 1] - \Pr[b' = 1|b = 0]|$$
$$= |p \cdot (\Pr[\mathsf{S}_2 \wedge \overline{\mathsf{D}_2}] - \Pr[\mathsf{S}_3 \wedge \overline{\mathsf{D}_3}]) + q \cdot (\Pr[\mathsf{D}_2] - \Pr[\mathsf{D}_3])|.$$

This completes the proof of Claim 3.   □

Claim 4. *There exists a PPTA $\mathcal{B}$ such that* $\mathsf{Adv}^{\mathsf{OT}}_{E_{\mathrm{in}},\mathcal{B}}(k) = 2 \cdot |\Pr[\mathsf{S}_3 \wedge \overline{\mathsf{D}_3}] + \frac{1}{2}\Pr[\mathsf{D}_3] - \frac{1}{2}|.$

Proof of Claim 4. Using $\mathcal{A}$ as a building block, we construct a PPTA adversary $\mathcal{B} = (\mathcal{B}_1, \mathcal{B}_2)$ that attacks the OT security of the SKE scheme E_{in} as follows:

$\mathcal{B}_1(1^k)$: $\mathcal{B}_1$ runs $(PK, SK) \leftarrow \overline{\mathsf{PKG}}(1^k)$ and $(m_0, m_1, \mathsf{st}) \leftarrow \mathcal{A}_1^{\overline{\mathsf{PDec}}(SK,\cdot)}(PK)$.
Then $\mathcal{B}_1$ sets state information $\mathsf{st}_{\mathcal{B}}$ as all the values known to $\mathcal{B}_1$, and terminates with output $(m_0, m_1, \mathsf{st}_{\mathcal{B}})$.

$\mathcal{B}_2(\tilde{c}^*_{\mathrm{in}}, \mathsf{st}_{\mathcal{B}})$: $\mathcal{B}_2$ runs $(c^*_{\mathrm{in}}, \beta') \leftarrow \mathsf{KEnc}_{\mathrm{in}}(pk_{\mathrm{in}})$, $(\tilde{c}^*_{\mathsf{A}}, \alpha^*) \leftarrow \mathsf{KEnc}_{\mathsf{A}}(pk_{\mathsf{A}})$, $\tilde{c}^*_{\mathsf{A}} \leftarrow$
$\mathsf{SEnc}_{\mathsf{A}}(\alpha^*, (c^*_{\mathrm{in}} \| \tilde{c}^*_{\mathrm{in}}))$, and $c^*_{\mathsf{B}} \leftarrow \mathsf{PEnc}_{\mathsf{B}}(pk_{\mathsf{B}}, \alpha^*)$. Then $\mathcal{B}_2$ sets $C^* \leftarrow (c^*_{\mathsf{A}}, \tilde{c}^*_{\mathsf{A}}, c^*_{\mathsf{B}})$,
and runs $\mathcal{A}_2(C^*, \mathsf{st})$. $\mathcal{B}_2$ answers to $\mathcal{A}_2$'s decryption queries C as specified in
Game 3 (using SK), except that if the query is dangerous (which can be detected by using SK and F), then $\mathcal{B}_2$ terminates with output a random bit
$b' \in \{0,1\}$. When $\mathcal{A}_2$ terminates with output b', $\mathcal{B}_2$ terminates with output
this bit b'.

The above completes the description of $\mathcal{B}$.

Let b be $\mathcal{B}$'s challenge bit, and $\mathsf{D}_{\mathcal{B}}$ be the event that $\mathcal{A}_2$ submits a dangerous query in the experiment simulated by $\mathcal{B}$. It is easy to see that unless $\mathcal{A}_2$ submits a dangerous query, $\mathcal{B}$ perfectly simulates Game 3 for $\mathcal{A}$ in which the challenge bit of $\mathcal{A}$ is that of $\mathcal{B}$'s, and K^* in Game 3 is the key chosen in $\mathcal{B}$'s OT experiment. Furthermore, if no dangerous query is made, $\mathcal{B}$ succeeds whenever $\mathcal{A}$ succeeds. Therefore, we have $\Pr[b' = b \wedge \overline{\mathsf{D}_{\mathcal{B}}}] = \Pr[\mathsf{S}_3 \wedge \overline{\mathsf{D}_3}]$ and $\Pr[\mathsf{D}_{\mathcal{B}}] = \Pr[\mathsf{D}_3]$. Furthermore, if $\mathcal{A}_2$ submits a dangerous query, $\mathcal{B}_2$ can detect it by using F and outputs a random bit, which implies $\Pr[b' = b | \mathsf{D}_{\mathcal{B}}] = 1/2$.

Using these, $\mathcal{B}$'s OT advantage can be calculated as follows:

$$\mathsf{Adv}^{\mathsf{OT}}_{E_{\mathrm{in}},\mathcal{B}}(k) = 2 \cdot |\Pr[b' = b] - \frac{1}{2}|$$

$$= 2 \cdot |\Pr[b' = b \wedge \overline{\mathsf{D}_{\mathcal{B}}}] + \Pr[b' = b | \mathsf{D}_{\mathcal{B}}] \cdot \Pr[\mathsf{D}_{\mathcal{B}}] - \frac{1}{2}|$$

$$= 2 \cdot |\Pr[\mathsf{S}_3 \wedge \overline{\mathsf{D}_3}] + \frac{1}{2}\Pr[\mathsf{D}_3] - \frac{1}{2}|.$$

This completes the proof of Claim 4.   □

Claim 5. *There exists a PPTA $\mathcal{B}$ such that* $\mathsf{Adv}^{\mathsf{CPA}}_{\Pi_{\mathsf{B}},\mathcal{B}}(k) = |\Pr[\mathsf{D}_3] - \Pr[\mathsf{D}_4]|.$

Proof of Claim 5. Using $\mathcal{A}$ as a building block, we construct a PPTA adversary $\mathcal{B} = (\mathcal{B}_1, \mathcal{B}_2)$ that attacks the CPA security of the PKE scheme Π_{B} as follows:

$\mathcal{B}_1(pk_{\mathsf{B}})$: $\mathcal{B}_1$ first runs $(pk_{\mathrm{in}}, sk_{\mathrm{in}}) \leftarrow \mathsf{KKG}_{\mathrm{in}}(1^k)$ and $(pk_{\mathsf{A}}, sk_{\mathsf{A}}) \leftarrow \mathsf{KKG}_{\mathsf{A}}(1^k)$.
Then $\mathcal{B}_1$ sets $PK \leftarrow (pk_{\mathrm{in}}, pk_{\mathsf{A}}, pk_{\mathsf{B}})$ and $SK \leftarrow (sk_{\mathrm{in}}, sk_{\mathsf{A}}, \perp)$, and runs
$(m_0, m_1, \mathsf{st}) \leftarrow \mathcal{A}_1^{\overline{\mathsf{PDec}}(SK,\cdot)}(PK)$ (note that sk_{B} is unnecessary to run $\overline{\mathsf{PDec}}$).
Next $\mathcal{B}_1$ runs $(c^*_{\mathsf{A}}, \alpha^*) \leftarrow \mathsf{KEnc}_{\mathsf{A}}(pk_{\mathsf{A}})$, and sets $M_1 \leftarrow \alpha^*$ and $M_0 \leftarrow 0^k$. Then
$\mathcal{B}_1$ sets state information $\mathsf{st}_{\mathcal{B}}$ as all the values known to $\mathcal{B}_1$, and terminates
with output $(M_0, M_1, \mathsf{st}_{\mathcal{B}})$.

$\mathcal{B}_2(c_{\mathsf{B}}^*, \mathsf{st}_{\mathcal{B}})$: $\mathcal{B}_2$ first picks $\gamma \in \{0,1\}$ and $K^* \in \{0,1\}^k$ uniformly, and then runs $(c_{\mathsf{in}}^*, \beta') \leftarrow \mathsf{KEnc}_{\mathsf{in}}(pk_{\mathsf{in}})$, $\widetilde{c}_{\mathsf{in}}^* \leftarrow \mathsf{SEnc}_{\mathsf{in}}(K^*, m_\gamma)$, and $\widetilde{c}_{\mathsf{A}}^* \leftarrow \mathsf{SEnc}_{\mathsf{A}}(\alpha^*, (c_{\mathsf{in}}^* \| \widetilde{c}_{\mathsf{in}}^*))$. Then $\mathcal{B}_2$ sets $C^* \leftarrow (c_{\mathsf{A}}^*, \widetilde{c}_{\mathsf{A}}^*, c_{\mathsf{B}}^*)$, and runs $\mathcal{A}_2(C^*, \mathsf{st})$. $\mathcal{B}_2$ answers to $\mathcal{A}_2$'s decryption queries C as specified in Game 3 (again, sk_{B} is unnecessary for this). When $\mathcal{A}_2$ terminates, $\mathcal{B}_2$ checks whether $\mathcal{A}_2$ has submitted a dangerous query (by using SK and F). If this is the case, then $\mathcal{B}_2$ sets $b' \leftarrow 1$, otherwise $\mathcal{B}_2$ sets $b' \leftarrow 0$, and terminates with output b'.

The above completes the description of $\mathcal{B}$. Let b be $\mathcal{B}$'s challenge bit.

It is easy to see that if $b = 1$ (resp. $b = 0$), then $\mathcal{B}$ simulates Game 3 (resp. Game 4) perfectly for $\mathcal{A}$ (with its challenge bit γ). In particular, c_{B}^* is an encryption of α^* if $b = 1$ and is an encryption of 0^k if $b = 0$, which is how c_{B}^* in Game 3 and that in Game 4 are generated, respectively, and all other values (pk, the remaining parts of C^*, and the answers to decryption queries) are distributed identically to those of Game 3 or Game 4, depending on b. Furthermore, $\mathcal{B}_2$ can detect whether $\mathcal{A}_2$'s queries contain a dangerous one by using F, in which case (and only then) $\mathcal{B}_2$ outputs $b' = 1$. These imply $\Pr[b' = 1 | b = 1] = \Pr[\mathsf{D}_3]$ and $\Pr[b' = 1 | b = 0] = \Pr[\mathsf{D}_4]$. Using them, $\mathcal{B}$'s CPA advantage is shown as follows:

$$\mathsf{Adv}_{\Pi_{\mathsf{B}}, \mathcal{B}}^{\mathsf{CPA}}(k) = |\Pr[b' = 1 | b = 1] - \Pr[b' = 1 | b = 0]| = |\Pr[\mathsf{D}_3] - \Pr[\mathsf{D}_4]|.$$

This completes the proof of Claim 5.   □

Claim 6. $\Pr[\mathsf{D}_4] = \Pr[\mathsf{D}_5]$.

Proof of Claim 6.   Recall that the difference between Game 4 and Game 5 is in how $\mathcal{A}$'s decryption queries $C = (c_{\mathsf{A}}, \widetilde{c}_{\mathsf{A}}, c_{\mathsf{B}})$ satisfying $(c_{\mathsf{A}}, \widetilde{c}_{\mathsf{A}}) \neq (c_{\mathsf{A}}^*, \widetilde{c}_{\mathsf{A}}^*)$ are answered. In Game 4, they are answered using $\overline{\mathsf{PDec}}$, while they are answered using $\mathsf{AltPDec}$ in Game 5. However, by Lemma 2, these algorithms behave identically for all queries made in both Game 4 and Game 5, and thus $\mathcal{A}$'s view is distributed identically in both games. This in particular implies $\Pr[\mathsf{D}_4] = \Pr[\mathsf{D}_5]$. This completes the proof of Claim 6.   □

Claim 7. Let $\Pi[\Gamma_{\mathsf{A}}, E_{\mathsf{A}}]$ be the hybrid encryption scheme obtained from the KEM Γ_{A} and the SKE scheme E_{A} (see Section 2.4). Then, there exists a PPTA $\mathcal{B}$ such that $\mathsf{Adv}_{\Pi[\Gamma_{\mathsf{A}}, E_{\mathsf{A}}], \mathcal{B}}^{\mathsf{1\text{-}CCA}}(k) = \frac{1}{Q} | \sum_{j \in [Q]} (\Pr[\mathsf{D}_5^{(j)}] - \Pr[\mathsf{D}_6^{(j)}]) |$.

Proof of Claim 7.   Using $\mathcal{A}$ as a building block, we construct a PPTA adversary $\mathcal{B} = (\mathcal{B}_1, \mathcal{B}_2)$ that attacks the 1-CCA security of the hybrid encryption scheme $\Pi[\Gamma_{\mathsf{A}}, E_{\mathsf{A}}]$ as follows:

$\mathcal{B}_1^{\mathcal{O}}(pk_{\mathsf{A}})$: $\mathcal{B}_1$ first runs $(pk_{\mathsf{in}}, sk_{\mathsf{in}}) \leftarrow \mathsf{KKG}_{\mathsf{in}}(1^k)$ and $(pk_{\mathsf{B}}, sk_{\mathsf{B}}) \leftarrow \mathsf{PKG}_{\mathsf{B}}(1^k)$. Then $\mathcal{B}_1$ sets $PK \leftarrow (pk_{\mathsf{in}}, pk_{\mathsf{A}}, pk_{\mathsf{B}})$ and $SK \leftarrow (sk_{\mathsf{in}}, \perp, sk_{\mathsf{B}})$, and runs $(m_0, m_1, \mathsf{st}) \leftarrow \mathcal{A}_1^{\mathsf{AltPDec}(SK, \cdot)}(PK)$ (note that sk_{A} is unnecessary to run $\mathsf{AltPDec}$). Next, $\mathcal{B}_1$ picks $\gamma \in \{0,1\}$ and $K^* \in \{0,1\}^k$ uniformly, and then runs $(c_{\mathsf{in}}^*, \beta') \leftarrow \mathsf{KEnc}_{\mathsf{in}}(pk_{\mathsf{in}})$ and $\widetilde{c}_{\mathsf{in}}^* \leftarrow \mathsf{SEnc}_{\mathsf{in}}(K^*, m_\gamma)$. Then, $\mathcal{B}_1$ sets $M_1 \leftarrow (c_{\mathsf{in}}^* \| \widetilde{c}_{\mathsf{in}}^*)$ and $M_0 \leftarrow 0^{|M_1|}$, sets state information $\mathsf{st}_{\mathcal{B}}$ as all the values known to $\mathcal{B}_1$, and terminates with output $(M_0, M_1, \mathsf{st}_{\mathcal{B}})$.

$\mathcal{B}_2^{\mathcal{O}}((c_A^*, \widetilde{c}_A^*), \mathsf{st}_\mathcal{B})$: $\mathcal{B}_2$ runs $c_B^* \leftarrow \mathsf{PEnc}(pk_B, 0^k)$, sets $C^* \leftarrow (c_A^*, \widetilde{c}_A^*, c_B^*)$, and then runs $\mathcal{A}_2(C^*, \mathsf{st})$. $\mathcal{B}_2$ answers to $\mathcal{A}_2$'s decryption queries C as specified in Game 5 (again, sk_A is unnecessary for this). When $\mathcal{A}_2$ terminates, $\mathcal{B}_2$ picks $\ell \in [Q]$ uniformly, and checks whether $\mathcal{A}_2$'s ℓ-th query $C_\ell = (c_A^{(\ell)}, \widetilde{c}_A^{(\ell)}, c_B^{(\ell)})$ is dangerous, by using $\mathcal{B}_2$'s decryption oracle $\mathcal{O}$ (if necessary). More specifically, if $(c_A^{(\ell)}, \widetilde{c}_A^{(\ell)}) = (c_A^*, \widetilde{c}_A^*)$, then C_ℓ is not dangerous. Otherwise, $\mathcal{B}_2$ submits $(c_A^{(\ell)}, \widetilde{c}_A^{(\ell)})$ to $\mathcal{O}$, and uses the returned value to further check whether C_ℓ is dangerous. If C_ℓ is dangerous, then $\mathcal{B}_2$ sets $b' \leftarrow 1$, otherwise $\mathcal{B}_2$ sets $b' \leftarrow 0$, and terminates with output b'.

The above completes the description of $\mathcal{B}$. Note that $\mathcal{B}$ uses the decryption oracle $\mathcal{O}$ at most once, and it never submits the prohibited query $(c_A^*, \widetilde{c}_A^*)$ to $\mathcal{O}$.

Let b be $\mathcal{B}$'s challenge bit. Furthermore, for $j \in [Q]$, let $\mathsf{D}_\mathcal{B}^{(j)}$ be the event that $\mathcal{A}_2$'s j-th query C_j is dangerous in the experiment simulated by $\mathcal{B}$. It is easy to see that if $b = 1$ (resp. $b = 0$), then $\mathcal{B}$ simulates Game 5 (resp. Game 6) perfectly for $\mathcal{A}$ (with its challenge bit γ). In particular, $(c_A^*, \widetilde{c}_A^*)$ is an encryption of $(c_{in}^*\|\widetilde{c}_{in}^*)$ if $b = 1$, and is an encryption of the zero-string if $b = 0$, which is how $(c_A^*, \widetilde{c}_A^*)$ in Game 5 and that in Game 6 are generated, respectively. Note that $\mathcal{B}_2$ can detect whether $\mathsf{D}_\mathcal{B}^{(\ell)}$ has occurred (for a randomly chosen index $\ell \in [Q]$) by using its own decryption oracle and using F, in which case (and only then) $\mathcal{B}_2$ outputs 1. Note also that $\ell \in [Q]$ is chosen uniformly and independently of b and $\mathcal{A}$'s behavior. These imply that for all $j \in [Q]$, we have $\Pr[\mathsf{D}_\mathcal{B}^{(j)}|\ell = j \wedge b = 1] = \Pr[\mathsf{D}_5^{(j)}]$, $\Pr[\mathsf{D}_\mathcal{B}^{(j)}|\ell = j \wedge b = 0] = \Pr[\mathsf{D}_6^{(j)}]$, and $\Pr[\ell = j|b = 1] = \Pr[\ell = j|b = 0] = 1/Q$.

Using these, $\mathcal{B}$'s 1-CCA advantage can be calculated as follows:

$$\mathsf{Adv}_{\Pi[\Gamma_A, E_A], \mathcal{B}}^{\mathsf{1\text{-}CCA}}(k) = |\Pr[b' = 1|b = 1] - \Pr[b' = 1|b = 0]|$$

$$= |\Pr[\mathsf{D}_\mathcal{B}^{(\ell)}|b = 1] - \Pr[\mathsf{D}_\mathcal{B}^{(\ell)}|b = 0]|$$

$$= |\sum_{j \in [Q]} \left(\Pr[\mathsf{D}_\mathcal{B}^{(j)}|\ell = j \wedge b = 1] \cdot \Pr[\ell = j|b = 1]\right.$$

$$\left. - \Pr[\mathsf{D}_\mathcal{B}^{(j)}|\ell = j \wedge b = 0] \cdot \Pr[\ell = j|b = 0]\right)|$$

$$= \frac{1}{Q}|\sum_{j \in [Q]} \left(\Pr[\mathsf{D}_5^{(j)}] - \Pr[\mathsf{D}_6^{(j)}]\right)|.$$

This completes the proof of Claim 7. $\square$

Claim 8. *There exists a PPTA $\mathcal{B}$ such that* $\mathsf{Adv}_{\Gamma_{in}, \mathcal{B}}^{\mathsf{UNP}}(k) = \frac{1}{Q}\sum_{j \in [Q]} \Pr[\mathsf{D}_6^{(j)}]$.

Proof of Claim 8. Using $\mathcal{A}$ as a building block, we construct a PPTA adversary $\mathcal{B}$ that attacks the unpredictability of the detectable KEM Γ_{in} as follows:

$\mathcal{B}^{\mathcal{O}}(pk_{in})$: $\mathcal{B}$ first runs $(pk_A, sk_A) \leftarrow \mathsf{KKG}_A(1^k)$ and $(pk_B, sk_B) \leftarrow \mathsf{PKG}_B(1^k)$. Then $\mathcal{B}$ sets $PK \leftarrow (pk_{in}, pk_A, pk_B)$ and $SK \leftarrow (\bot, sk_A, sk_B)$, and runs $(m_0, m_1, \mathsf{st}) \leftarrow \mathcal{A}_1^{\mathsf{AltPDec}(SK, \cdot)}(PK)$, where $\mathcal{B}$ uses its own decapsulation oracle $\mathcal{O}$ as a substitute for $\mathsf{KDec}_{in}(sk_{in}, \cdot)$. Next, $\mathcal{B}$ runs $(c_A^*, \alpha^*) \leftarrow \mathsf{KEnc}_A(pk_A)$,

$\widetilde{c}_A^* \leftarrow \mathsf{SEnc}_A(\alpha^*, 0^t)$ (where $t = |c_{in}^*| + |\widetilde{c}_{in}^*|$ which we assume can be computed efficiently from k and $|m_0|$) and $c_B^* \leftarrow \mathsf{PEnc}_B(pk_B, 0^k)$. Then $\mathcal{B}$ sets $C^* \leftarrow (c_A^*, \widetilde{c}_A^*, c_B^*)$, and runs $\mathcal{A}_2(C^*, \mathsf{st})$. $\mathcal{B}$ answers to $\mathcal{A}_2$'s queries as specified in Game 6, again using $\mathcal{B}$'s oracle $\mathcal{O}$ as a substitute for $\mathsf{KDec}_{in}(sk_{in}, \cdot)$. When $\mathcal{A}$ terminates, $\mathcal{B}$ picks $\ell \in [Q]$ uniformly, and proceeds as follows: Let $C_\ell = (c_A^{(\ell)}, \widetilde{c}_A^{(\ell)}, c_B^{(\ell)})$ be the ℓ-th query submitted by $\mathcal{A}_2$. If $(c_A^{(\ell)}, \widetilde{c}_A^{(\ell)}) \neq (c_A^*, \widetilde{c}_A^*)$, $\mathsf{KDec}_A(sk_A, c_A^{(\ell)}) = \alpha_\ell \neq \bot$, and $\mathsf{SDec}_A(\alpha_\ell, \widetilde{c}_A^{(\ell)}) = (c_{in}^{(\ell)} \| \widetilde{c}_{in}^{(\ell)}) \neq \bot$ hold, then $\mathcal{B}$ terminates with output $c_{in}^{(\ell)}$. Otherwise, $\mathcal{B}$ gives up and aborts.

The above completes the description of $\mathcal{B}$.

It is easy to see that $\mathcal{B}$ perfectly simulates Game 6 for $\mathcal{A}$. In particular, the decryption queries from $\mathcal{A}$ are answered using $\mathsf{AltPDec}$ where $\mathsf{KDec}_{in}(sk_{in}, \cdot)$ is performed perfectly using $\mathcal{B}$'s decapsulation oracle $\mathcal{O}$.

Let c_{in}^* be the ciphertext generated by $\mathcal{B}$'s unpredictability experiment, and for each $j \in [Q]$, let $\mathsf{D}_\mathcal{B}^{(j)}$ be the event that $\mathcal{A}_2$'s j-th query is dangerous in the experiment simulated by $\mathcal{B}$, where the notion "dangerous" is with respect to c_{in}^* generated by $\mathcal{B}$'s unpredictability experiment. Recall that in Game 6, $\mathcal{A}$'s view is independent of c_{in}^*. Note also that the distribution of c_{in}^* in Game 6 and that of c_{in}^* generated in $\mathcal{B}$'s unpredictability experiment are identical. Therefore, for all $j \in [Q]$, the probability that $\mathsf{D}_\mathcal{B}^{(j)}$ occurs in $\mathcal{B}$'s unpredictability experiment is exactly the same as the probability that the event $\mathsf{D}_6^{(j)}$ occurs in Game 6. Note also that the index $\ell \in [Q]$ is chosen uniformly, independently of $\mathcal{A}$'s behavior. Hence, for all $j \in [Q]$, we have $\Pr[\mathsf{D}_\mathcal{B}^{(j)} | \ell = j] = \Pr[\mathsf{D}_6^{(j)}]$ and $\Pr[\ell = j] = 1/Q$.

Using these, $\mathcal{A}$'s unpredictability advantage can be calculated as follows:

$$\mathsf{Adv}_{\Gamma_{in}, \mathcal{B}}^{\mathsf{UNP}}(k) = \Pr[\mathsf{D}_\mathcal{B}^{(\ell)}] = \sum_{j \in [Q]} \Pr[\mathsf{D}_\mathcal{B}^{(j)} | \ell = j] \cdot \Pr[\ell = j] = \frac{1}{Q} \sum_{j \in [Q]} \Pr[\mathsf{D}_6^{(j)}].$$

This completes the proof of Claim 8.   □

According to Claims 1 to 8, there exist PPTAs $\mathcal{B}_{in}$, $\mathcal{B}_{in}'$, $\widetilde{\mathcal{B}}_{in}$, $\widehat{\mathcal{B}}_{in}$, $\mathcal{B}_A$, and $\mathcal{B}_B$, such that

$$\mathsf{Adv}_{\overline{\Pi}, \mathcal{A}}^{\mathsf{CCA}}(k) \leq 2 \cdot \mathsf{Adv}_{\Gamma_{in}, \mathcal{B}_{in}}^{\mathsf{DCCA}}(k) + \mathsf{Adv}_{\Gamma_{in}, \mathcal{B}_{in}'}^{\mathsf{DCCA}}(k) + \mathsf{Adv}_{E_{in}, \widetilde{\mathcal{B}}_{in}}^{\mathsf{OT}}(k)$$
$$+ \mathsf{Adv}_{\Pi_B, \mathcal{B}_B}^{\mathsf{CPA}}(k) + Q \cdot \left(\mathsf{Adv}_{\Pi[\Gamma_A, E_A], \mathcal{B}_A}^{\mathsf{1\text{-}CCA}}(k) + \mathsf{Adv}_{\Gamma_{in}, \widehat{\mathcal{B}}_{in}}^{\mathsf{UNP}}(k) \right),$$

where the first and the second terms are from Claim 3 with $(p, q) = (1, 1/2)$ and $(p, q) = (0, 1)$, respectively. By the assumptions on the building blocks and Lemma 1, we conclude that $\mathsf{Adv}_{\overline{\Pi}, \mathcal{A}}^{\mathsf{CCA}}(k)$ is negligible. Recall that the choice of $\mathcal{A}$ was arbitrarily, and thus for any PPTA $\mathcal{A}$, we can show a negligible upperbound on $\mathsf{Adv}_{\overline{\Pi}, \mathcal{A}}^{\mathsf{CCA}}(k)$. This completes the proof of Theorem 1.   □

Direct Construction of a CCA *Secure KEM.* If one's purpose is to construct a CCA secure KEM (rather than a PKE scheme), then we can omit all operations regarding the building block SKE scheme E_{in} and directly use K as a session-key. (That is, in the encapsulation algorithm, SEnc_A only encrypts c_{in}, and the

Table 1. Comparison of the ciphertext overhead of the proposed scheme with the HLW construction. (Notations in the right column are explained in Section 5.)

Scheme	Ciphertext Size
HLW [7] (see also Fig. 2)	$OH_{1\text{-}CCA} + OH_{CPA} + 2OH_{DCCA} + 4k + 2|m|$
HLW (used as a KEM) & DEM	$OH_{1\text{-}CCA} + OH_{CPA} + 2OH_{DCCA} + 6k + |m|$
Proposed Scheme (§ 4)	$OH_{1\text{-}CCA} + OH_{CPA} + OH_{DCCA} + k + |m|$

decapsulation algorithm outputs K rather than $SDec_{in}(K, \tilde{c}_{in})$.) The security proof is essentially the same as that of our proposed PKE scheme (only slightly simpler).

5 Comparison

Table 1 shows a comparison of the ciphertext size between our proposed PKE scheme and the original HLW construction [7]. For a fair comparison, we consider two versions for the HLW construction: the original scheme (as described in Fig. 2), and the version in which the original scheme is used as a KEM by encrypting a k-bit randomness K and combined with a CCA secure SKE scheme (DEM). We assume that all building block PKE schemes (both in our scheme and in the HLW construction) are implemented by hybrid encryption so that the ciphertext overhead from these components become small. Since we can implement a CCA secure SKE scheme with zero ciphertext overhead [16], we assume that a ciphertext overhead of a PKE scheme equals the size of its KEM ciphertext. In the right column of the table, let OH_X (which stands for "overhead") represent the size of the "KEM-part" of the X secure PKE scheme or that of the X secure KEM used in the constructions. For example, if we encrypt a message m and obtain c with a CPA secure PKE scheme, then $|c| = OH_{CPA} + |m|$.

As is obvious from Table 1, our proposed scheme is more space-efficient. Due to its design of doubly-encrypting c_{in} by the "outer" schemes Π_A and Π_B, the ciphertext size of the HLW constructions counts OH_{DCCA} twice (in the original construction case it further counts $|m|$ and two k-bit strings twice). On the other hand, in our proposed scheme OH_{DCCA} is counted only once, since c_{in} is encrypted only once by the "outer" SKE scheme E_A.

Acknowledgement. The authors would like to thank the anonymous reviewers and the members of Shin-Akarui-Angou-Benkyou-Kai for their constructive comments and suggestions.

References

1. Bleichenbacher, D.: Chosen ciphertext attacks against protocols based on the RSA encryption standard PKCS #1. In: Krawczyk, H. (ed.) CRYPTO 1998. LNCS, vol. 1462, pp. 1–12. Springer, Heidelberg (1998)

2. Canetti, R.: Universally composable security: A new paradigm for cryptographic protocols. In: FOCS 2001, pp. 136–145 (2001)
3. Canetti, R., Halevi, S., Katz, J.: Chosen-ciphertext security from identity-based encryption. In: Cachin, C., Camenisch, J.L. (eds.) EUROCRYPT 2004. LNCS, vol. 3027, pp. 207–222. Springer, Heidelberg (2004)
4. Cramer, R., Hanaoka, G., Hofheinz, D., Imai, H., Kiltz, E., Pass, R., Shelat, A., Vaikuntanathan, V.: Bounded CCA2-secure encryption. In: Kurosawa, K. (ed.) ASIACRYPT 2007. LNCS, vol. 4833, pp. 502–518. Springer, Heidelberg (2007)
5. Cramer, R., Shoup, V.: Design and analysis of practical public-key encryption schemes secure against adaptive chosen ciphertext attack. SIAM J. Computing 33(1), 167–226 (2003)
6. Dolev, D., Dwork, C., Naor, M.: Non-malleable cryptography. In: STOC 1991, pp. 542–552 (1991)
7. Hohenberger, S., Lewko, A., Waters, B.: Detecting dangerous queries: A new approach for chosen ciphertext security. In: Pointcheval, D., Johansson, T. (eds.) EUROCRYPT 2012. LNCS, vol. 7237, pp. 663–681. Springer, Heidelberg (2012)
8. Kiltz, E.: Chosen-ciphertext security from tag-based encryption. In: Halevi, S., Rabin, T. (eds.) TCC 2006. LNCS, vol. 3876, pp. 581–600. Springer, Heidelberg (2006)
9. Kiltz, E., Mohassel, P., O'Neill, A.: Adaptive trapdoor functions and chosen-ciphertext security. In: Gilbert, H. (ed.) EUROCRYPT 2010. LNCS, vol. 6110, pp. 673–692. Springer, Heidelberg (2010)
10. Lin, H., Tessaro, S.: Amplification of chosen-ciphertext security. In: Johansson, T., Nguyen, P.Q. (eds.) EUROCRYPT 2013. LNCS, vol. 7881, pp. 503–519. Springer, Heidelberg (2013)
11. Lindell, Y.: A simpler construction of CCA2-secure public-key encryption under general assumptions. In: Biham, E. (ed.) EUROCRYPT 2003. LNCS, vol. 2656, pp. 241–254. Springer, Heidelberg (2003)
12. Mol, P., Yilek, S.: Chosen-ciphertext security from slightly lossy trapdoor functions. In: Nguyen, P.Q., Pointcheval, D. (eds.) PKC 2010. LNCS, vol. 6056, pp. 296–311. Springer, Heidelberg (2010)
13. Myers, S., Shelat, A.: Bit encryption is complete. In: FOCS 2009, pp. 607–616 (2009)
14. Naor, M., Yung, M.: Public-key cryptosystems provably secure against chosen ciphertext attacks. In: STOC 1990, pp. 427–437 (1990)
15. Peikert, C., Waters, B.: Lossy trapdoor functions and their applications. In: STOC 2008, pp. 187–196 (2008)
16. Phan, D.H., Pointcheval, D.: About the security of ciphers (Semantic security and pseudo-random permutations). In: Handschuh, H., Hasan, M.A. (eds.) SAC 2004. LNCS, vol. 3357, pp. 182–197. Springer, Heidelberg (2004)
17. Rackoff, C., Simon, D.R.: Non-interactive zero-knowledge proof of knowledge and chosen ciphertext attack. In: Feigenbaum, J. (ed.) CRYPTO 1991. LNCS, vol. 576, pp. 433–444. Springer, Heidelberg (1992)
18. Rosen, A., Segev, G.: Chosen-ciphertext security via correlated products. In: Reingold, O. (ed.) TCC 2009. LNCS, vol. 5444, pp. 419–436. Springer, Heidelberg (2009)
19. Sahai, A.: Non-malleable non-interactive zero knowledge and adaptive chosen-ciphertext security. In: FOCS 1999, pp. 543–553 (1999)
20. Shamir, A.: Identity-based cryptosystems and signature schemes. In: Blakely, G.R., Chaum, D. (eds.) CRYPTO 1984. LNCS, vol. 196, pp. 47–53. Springer, Heidelberg (1985)

21. Wee, H.: Efficient chosen-ciphertext security via extractable hash proofs. In: Rabin, T. (ed.) CRYPTO 2010. LNCS, vol. 6223, pp. 314–332. Springer, Heidelberg (2010)

A   Proof of Claim 1

By definition of the games and events, $\mathcal{A}$'s CCA advantage is calculated as follows:

$$\mathsf{Adv}^{\mathsf{CCA}}_{\Pi,\mathcal{A}}(k) = 2 \cdot |\Pr[\mathsf{S}_1] - \frac{1}{2}| \leq 2 \cdot |\Pr[\mathsf{S}_1] - \Pr[\mathsf{S}_2]| + 2 \cdot |\Pr[\mathsf{S}_2] - \frac{1}{2}|$$

The second term in the right hand side can be further calculated as follows:

$$|\Pr[\mathsf{S}_2] - \frac{1}{2}| = |\Pr[\mathsf{S}_2 \wedge \overline{\mathsf{D}_2}] + \Pr[\mathsf{S}_2|\mathsf{D}_2] \cdot \Pr[\mathsf{D}_2] - \frac{1}{2}|$$

$$\leq |\Pr[\mathsf{S}_2 \wedge \overline{\mathsf{D}_2}] + \frac{1}{2}\Pr[\mathsf{D}_2] - \frac{1}{2}| + |\Pr[\mathsf{S}_2|\mathsf{D}_2] - \frac{1}{2}| \cdot \Pr[\mathsf{D}_2]$$

$$\overset{(*)}{\leq} |\Pr[\mathsf{S}_2 \wedge \overline{\mathsf{D}_2}] + \frac{1}{2}\Pr[\mathsf{D}_2] - \frac{1}{2}| + \frac{1}{2}\Pr[\mathsf{D}_2]$$

$$\leq |\Pr[\mathsf{S}_2 \wedge \overline{\mathsf{D}_2}] - \Pr[\mathsf{S}_3 \wedge \overline{\mathsf{D}_3}] + \frac{1}{2}(\Pr[\mathsf{D}_2] - \Pr[\mathsf{D}_3])|$$

$$+ |\Pr[\mathsf{S}_3 \wedge \overline{\mathsf{D}_3}] + \frac{1}{2}\Pr[\mathsf{D}_3] - \frac{1}{2}| + \frac{1}{2}\Pr[\mathsf{D}_2]$$

$$\leq |\Pr[\mathsf{S}_2 \wedge \overline{\mathsf{D}_2}] - \Pr[\mathsf{S}_3 \wedge \overline{\mathsf{D}_3}] + \frac{1}{2}(\Pr[\mathsf{D}_2] - \Pr[\mathsf{D}_3])|$$

$$+ |\Pr[\mathsf{S}_3 \wedge \overline{\mathsf{D}_3}] + \frac{1}{2}\Pr[\mathsf{D}_3] - \frac{1}{2}| + \frac{1}{2}\Big(\sum_{i\in\{2,3,4\}} |\Pr[\mathsf{D}_i] - \Pr[\mathsf{D}_{i+1}]| + \Pr[\mathsf{D}_5]\Big)$$

where in the inequality (*) we used $|\Pr[\mathsf{S}_2|\mathsf{D}_2] - 1/2| \leq 1/2$, and other inequalities are due to the triangle inequality.

Finally, we estimate the upperbound of $\Pr[\mathsf{D}_5]$ by

$$\Pr[\mathsf{D}_5] = \Pr[\bigvee_{j\in[Q]} \mathsf{D}_5^{(j)}] \leq \sum_{j\in[Q]} \Pr[\mathsf{D}_5^{(j)}]$$

$$\leq |\sum_{j\in[Q]} (\Pr[\mathsf{D}_5^{(j)}] - \Pr[\mathsf{D}_6^{(j)}])| + \sum_{j\in[Q]} \Pr[\mathsf{D}_6^{(j)}]$$

Combining all the inequalities yields the claim.   □

Cryptanalysis of the Quaternion Rainbow

Yasufumi Hashimoto

Department of Mathematical Sciences, University of the Ryukyus

Abstract. Rainbow is one of the signature schemes based on multivariate problems. While its signature generation and verification are fast and the security is presently sufficient under suitable parameter selections, the key size is relatively large. Recently, Quaternion Rainbow – Rainbow over quaternion ring – was proposed by Yasuda, Sakurai and Takagi (CT-RSA'12) to reduce the key size of Rainbow without impairing the security. However, a new vulnerability emerges from the structure of quaternion ring; in fact, Thomae (SCN'12) found that Quaternion Rainbow is less secure than the same-size original Rainbow. In the present paper, we further study the security of Quaternion Rainbow and get better security results than Thomae's ones. Especially, we find that Quaternion Rainbow over even characteristic field, whose security level is estimated as about the original Rainbow of at most 3/4 by Thomae's analysis, is almost as secure as the original Rainbow of at most 1/4-size.

Keywords: post-quantum cryptography, multivariate public-key cryptosystems, Rainbow, quaternion ring.

1 Introduction

The multivariate public key cryptosystem (MPKC) is a family of cryptosystems based on the problem of solving a set of multivariate quadratic equations, and is expected to be a post-quantum cryptology. Rainbow [4] is one of the signature schemes consisting in MPKC. This is known as a nice scheme in the sense that the signature generations and verifications are faster than RSA and ECC [2] and the security is presently sufficient under suitable parameter selections. However, the key size of Rainbow is relatively large and then reducing it is required for implementations in practice.

TTS [12] and Cyclic Rainbow [9] are famous variations of Rainbow whose key sizes are smaller than those in the original Rainbow; the secret keys are smaller in the former scheme and the public keys are smaller in the latter scheme.

Recently, a new Rainbow variant was proposed by Yasuda, Sakurai and Takagi [13]. Their idea is to construct Rainbow on the quaternion ring; thus we call it Quaternion Rainbow. They claimed that the size of secret keys is about 75% of the same-size original Rainbow under the same security level. However, a new vulnerability emerges from the structure of the quaternion ring; in fact, Thomae [11] showed that the security of Quaternion Rainbow against rank attacks [12] is less than that expected by the authors of [13].

K. Sakiyama and M. Terada (Eds.): IWSEC 2013, LNCS 8231, pp. 244–257, 2013.
© Springer-Verlag Berlin Heidelberg 2013

In the present paper, we further study vulnerabilities of Quaternion Rainbow emerging from the structure of the quaternion ring. It is well known that there are nontrivial zero divisors in the quaternion ring over a finite field. Taking such zero divisors with several conditions as a basis of the quaternion ring, we find a sparseness of the quadratic forms in Quaternion Rainbow. Its sparseness causes a vulnerability of Quaternion Rainbow. Especially, when the field is of even characteristic, the quadratic forms in Quaternion Rainbow are described by balanced Oil and Vinegar type quadratic forms [8,7,3]. Thus the problem of recovering the secret keys of Quaternion Rainbow can be reduced to that of recovering them of the original Rainbow of at most 1/4-size by the Kipnis-Shamir attack [8,7,3] in polynomial time. This means that the security level of the Quaternion Rainbow over even characteristic field is almost 1/4 of that expected in [13] and about 1/3 of that estimated by Thomae [11].

2 Rainbow

The Rainbow [4] is a signature scheme consisting in MPKC. Throughout this paper, we study the double-layer version of Rainbow for simplicity.

2.1 Scheme

Let q be a power of prime and k a finite field of order q. For integers $o_1, o_2, v \geq 1$, $m := o_1 + o_2$ and $n := m + v$, the quadratic map $G : k^n \to k^m$ for Rainbow is given as follows.

$$G(x) = (g_1(x), \cdots, g_m(x)), \qquad x = (x_1, \cdots, x_n)^t \in k^n,$$

where

$$g_l(x) := \sum_{(i,j) \in L_{l,1}} \alpha_{i,j}^{(l)} x_i x_j + \sum_{i \in L_{l,2}} \beta_i^{(l)} x_i + \gamma^{(l)}$$

with $\alpha_{i,j}^{(l)}, \beta_i^{(l)}, \gamma^{(l)} \in k$ and

$$L_{l,1} := \begin{cases} \{o_1 + 1 \leq i, j \leq n\} \backslash \{o_1 + 1 \leq i, j \leq m\}, & (1 \leq l \leq o_2), \\ \{1 \leq i, j \leq n\} \backslash \{1 \leq i, j \leq o_1\}, & (o_2 + 1 \leq l \leq m), \end{cases}$$

$$L_{l,2} := \begin{cases} \{o_1 + 1 \leq i \leq n\}, & (1 \leq l \leq o_2), \\ \{1 \leq i \leq n\}, & (o_2 + 1 \leq l \leq m). \end{cases}$$

Note that g_l's are described by

$$g_l(x) = \begin{cases} x^t \begin{pmatrix} 0_{o_1} & 0 & 0 \\ 0 & 0_{o_2} & * \\ 0 & * & *_v \end{pmatrix} x + (\text{linear form of } x_{o_1+1}, \ldots, x_n), & (1 \leq l \leq o_2), \\[4mm] x^t \begin{pmatrix} 0_{o_1} & * \\ * & *_{o_2+v} \end{pmatrix} x + (\text{linear form of } x_1, \ldots, x_n), & (o_2 + 1 \leq l \leq m). \end{cases}$$

The (o_1, o_2, v)-Rainbow is constructed as follows.

Secret Key: The secret keys consists of two invertible affine maps $S : k^n \to k^n$, $T : k^m \to k^m$ and the quadratic map $G : k^n \to k^m$ given above.

Public Key: The public key is the convolution $F := T \circ G \circ S : k^n \to k^m$ of three maps S, G, T, namely

$$F : k^n \xrightarrow{S} k^n \xrightarrow{G} k^m \xrightarrow{T} k^m.$$

Signature Generation: For a message $y \in k^m$, the signature is given as follows.
Step 1. Compute $z := T^{-1}(y) = (z_1, \cdots, z_m)^t \in k^m$.
Step 2. Choose $r_1, \ldots, r_v \in k$ randomly.
Step 3. Find $x_{o_1+1}, \ldots, x_m \in k$ such that

$$g_1(x_1, \cdots, x_m, r_1, \cdots, r_v) = z_1,$$

$$\vdots$$

$$g_{o_2}(x_1, \cdots, x_m, r_1, \cdots, r_v) = z_{o_2}.$$

By the definition of G, the equations above are linear equations of $x_{o_1+1}, \ldots, x_m$. Therefore $x_{o_1+1}, \ldots, x_m$ are found by the Gaussian elimination and are independent on the choice of $x_1, \cdots, x_{o_1}$.
Step 4. For $x_{o_1+1}, \cdots, x_m$ given in Step 3, find $x_1, \cdots, x_{o_1} \in k$ such that

$$g_{o_2+1}(x_1, \cdots, x_m, r_1, \cdots, r_v) = z_{o_2+1},$$

$$\vdots$$

$$g_m(x_1, \cdots, x_m, r_1, \cdots, r_v) = z_m.$$

Similar to Step 3, they can be found by the Gaussian eliminations.
Step 5. The signature for $y \in k^m$ is $w := S^{-1}((x_1, \cdots, x_m, r_1, \cdots, r_v)^t) \in k^n$.
Signature Verification: Check whether $F(w) = y$.

2.2   Major Attacks

The following attacks are applicable to Rainbow.

1. Kipnis-Shamir's Attack on UOV. It was proposed by Kipnis and Shamir [8,7,3] against the (unbalanced) Oil and Vinegar signature scheme. If the coefficient matrices of g_l's are in the form $\begin{pmatrix} 0_N & * \\ * & *_M \end{pmatrix}$, this attack recovers S partially with the complexity $O(q^{\max(0, M-N)} \cdot (\text{polyn.}))$. On the (o_1, o_2, v)-Rainbow, the complexity of Kipnis-Shamir's attack is $O(q^{v+o_2-o_1} \cdot (\text{polyn.}))$ [8,7,3].

2. The High-Rank Attack. If there are gaps among the ranks of the coefficient matrices of g_l's, the high-rank attack recovers T partially. On the (o_1, o_2, v)-Rainbow, the complexity of the high-rank attack is $O(q^{o_1} \cdot (\text{polyn.}))$ [12,10].

3. The Min-Rank Attack. If there are g_l's whose coefficient matrices are of small ranks, the min-rank attack recovers T partially. On the (o_1, o_2, v)-Rainbow, the complexity of the min-rank attack is $O(q^{v+o_2} \cdot (\text{polyn.}))$ [12,10].

Other than the attacks above, the security against the Gröbner basis attacks [6,1], the UOV-Reconciliation attacks and the Rainbow Band Separation attacks [5] have been studied. See [10] for experiments of these attacks on Rainbow with smaller n and m.

3   Quaternion Rainbow

In this section, we survey Quaternion Rainbow [13].

3.1   Scheme

Let q be a power of prime and k a finite field of order q. The quaternion ring $Q(k)$ over k is defined by

$$Q(k) := k + ki + kj + kij$$
$$= \{a_1 + a_2 i + a_3 j + a_4 ij \mid a_1, a_2, a_3, a_4 \in k\},$$

where i, j satisfy $i^2 = j^2 = -1$ and $ij = -ji$. Notice that $Q(k)$ is non-commutative when q is odd.

For integers $\tilde{o}_1, \tilde{o}_2, \tilde{v} \geq 1$, put $\tilde{m} := \tilde{o}_1 + \tilde{o}_2$ and $\tilde{n} := \tilde{m} + \tilde{v}$. The secret quadratic map $\tilde{G} : Q(k)^{\tilde{n}} \to Q(k)^{\tilde{m}}$ of Quaternion Rainbow is defined as follows [13].

$$\tilde{g}_l(x) := \sum_{(i,j) \in \tilde{L}_{l,1}} \tilde{x}_i \tilde{\alpha}_{i,j}^{(l)} \tilde{x}_j + \sum_{i \in \tilde{L}_{l,2}} (\tilde{\beta}_{i,1}^{(l)} \tilde{x}_i + \tilde{x}_i \tilde{\beta}_{i,2}^{(l)}) + \tilde{\gamma}^{(l)}$$

where $\tilde{\alpha}_{i,j}^{(l)}, \tilde{\beta}_i^{(l,1)}, \tilde{\beta}_{i,2}^{(l)}, \tilde{\gamma}^{(l)} \in k$ and

$$\tilde{L}_{l,1} := \begin{cases} \{\tilde{o}_1 + 1 \leq i, j \leq \tilde{n}\} \backslash \{\tilde{o}_1 + 1 \leq i, j \leq \tilde{m}\}, & (1 \leq l \leq \tilde{o}_2), \\ \{1 \leq i, j \leq \tilde{n}\} \backslash \{1 \leq i, j \leq \tilde{o}_1\}, & (\tilde{o}_2 + 1 \leq l \leq \tilde{m}), \end{cases}$$

$$\tilde{L}_{l,2} := \begin{cases} \{\tilde{o}_1 + 1 \leq i \leq \tilde{n}\}, & (1 \leq l \leq \tilde{o}_2), \\ \{1 \leq i \leq \tilde{n}\}, & (\tilde{o}_2 + 1 \leq l \leq \tilde{m}). \end{cases}$$

Denoting $\tilde{x} = x_1 + x_2 i + x_3 j + x_4 ij$ and $\tilde{x}' = x_1^t + x_2^t i + x_3^t j + x_4^t ij$ with $x_1, x_2, x_3, x_4 \in k^{\tilde{n}}$, we can rewrite $\tilde{g}_l$ in the form

$$\tilde{g}_l(\tilde{x}) = \tilde{x}' \tilde{G}_l \tilde{x} + \text{(linear)}, \tag{1}$$

where $\tilde{G}_l$ is an $\tilde{n} \times \tilde{n}$ matrix with $Q(k)$-entries given as follows.

$$\tilde{G}_l = \begin{cases} \begin{pmatrix} 0_{\tilde{o}_1} & 0 & 0 \\ 0 & 0_{\tilde{o}_2} & * \\ 0 & * & *_{\tilde{v}} \end{pmatrix}, & (1 \leq l \leq \tilde{o}_2) \\ \begin{pmatrix} 0_{\tilde{o}_1} & * \\ * & *_{\tilde{o}_2 + \tilde{v}} \end{pmatrix}, & (\tilde{o}_2 + 1 \leq l \leq \tilde{m}). \end{cases} \tag{2}$$

Let $\psi : k^{4\tilde{n}} \to Q(k)^{\tilde{n}}$ and $\varphi : Q(k)^{\tilde{m}} \to k^{4\tilde{m}}$ be one-to-one maps given by

$$\psi(a_1, a_2, a_3, a_4) = a_1 + a_2 i + a_3 j + a_4 ij,$$
$$\varphi(b_1 + b_2 i + b_3 j + b_4 ij) = (b_1, b_2, b_3, b_4)$$

for $a_1, a_2, a_3, a_4 \in k^{\tilde{n}}$ and $b_1, b_2, b_3, b_4 \in k^{\tilde{m}}$, and put $G := \varphi \circ \tilde{G} \circ \psi$.

$$G : k^{4\tilde{n}} \xrightarrow{\psi} Q(k)^{\tilde{n}} \xrightarrow{\tilde{G}} Q(k)^{\tilde{m}} \xrightarrow{\varphi} k^{4\tilde{m}}.$$

We now study the quadratic map G. Let $\tilde{G}_{l,1}, \tilde{G}_{l,2}, \tilde{G}_{l,3}, \tilde{G}_{l,4}$ $(1 \le l \le \tilde{m})$ be the $\tilde{n} \times \tilde{n}$ matrices with k entries such that

$$\tilde{G}_l = \tilde{G}_{l,1} + \tilde{G}_{l,2} i + \tilde{G}_{l,3} j + \tilde{G}_{l,4} ij$$

and $x := \begin{pmatrix} x_1 \\ x_2 \\ x_3 \\ x_4 \end{pmatrix}$ for $\tilde{x} = x_1 + x_2 i + x_3 j + x_4 ij$. Then we have

$$\tilde{x}' \tilde{G}_l x = (x_1^t + x_2^t i + x_3^t j + x_4^t ij)(\tilde{G}_{l,1} + \tilde{G}_{l,2} i + \tilde{G}_{l,3} j + \tilde{G}_{l,4} ij)$$
$$\cdot (x_1 + x_2 i + x_3 j + x_4 ij)$$

$$= x^t \begin{pmatrix} \tilde{G}_{l,1} & -\tilde{G}_{l,2} & -\tilde{G}_{l,3} & -\tilde{G}_{l,4} \\ -\tilde{G}_{l,2} & -\tilde{G}_{l,1} & -\tilde{G}_{l,4} & \tilde{G}_{l,3} \\ -\tilde{G}_{l,3} & \tilde{G}_{l,4} & -\tilde{G}_{l,1} & -\tilde{G}_{l,2} \\ -\tilde{G}_{l,4} & -\tilde{G}_{l,3} & \tilde{G}_{l,2} & -\tilde{G}_{l,1} \end{pmatrix} x \cdot 1 + x^t \begin{pmatrix} \tilde{G}_{l,2} & \tilde{G}_{l,1} & \tilde{G}_{l,4} & -\tilde{G}_{l,3} \\ \tilde{G}_{l,1} & -\tilde{G}_{l,2} & -\tilde{G}_{l,3} & -\tilde{G}_{l,4} \\ \tilde{G}_{l,4} & -\tilde{G}_{l,3} & \tilde{G}_{l,2} & -\tilde{G}_{l,1} \\ \tilde{G}_{l,3} & \tilde{G}_{l,4} & \tilde{G}_{l,1} & -\tilde{G}_{l,2} \end{pmatrix} x \cdot i$$

$$+ x^t \begin{pmatrix} \tilde{G}_{l,3} & -\tilde{G}_{l,4} & \tilde{G}_{l,1} & \tilde{G}_{l,2} \\ \tilde{G}_{l,4} & \tilde{G}_{l,3} & -\tilde{G}_{l,2} & \tilde{G}_{l,1} \\ \tilde{G}_{l,1} & -\tilde{G}_{l,2} & -\tilde{G}_{l,3} & -\tilde{G}_{l,4} \\ -\tilde{G}_{l,2} & -\tilde{G}_{l,1} & -\tilde{G}_{l,4} & -\tilde{G}_{l,3} \end{pmatrix} x \cdot j + x^t \begin{pmatrix} \tilde{G}_{l,4} & \tilde{G}_{l,3} & -\tilde{G}_{l,2} & \tilde{G}_{l,1} \\ -\tilde{G}_{l,3} & \tilde{G}_{l,4} & -\tilde{G}_{l,1} & -\tilde{G}_{l,2} \\ \tilde{G}_{l,2} & \tilde{G}_{l,1} & -\tilde{G}_{l,4} & -\tilde{G}_{l,3} \\ \tilde{G}_{l,1} & -\tilde{G}_{l,2} & -\tilde{G}_{l,3} & -\tilde{G}_{l,4} \end{pmatrix} x \cdot ij.$$

Since $\tilde{G}_{l,1}, \tilde{G}_{l,2}, \tilde{G}_{l,3}, \tilde{G}_{l,4}$ are in the forms (2), it is easy to see that the quadratic map $G : k^{4\tilde{n}} \to k^{4\tilde{m}}$ in the $(\tilde{o}_1, \tilde{o}_2, \tilde{v})$-Quaternion Rainbow is written as G in the $(4\tilde{o}_1, 4\tilde{o}_2, 4\tilde{v})$-Rainbow. Thus, setting two invertible affine maps $S : k^{4\tilde{n}} \to k^{4\tilde{n}}$, $T : k^{4\tilde{m}} \to k^{4\tilde{m}}$ as the secret keys and $F := T \circ G \circ S : k^{4\tilde{n}} \to k^{4\tilde{m}}$ as the public key in the $(\tilde{o}_1, \tilde{o}_2 4\tilde{v})$-Quaternion Rainbow, we can interpret $(\tilde{o}_1, \tilde{o}_2, \tilde{v})$-Quaternion Rainbow as a special version of the $(4\tilde{o}_1, 4\tilde{o}_2, 4\tilde{v})$-Rainbow.

3.2   Previous Security Analysis

The authors in [13] claimed that the security levels of the $(\tilde{o}_1, \tilde{o}_2, \tilde{v})$-Quaternion Rainbow and the $(4\tilde{o}_1, 4\tilde{o}_2, 4\tilde{v})$-Rainbow are same. However, Thomae [11] found that, taking linear sums of the coefficient matrices in G, one gets matrices of smaller ranks, and then the security of Quaternion Rainbow against the rank attacks is weaker than the same size original Rainbow. Table 1 summarizes the security levels estimated in [13] and [11].

Table 1. Previous security analysis on Quaternion Rainbow

	Kipnis-Shamir	Min-Rank	High-Rank
Original Rainbow	$q^{4\tilde{v}+4\tilde{o}_2-4\tilde{o}_1}$	$q^{4\tilde{v}+4\tilde{o}_2}$	$q^{4\tilde{o}_1}$
Yasuda-Sakurai-Takagi [13]	$q^{4\tilde{v}+4\tilde{o}_2-4\tilde{o}_1}$	$q^{4\tilde{v}+4\tilde{o}_2}$	$q^{4\tilde{o}_1}$
Thomae [11] $(2 \nmid q)$	$q^{4\tilde{v}+4\tilde{o}_2-4\tilde{o}_1}$	$q^{4\tilde{v}+3\tilde{o}_2}$	$q^{3\tilde{o}_1}$
Thomae [11] $(2 \mid q)$	$q^{4\tilde{v}+4\tilde{o}_2-4\tilde{o}_1}$	$q^{4\tilde{v}+\tilde{o}_2}$	$q^{3\tilde{o}_1}$

4   Proposed Security Analysis

In this section, we analyze the security of Quaternion Rainbow.

4.1   The Case of Odd Characteristic

Study the case that q is odd. Suppose that $a, b \in k$ satisfy

$$a^2 + b^2 = -4^{-1} \tag{3}$$

and put

$$\alpha := 2^{-1} + ai + bij, \qquad \bar{\alpha} := 2^{-1} - ai - bij. \tag{4}$$

There always exist $a, b \in k$ with (3) due to Lemma 2 in [11], . The following properties of $\alpha, \bar{\alpha} \in Q(k)$ are given by the equation (3):

$$\alpha^2 = \alpha, \quad \bar{\alpha}^2 = \bar{\alpha}, \quad \alpha\bar{\alpha} = \bar{\alpha}\alpha = 0, \quad \alpha j = j\bar{\alpha}, \quad \bar{\alpha}j = j\alpha. \tag{5}$$

We now state the following lemma.

Lemma 1. *Let k be a finite field of odd characteristic. Then, for any $(a_1, a_2, a_3, a_4) \in k^4$, there exists a unique $(b_1, b_2, b_3, b_4) \in k^4$ such that*

$$a_1 + a_2i + a_3j + a_4ij = b_1\alpha + b_2\bar{\alpha} + b_3\alpha j + b_4\bar{\alpha}j, \tag{6}$$

namely $\{\alpha, \bar{\alpha}, \alpha j, \bar{\alpha}j\}$ is a basis of $Q(k)$ over k.

Proof. Equation (6) holds if

$$\begin{pmatrix} a_1 \\ a_2 \\ a_3 \\ a_4 \end{pmatrix} = \begin{pmatrix} 2^{-1} & 2^{-1} & 0 & 0 \\ a & -a & -b & b \\ 0 & 0 & 2^{-1} & 2^{-1} \\ b & -b & a & -a \end{pmatrix} \begin{pmatrix} b_1 \\ b_2 \\ b_3 \\ b_4 \end{pmatrix}. \tag{7}$$

Due to (3), we see that the square matrix in the right hand side of the equation above is invertible. Thus the claim in this lemma holds. $\qquad\square$

According to the properties (5) of $\alpha, \bar{\alpha}$, we can find the following multiplicative properties among the elements in the basis:

$$
\begin{array}{llll}
\alpha \cdot \alpha = \alpha, & \alpha \cdot \bar{\alpha} = 0, & \alpha \cdot \alpha j = \alpha j, & \alpha \cdot \bar{\alpha} j = 0, \\
\bar{\alpha} \cdot \alpha = 0, & \bar{\alpha} \cdot \bar{\alpha} = \bar{\alpha}, & \bar{\alpha} \cdot \alpha j = 0, & \bar{\alpha} \cdot \bar{\alpha} j = \bar{\alpha} j, \\
\alpha j \cdot \alpha = 0, & \alpha j \cdot \bar{\alpha} = \alpha j, & \alpha j \cdot \alpha j = 0, & \alpha j \cdot \bar{\alpha} j = -\alpha, \\
\bar{\alpha} j \cdot \alpha = \bar{\alpha} j, & \bar{\alpha} j \cdot \bar{\alpha} = 0, & \bar{\alpha} j \cdot \alpha j = -\bar{\alpha}, & \bar{\alpha} j \cdot \bar{\alpha} j = 0.
\end{array}
$$

By using the basis $\{\alpha, \bar{\alpha}, \alpha j, \bar{\alpha} j\}$, we can rewrite the quadratic map $\tilde{G}$ in Quaternion Rainbow as follows:

$$
\tilde{x} = y_1 \alpha + y_2 \bar{\alpha} + y_3 \alpha j + y_4 \bar{\alpha} j, \quad \tilde{x}' = y_1^t \alpha + y_2^t \bar{\alpha} + y_3^t \alpha j + y_4^t \bar{\alpha} j,
$$
$$
\tilde{g}_l(\tilde{x}) = \tilde{x}' \tilde{G}_l \tilde{x} + (\text{linear form}),
$$

where y_1, y_2, y_3, y_4 are unknowns in $k^{\tilde{n}}$ and $\tilde{G}_l$ is an $\tilde{n} \times \tilde{n}$-matrix with $Q(k)$-entries given in (2). Note that, since the entries in $\tilde{G}_l$ are linear sums of $\{\alpha, \bar{\alpha}, \alpha j, \bar{\alpha} j\}$ over k, $\tilde{G}_l$ is given by

$$
\tilde{G}_l = H_{l,1} \alpha + H_{l,2} \bar{\alpha} + H_{l,3} \alpha j + H_{l,4} \bar{\alpha} j,
$$

where $H_{l,1}, H_{l,2}, H_{l,3}, H_{l,4}$ are $\tilde{n} \times \tilde{n}$-matrices with k-entries in the forms

$$
H_{l,1}, H_{l,2}, H_{l,3}, H_{l,4} = \begin{cases} \begin{pmatrix} 0_{\tilde{o}_1} & 0 & 0 \\ 0 & 0_{\tilde{o}_2} & * \\ 0 & * & *_{\tilde{v}} \end{pmatrix}, & (1 \le l \le \tilde{o}_2) \\ \begin{pmatrix} 0_{\tilde{o}_1} & * \\ * & *_{\tilde{o}_2 + \tilde{v}} \end{pmatrix}, & (\tilde{o}_2 + 1 \le l \le \tilde{m}). \end{cases} \tag{8}
$$

Thus $\tilde{g}_l(\tilde{x})$ is written by

$$
\begin{aligned}
\tilde{x}' \tilde{G}_l \tilde{x} =& (y_1^t \alpha + y_2^t \bar{\alpha} + y_3^t \alpha j + y_4^t \bar{\alpha} j)(H_{l,1} \alpha + H_{l,2} \bar{\alpha} + H_{l,3} \alpha j + H_{l,4} \bar{\alpha} j) \\
& \cdot (y_1 \alpha + y_2 \bar{\alpha} + y_3 \alpha j + y_4 \bar{\alpha} j) \\
=& (y_1^t H_{l,1} y_1 - y_4^t H_{l,2} y_3 - y_1^t H_{l,3} y_4 - y_3^t H_{l,4} y_1) \alpha \\
& + (-y_4^t H_{l,1} y_3 + y_2^t H_{l,2} y_2 - y_2^t H_{l,3} y_4 - y_3^t H_{l,4} y_2) \bar{\alpha} \\
& + (y_2^t H_{l,1} y_3 + y_3^t H_{l,2} y_1 + y_1^t H_{l,3} y_2 - y_3^t H_{l,4} y_3) \alpha j \\
& + (y_1^t H_{l,1} y_4 + y_4^t H_{l,2} y_2 - y_4^t H_{l,3} y_4 + y_2^t H_{l,4} y_1) \bar{\alpha} j.
\end{aligned} \tag{9}
$$

Putting $y := \begin{pmatrix} y_1 \\ y_2 \\ y_3 \\ y_4 \end{pmatrix}$, we have

$$
\tilde{x}'\tilde{G}_l\tilde{x} = y^t \begin{pmatrix} H_{l,1} & 0_{\tilde{n}} & -\frac{1}{2}H_{l,4} & -\frac{1}{2}H_{l,3} \\ 0_{\tilde{n}} & 0_{\tilde{n}} & 0_{\tilde{n}} & 0_{\tilde{n}} \\ -\frac{1}{2}H_{l,4} & 0_{\tilde{n}} & 0_{\tilde{n}} & -\frac{1}{2}H_{l,2} \\ -\frac{1}{2}H_{l,3} & 0_{\tilde{n}} & -\frac{1}{2}H_{l,2} & 0_{\tilde{n}} \end{pmatrix} y \cdot \alpha + y^t \begin{pmatrix} 0_{\tilde{n}} & 0_{\tilde{n}} & 0_{\tilde{n}} & 0_{\tilde{n}} \\ 0_{\tilde{n}} & H_{l,2} & -\frac{1}{2}H_{l,4} & -\frac{1}{2}H_{l,3} \\ 0_{\tilde{n}} & -\frac{1}{2}H_{l,4} & 0_{\tilde{n}} & -\frac{1}{2}H_{l,1} \\ 0_{\tilde{n}} & -\frac{1}{2}H_{l,3} & -\frac{1}{2}H_{l,1} & 0_{\tilde{n}} \end{pmatrix} y \cdot \bar{\alpha}
$$

$$
+ y^t \begin{pmatrix} 0_{\tilde{n}} & \frac{1}{2}H_{l,3} & \frac{1}{2}H_{l,1} & 0_{\tilde{n}} \\ \frac{1}{2}H_{l,3} & 0_{\tilde{n}} & \frac{1}{2}H_{l,2} & 0_{\tilde{n}} \\ \frac{1}{2}H_{l,1} & \frac{1}{2}H_{l,2} & -H_{l,4} & 0_{\tilde{n}} \\ 0_{\tilde{n}} & 0_{\tilde{n}} & 0_{\tilde{n}} & 0_{\tilde{n}} \end{pmatrix} y \cdot \alpha j + y^t \begin{pmatrix} 0_{\tilde{n}} & \frac{1}{2}H_{l,4} & 0_{\tilde{n}} & \frac{1}{2}H_{l,1} \\ \frac{1}{2}H_{l,4} & 0_{\tilde{n}} & 0_{\tilde{n}} & \frac{1}{2}H_{l,2} \\ 0_{\tilde{n}} & 0_{\tilde{n}} & 0_{\tilde{n}} & 0_{\tilde{n}} \\ \frac{1}{2}H_{l,1} & \frac{1}{2}H_{l,2} & 0_{\tilde{n}} & -H_{l,3} \end{pmatrix} y \cdot \bar{\alpha} j
$$

$$
=: y^t \tilde{H}_{l,1} y \cdot \alpha + y^t \tilde{H}_{l,2} y \cdot \bar{\alpha} + y^t \tilde{H}_{l,3} y \cdot \alpha j + y^t \tilde{H}_{l,4} y \cdot \bar{\alpha} j. \tag{10}
$$

For $a_1, a_2, a_3, a_4 \in k^{\tilde{n}}$ and $b_1, b_2, b_3, b_4 \in k^{\tilde{m}}$, let $\psi, \psi_1 : k^{4\tilde{n}} \to Q(k)^{\tilde{n}}$ and $\varphi, \varphi_1 : Q(k)^{\tilde{m}} \to k^{4\tilde{m}}$ be the one-to-one maps as follows:

$$
\psi(a_1, a_2, a_3, a_4) = a_1 + a_2 i + a_3 j + a_4 i j,
$$
$$
\psi_1(a_1, a_2, a_3, a_4) = a_1 \alpha + a_2 \bar{\alpha} + a_3 \alpha j + a_4 \bar{\alpha} j,
$$
$$
\varphi(b_1 + b_2 i + b_3 j + b_4 i j) = (b_1, b_2, b_3, b_4),
$$
$$
\varphi_1(b_1 \alpha + b_2 \bar{\alpha} + b_3 \alpha j + b_4 \bar{\alpha} j) = (b_1, b_2, b_3, b_4).
$$

Due to Lemma 1 and its proof, we see that there exist invertible linear transformations $U : k^{4\tilde{n}} \to k^{4\tilde{n}}$ and $V : k^{4\tilde{m}} \to k^{4\tilde{m}}$ such that

$$
\psi = \psi_1 \circ U, \qquad \varphi = V \circ \varphi_1, \tag{11}
$$

and U, V are explicitly described by the matrix in (7). According to (10) and (11), we have

$$
T \circ \varphi \circ \tilde{G} \circ \psi \circ S = \tilde{T} \circ \tilde{H} \circ \tilde{S},
$$

where

$$
\tilde{S} := U \circ S : k^{4\tilde{n}} \to k^{4\tilde{n}},
$$
$$
\tilde{T} := T \circ V : k^{4\tilde{m}} \to k^{4\tilde{m}}
$$

are invertible affine maps and

$$
\tilde{H} := \varphi_1 \circ \tilde{G} \circ \psi_1 : k^{4\tilde{n}} \to k^{4\tilde{m}}
$$

is a quadratic map whose coefficient matrices in the quadratic forms are given by $\tilde{H}_{l,1}, \tilde{H}_{l,2}, \tilde{H}_{l,3}, \tilde{H}_{l,4}$ ($1 \le l \le \tilde{m}$). This means that the $(\tilde{o}_1, \tilde{o}_2, \tilde{v})$-Quaternion Rainbow proposed in §3.1 is interpreted by an MPKC scheme such that $\tilde{S}, \tilde{T}$ and $\tilde{H}$ are the secret keys and $F = \tilde{T} \circ \tilde{H} \circ \tilde{S}$ is the public key.

Based on this fact, we now explain how (partial information of) the secret keys $\tilde{S}, \tilde{T}$ are recovered by the rank attacks.

The Min-Rank Attack. Let $F_1, \ldots, F_{4\tilde{n}}$ be the coefficient matrices of the quadratic forms in the public key F. Recall that any F_l $(1 \leq l \leq 4\tilde{m})$ is a linear sum of $\tilde{S}^t \tilde{H}_{l,1} \tilde{S}, \tilde{S}^t \tilde{H}_{l,2} \tilde{S}, \tilde{S}^t \tilde{H}_{l,3} \tilde{S}, \tilde{S}^t \tilde{H}_{l,4} \tilde{S}$ $(1 \leq l \leq \tilde{m})$. Due to (8) and (10), we see that the minimum of the ranks of $\tilde{H}_{l,1}, \tilde{H}_{l,2}, \tilde{H}_{l,3}, \tilde{H}_{l,4}$ is $3\tilde{v} + 3\tilde{o}_2$. Then the min-rank attack [12] finds $c_1, \ldots, c_{4\tilde{m}} \in k$ such that the rank of $\hat{F} := c_1 F_1 + \cdots + c_{4\tilde{m}} F_{4\tilde{m}}$ is $3\tilde{v} + 3\tilde{o}_2$ with the complexity $q^{3\tilde{v}+3\tilde{o}_2} \cdot$ (polyn.). Note that $c_1, \ldots, c_{4\tilde{m}}$ are partial information of $\tilde{T}$. Once such $c_1, \ldots, c_{4\tilde{m}}$ are given, partial information of $\tilde{S}$ and further information of $\tilde{T}$ are recovered from $\hat{F}$.

The High-Rank Attack. Recall again that any F_l $(1 \leq l \leq 4\tilde{m})$ is a linear sum of $\tilde{S}^t \tilde{H}_{l,1} \tilde{S}, \tilde{S}^t \tilde{H}_{l,2} \tilde{S}, \tilde{S}^t \tilde{H}_{l,3} \tilde{S}, \tilde{S}^t \tilde{H}_{l,4} \tilde{S}$ $(1 \leq l \leq \tilde{m})$. Due to (8) and (10), we see that, removing the contributions of $3\tilde{o}_1$ matrices $\{\tilde{S}^t \tilde{H}_{l,1} \tilde{S}, \tilde{S}^t \tilde{H}_{l,2} \tilde{S}, \tilde{S}^t \tilde{H}_{l,3} \tilde{S}\}_{\tilde{o}_2+1 \leq l \leq \tilde{m}}$, $\{\tilde{S}^t \tilde{H}_{l,1} \tilde{S}, \tilde{S}^t \tilde{H}_{l,2} \tilde{S}, \tilde{S}^t \tilde{H}_{l,4} \tilde{S}\}_{\tilde{o}_2+1 \leq l \leq \tilde{m}}$, $\{\tilde{S}^t \tilde{H}_{l,1} \tilde{S}, \tilde{S}^t \tilde{H}_{l,3} \tilde{S}, \tilde{S}^t \tilde{H}_{l,4} \tilde{S}\}_{\tilde{o}_2+1 \leq l \leq \tilde{m}}$ or $\{\tilde{S}^t \tilde{H}_{l,2} \tilde{S}, \tilde{S}^t \tilde{H}_{l,3} \tilde{S}, \tilde{S}^t \tilde{H}_{l,4} \tilde{S}\}_{\tilde{o}_2+1 \leq l \leq \tilde{m}}$, we get a matrix of rank $4\tilde{n} - \tilde{o}_1$. Then the high-rank attack [12] finds $c_1, \ldots, c_{4\tilde{m}} \in k$ such that the rank of $\hat{F} := c_1 F_1 + \cdots + c_{4\tilde{m}} F_{4\tilde{m}}$ is $4\tilde{n} - \tilde{o}_1$ with the complexity $q^{3\tilde{o}_1} \cdot$ (polyn.). Note that $c_1, \ldots, c_{4\tilde{m}}$ are partial information of $\tilde{T}$. Once such $c_1, \ldots, c_{4\tilde{m}}$ are given, partial information of $\tilde{S}$ and further information of $\tilde{T}$ are recovered from $\hat{F}$.

4.2   The Case of Even Characteristic

Study the case of even characteristic. Note that $Q(k)$ is commutative since $-1 = 1$ in even characteristic k. Let

$$\alpha := 1 + i, \qquad \beta := 1 + j.$$

Similar to the case of odd characteristic, the following lemma holds.

Lemma 2. Let k be a finite field of even characteristic. Then, for any $(a_1, a_2, a_3, a_4) \in k^4$, there exists a unique $(b_1, b_2, b_3, b_4) \in k^4$ such that

$$a_1 + a_2 i + a_3 j + a_4 ij = b_1 + b_2 \alpha + b_3 \beta + b_4 \alpha\beta, \tag{12}$$

namely $\{1, \alpha, \beta, \alpha\beta\}$ is a basis of $Q(k)$ over k.   □

The multiplicative relations among $1, \alpha, \beta, \alpha\beta$ are as follows:

$$
\begin{array}{llll}
1 \cdot 1 = 1, & 1 \cdot \alpha = \alpha, & 1 \cdot \beta = \beta, & 1 \cdot \alpha\beta = \alpha\beta, \\
\alpha \cdot 1 = \alpha, & \alpha \cdot \alpha = 0, & \alpha \cdot \beta = \alpha\beta, & \alpha \cdot \alpha\beta = 0, \\
\beta \cdot 1 = \beta, & \beta \cdot \alpha = \alpha\beta, & \beta \cdot \beta = 0, & \beta \cdot \alpha\beta = 0, \\
\alpha\beta \cdot 1 = \alpha\beta, & \alpha\beta \cdot \alpha = 0, & \alpha\beta \cdot \beta = 0, & \alpha\beta \cdot \alpha\beta = 0.
\end{array}
$$

By using the basis $\{1, \alpha, \beta, \alpha\beta\}$, we can rewrite the quadratic form $\tilde{G}$ as follows:

$$\tilde{x} = y_1 + y_2 \alpha + y_3 \beta + y_4 \alpha\beta, \quad \tilde{g}_l(\tilde{x}) = \tilde{x}^t \tilde{G}_l \tilde{x} + \text{(linear form)},$$

where y_1, y_1, y_3, y_4 are unknowns in $k^{\tilde{n}}$ and $\tilde{G}_l$ is an $\tilde{n} \times \tilde{n}$-matrix with $Q(k)$-entries given in (2). Note that, since the entries in $\tilde{G}_l$ are linear sums of $\{1, \alpha, \beta, \alpha\beta\}$ over k, $\tilde{G}_l$ is given by

$$\tilde{G}_l = H_{l,1} + H_{l,2}\alpha + H_{l,3}\beta + H_{l,4}\alpha\beta,$$

where $H_{l,1}, H_{l,2}, H_{l,3}, H_{l,4}$ are $\tilde{n} \times \tilde{n}$-matrices with k-entries in the same forms of (8). Thus $\tilde{g}_l(\tilde{x})$ is written by

$$
\begin{aligned}
\tilde{x}^t \tilde{G}_l \tilde{x} =& (y_1^t + y_2^t\alpha + y_3^t\beta + y_4^t\alpha\beta)(H_{l,1} + H_{l,2}\alpha + H_{l,3}\beta + H_{l,4}\alpha\beta) \\
& \cdot (y_1 + y_2\alpha + y_3\beta + y_4\alpha\beta) \\
=& y_1^t H_{l,1} y_1 \cdot 1 + (y_1^t H_{l,1} y_2 + y_2^t H_{l,1} y_1 + y_1^t H_{l,2} y_1) \cdot \alpha \\
& + (y_1^t H_{l,1} y_3 + y_3^t H_{l,1} y_1 + y_1^t H_{l,3} y_1) \cdot \beta \\
& + (y_1^t H_{l,1} y_4 + y_2^t H_{l,1} y_3 + y_3^t H_{l,1} y_2 + y_4^t H_{l,1} y_1 \\
& \quad + y_1^t H_{l,2} y_3 + y_3^t H_{l,2} y_1 + y_2^t H_{l,3} y_1 + y_1^t H_{l,3} y_2 + y_1^t H_{l,4} y_1) \cdot \alpha\beta \quad (13)
\end{aligned}
$$

Putting $y := \begin{pmatrix} y_1 \\ y_2 \\ y_3 \\ y_4 \end{pmatrix}$, we have

$$
\begin{aligned}
\tilde{x}^t \tilde{G}_l \tilde{x} =& y^t \begin{pmatrix} H_{l,1} & 0_{\tilde{n}} & 0_{\tilde{n}} & 0_{\tilde{n}} \\ 0_{\tilde{n}} & 0_{\tilde{n}} & 0_{\tilde{n}} & 0_{\tilde{n}} \\ 0_{\tilde{n}} & 0_{\tilde{n}} & 0_{\tilde{n}} & 0_{\tilde{n}} \\ 0_{\tilde{n}} & 0_{\tilde{n}} & 0_{\tilde{n}} & 0_{\tilde{n}} \end{pmatrix} y \cdot 1 + y^t \begin{pmatrix} H_{l,2} & H_{l,1} & 0_{\tilde{n}} & 0_{\tilde{n}} \\ H_{l,1} & 0_{\tilde{n}} & 0_{\tilde{n}} & 0_{\tilde{n}} \\ 0_{\tilde{n}} & 0_{\tilde{n}} & 0_{\tilde{n}} & 0_{\tilde{n}} \\ 0_{\tilde{n}} & 0_{\tilde{n}} & 0_{\tilde{n}} & 0_{\tilde{n}} \end{pmatrix} y \cdot \alpha \\
& + y^t \begin{pmatrix} H_{l,3} & 0_{\tilde{n}} & H_{l,1} & 0_{\tilde{n}} \\ 0_{\tilde{n}} & 0_{\tilde{n}} & 0_{\tilde{n}} & 0_{\tilde{n}} \\ H_{l,1} & 0_{\tilde{n}} & 0_{\tilde{n}} & 0_{\tilde{n}} \\ 0_{\tilde{n}} & 0_{\tilde{n}} & 0_{\tilde{n}} & 0_{\tilde{n}} \end{pmatrix} y \cdot \beta + y^t \begin{pmatrix} H_{l,4} & H_{l,3} & H_{l,2} & H_{l,1} \\ H_{l,3} & 0_{\tilde{n}} & H_{l,1} & 0_{\tilde{n}} \\ H_{l,2} & H_{l,1} & 0_{\tilde{n}} & 0_{\tilde{n}} \\ H_{l,1} & 0_{\tilde{n}} & 0_{\tilde{n}} & 0_{\tilde{n}} \end{pmatrix} y \cdot \alpha\beta \\
=:& y^t \tilde{H}_{l,1} y \cdot 1 + y^t \tilde{H}_{l,2} y \cdot \alpha + y^t \tilde{H}_{l,3} y \cdot \beta + y^t \tilde{H}_{l,4} y \cdot \alpha\beta. \quad (14)
\end{aligned}
$$

For $a_1, a_2, a_3, a_4 \in k^{\tilde{n}}$ and $b_1, b_2, b_3, b_4 \in k^{\tilde{m}}$, let $\psi, \psi_1 : k^{4\tilde{n}} \to Q(k)^{\tilde{n}}$ and $\varphi, \varphi_1 : Q(k)^{\tilde{m}} \to k^{4\tilde{m}}$ be the one-to-one maps as follows:

$$
\begin{aligned}
\psi(a_1, a_2, a_3, a_4) =& a_1 + a_2 i + a_3 j + a_4 ij, \\
\psi_1(a_1, a_2, a_3, a_4) =& a_1 + a_2\alpha + a_3\beta + a_4\alpha\beta, \\
\varphi(b_1 + b_2 i + b_3 j + b_4 ij) =& (b_1, b_2, b_3, b_4), \\
\varphi_1(b_1 + b_2\alpha + b_3\beta + b_4\alpha\beta) =& (b_1, b_2, b_3, b_4).
\end{aligned}
$$

Due to Lemma 2, we see that there exist invertible linear transformations $U : k^{4\tilde{n}} \to k^{4\tilde{n}}$ and $V : k^{4\tilde{m}} \to k^{4\tilde{m}}$ such that

$$\psi = \psi_1 \circ U, \qquad \varphi = V \circ \varphi_1, \quad (15)$$

and U, V are explicitly given. According to (14) and (15), we have

$$T \circ \varphi \circ \tilde{G} \circ \psi \circ S = \tilde{T} \circ \tilde{H} \circ \tilde{S},$$

where

$$\tilde{S} := U \circ S : k^{4\tilde{n}} \to k^{4\tilde{n}},$$

$$\tilde{T} := T \circ V : k^{4\tilde{m}} \to k^{4\tilde{m}}$$

are invertible affine maps and

$$\tilde{H} := \varphi_1 \circ \tilde{G} \circ \psi_1 : k^{4\tilde{n}} \to k^{4\tilde{m}}$$

is a quadratic map whose coefficient matrices in the quadratic forms are given by $\tilde{H}_{l,1}, \tilde{H}_{l,2}, \tilde{H}_{l,3}, \tilde{H}_{l,4}$ $(1 \le l \le \tilde{m})$. This means that the $(\tilde{o}_1, \tilde{o}_2, \tilde{v})$-Quaternion Rainbow proposed in §3.1 is interpreted by an MPKC scheme such that $\tilde{S}, \tilde{T}$ and $\tilde{H}$ are the secret keys and $F = \tilde{T} \circ \tilde{H} \circ \tilde{S}$ is the public key.

Based on this fact, we now explain how (partial information of) the secret keys $\tilde{S}, \tilde{T}$ are recovered by the rank attacks and the Kipnis-Shamir attack.

The Min-Rank Attack. Let $F_1, \ldots, F_{4\tilde{n}}$ be the coefficient matrices of the quadratic forms in the public key F. Due to (8) and (14), we see that the minimum of the ranks of $\tilde{H}_{l,1}, \tilde{H}_{l,2}, \tilde{H}_{l,3}, \tilde{H}_{l,4}$ is $\tilde{v} + \tilde{o}_2$. Then the min-rank attack [12] finds $c_1, \ldots, c_{4\tilde{m}} \in k$ such that the rank of $\hat{F} := c_1 F_1 + \cdots + c_{4\tilde{m}} F_{4\tilde{m}}$ is $\tilde{v} + \tilde{o}_2$ with the complexity $q^{\tilde{v}+\tilde{o}_2} \cdot$ (polyn.). Note that $c_1, \ldots, c_{4\tilde{m}}$ are partial information of $\tilde{T}$. Once such $c_1, \ldots, c_{4\tilde{m}}$ are given, partial information of $\tilde{S}$ and further information of $\tilde{T}$ are recovered from $\hat{F}$.

The High-Rank Attack. Due to (8) and (14), we see that, removing the contributions of $\tilde{o}_1$ matrices $\{\tilde{S}^t \tilde{H}_{l,4} \tilde{S}\}_{\tilde{o}_2+1 \le l \le \tilde{m}}$, we get a matrix of rank $4\tilde{n} - \tilde{o}_1$. Then the high-rank attack [12] finds $c_1, \ldots, c_{4\tilde{m}} \in k$ such that the rank of $\hat{F} := c_1 F_1 + \cdots + c_{4\tilde{m}} F_{4\tilde{m}}$ is $4\tilde{n} - \tilde{o}_1$ with the complexity $q^{\tilde{o}_1} \cdot$ (polyn.). Note that $c_1, \ldots, c_{4\tilde{m}}$ are partial information of $\tilde{T}$. Once such $c_1, \ldots, c_{4\tilde{m}}$ are given, partial information of $\tilde{S}$ and further information of $\tilde{T}$ are recovered from $\hat{F}$.

Kipnis-Shamir's Attack. We now use the following lemma.

Lemma 3. *([8,7,3]) Let $m, n \ge 1$ be integers and $P_1, \ldots, P_m$ be $2n \times 2n$-matrices with k-entries. If $P_1, \ldots, P_m$ are public and are given in the forms*

$$P_l = A^t \begin{pmatrix} *_n & * \\ * & 0_n \end{pmatrix} A \qquad (1 \le l \le m)$$

where A is an invertible $2n \times 2n$-matrix, then Kipnis-Shamir's attack [8,7,3] recovers an $n \times n$-matrix A_1 such that

$$A \begin{pmatrix} I_n & A_1 \\ 0 & I_n \end{pmatrix} = \begin{pmatrix} *_n & 0 \\ * & *_n \end{pmatrix}$$

in polynomial time of n, m and $\log q$.

Equation (14) tells that any linear sum of $\tilde{H}_{l,1}, \tilde{H}_{l,2}, \tilde{H}_{l,3}, \tilde{H}_{l,4}$ $(1 \le l \le \tilde{m})$ is in the form $\begin{pmatrix} *_{2\tilde{n}} & * \\ * & 0_{2\tilde{n}} \end{pmatrix}$. Then, due to Lemma 3, Kipnis-Shamir's attack [8,7,3] recovers a $2\tilde{n} \times 2\tilde{n}$ matrix M_1 such that

$$\tilde{S}\begin{pmatrix} I_{2\tilde{n}} & M_1 \\ 0 & I_{2\tilde{n}} \end{pmatrix} = \begin{pmatrix} *_{2\tilde{n}} & 0 \\ * & *_{2\tilde{n}} \end{pmatrix}$$

in polynomial time. Put $F_l^{(1)} := \begin{pmatrix} I_{2\tilde{n}} & 0 \\ M_1^t & I_{2\tilde{n}} \end{pmatrix} F_l \begin{pmatrix} I_{2\tilde{n}} & M_1 \\ 0 & I_{2\tilde{n}} \end{pmatrix}$. Since

$$\begin{pmatrix} * & * \\ 0 & * \end{pmatrix}\begin{pmatrix} * & * \\ * & 0 \end{pmatrix}\begin{pmatrix} * & 0 \\ * & * \end{pmatrix} = \begin{pmatrix} * & * \\ * & 0 \end{pmatrix}, \qquad \begin{pmatrix} * & * \\ 0 & * \end{pmatrix}\begin{pmatrix} * & 0 \\ 0 & 0 \end{pmatrix}\begin{pmatrix} * & 0 \\ * & * \end{pmatrix} = \begin{pmatrix} * & 0 \\ 0 & 0 \end{pmatrix}, \qquad (16)$$

any $F_l^{(1)}$ is in the form $\begin{pmatrix} *_{2\tilde{n}} & * \\ * & 0_{2\tilde{n}} \end{pmatrix}$ and we can find $2\tilde{m}$ linear sums $F_1^{(2)}, \ldots, F_{2\tilde{m}}^{(2)}$ of $F_1^{(1)}, \ldots, F_{4\tilde{n}}^{(1)}$ in the forms $\begin{pmatrix} *_{2\tilde{n}} & 0 \\ 0 & 0_{2\tilde{n}} \end{pmatrix}$ by the Gaussian eliminations. Note that $F_l^{(2)}$ $(1 \le l \le 2\tilde{m})$ is a linear sum of $\begin{pmatrix} I_{2\tilde{n}} & 0 \\ M_1^t & I_{2\tilde{n}} \end{pmatrix} \tilde{S}^t \tilde{H}_{l,1} \tilde{S} \begin{pmatrix} I_{2\tilde{n}} & M_1 \\ 0 & I_{2\tilde{n}} \end{pmatrix}$ and $\begin{pmatrix} I_{2\tilde{n}} & 0 \\ M_1^t & I_{2\tilde{n}} \end{pmatrix} \tilde{S}^t \tilde{H}_{l,2} \tilde{S} \begin{pmatrix} I_{2\tilde{n}} & M_1 \\ 0 & I_{2\tilde{n}} \end{pmatrix}$ $(1 \le l \le \tilde{m})$.

Equation (14) tells that the upper left block of any linear sum of $\tilde{H}_{l,1}, \tilde{H}_{l,2}$ $(1 \le l \le \tilde{m})$ is in the form $\begin{pmatrix} *_{\tilde{n}} & * \\ * & 0_{\tilde{n}} \end{pmatrix}$. Then, due to Lemma 3, Kipnis-Shamir's attack recovers an $\tilde{n} \times \tilde{n}$ matrix M_2 such that

$$\tilde{S}\begin{pmatrix} I_{2\tilde{n}} & M_1 \\ 0 & I_{2\tilde{n}} \end{pmatrix}\begin{pmatrix} I_{\tilde{n}} & M_2 & 0 \\ 0 & I_{\tilde{n}} & 0 \\ 0 & 0 & I_{2\tilde{n}} \end{pmatrix} = \begin{pmatrix} *_{\tilde{n}} & 0 & 0 \\ * & *_{\tilde{n}} & 0 \\ * & * & *_{2\tilde{n}} \end{pmatrix} \qquad (17)$$

in polynomial time. Put $F_l^{(3)} := \begin{pmatrix} I_{\tilde{n}} & 0 & 0 \\ M_2^t & I_{\tilde{n}} & 0 \\ 0 & 0 & I_{2\tilde{n}} \end{pmatrix}\begin{pmatrix} I_{2\tilde{n}} & 0 \\ M_1^t & I_{2\tilde{n}} \end{pmatrix} F_l^{(2)} \begin{pmatrix} I_{2\tilde{n}} & M_1 \\ 0 & I_{2\tilde{n}} \end{pmatrix}\begin{pmatrix} I_{\tilde{n}} & M_2 & 0 \\ 0 & I_{\tilde{n}} & 0 \\ 0 & 0 & I_{2\tilde{n}} \end{pmatrix}$

$(1 \le l \le 2\tilde{m})$. We see that any $F_l^{(3)}$ is in the form $\begin{pmatrix} *_{\tilde{n}} & * & 0 \\ * & 0_{\tilde{n}} & 0 \\ 0 & 0 & 0_{2\tilde{n}} \end{pmatrix}$ and we can find $\tilde{m}$ linear sums $F_1^{(4)}, \ldots, F_{\tilde{m}}^{(4)}$ of $F_1^{(3)}, \ldots, F_{2\tilde{m}}^{(3)}$ in the forms $\begin{pmatrix} *_{\tilde{n}} & 0 \\ 0 & 0_{3\tilde{n}} \end{pmatrix}$ by the Gaussian eliminations. Since the upper left blocks of $F_l^{(4)}$ $(1 \le l \le \tilde{m})$ is a linear sums of $\tilde{S}_1^t H_{l,1} \tilde{S}_1$ $(1 \le l \le \tilde{m})$ with an $\tilde{n} \times \tilde{n}$ invertible matrix $\tilde{S}_1$, the quadratic forms derived from $F_1^{(4)}, \ldots, F_{\tilde{n}}^{(4)}$ correspond to those in the $(\tilde{o}_1, \tilde{o}_2, \tilde{v})$-Rainbow.

Now let $\tilde{F}(x) = (\tilde{f}_1(\tilde{x}), \ldots, \tilde{f}_{4\tilde{n}}(\tilde{x}))$ be the quadratic map given by

$$\tilde{f}_l(x) := \begin{cases} x^t F_l^{(4)} x + (\text{linear}), & (1 \le l \le \tilde{m}), \\ x^t F_l^{(3)} x + (\text{linear}), & (\tilde{m}+1 \le l \le 2\tilde{m}), \\ x^t F_l^{(1)} x + (\text{linear}), & (2\tilde{m}+1 \le l \le 4\tilde{m}) \end{cases}$$

$$= \begin{cases} x^t \begin{pmatrix} *_{\tilde{n}} & 0 \\ 0 & 0_{3\tilde{n}} \end{pmatrix} x + (\text{linear}), & (1 \le l \le \tilde{m}), \\[2em] x^t \begin{pmatrix} *_{\tilde{n}} & * & 0 \\ * & 0_{\tilde{n}} & 0 \\ 0 & 0 & 0_{2\tilde{n}} \end{pmatrix} x + (\text{linear}), & (\tilde{m}+1 \le l \le 2\tilde{m}), \\[3em] x^t \begin{pmatrix} *_{2\tilde{n}} & * \\ * & 0_{2\tilde{n}} \end{pmatrix} x + (\text{linear}), & (2\tilde{m}+1 \le l \le 4\tilde{m}). \end{cases}$$

Table 2. Comparisons of the security analysis on Quaternion Rainbow

	Kipnis-Shamir	Min-Rank	High-Rank
Original Rainbow	$q^{4\tilde{v}+4\tilde{o}_2-4\tilde{o}_1}$	$q^{4\tilde{v}+4\tilde{o}_2}$	$q^{4\tilde{o}_1}$
Yasuda-Sakurai-Takagi [13]	$q^{4\tilde{v}+4\tilde{o}_2-4\tilde{o}_1}$	$q^{4\tilde{v}+4\tilde{o}_2}$	$q^{4\tilde{o}_1}$
Thomae [11] $(2 \nmid q)$	$q^{4\tilde{v}+4\tilde{o}_2-4\tilde{o}_1}$	$q^{4\tilde{v}+3\tilde{o}_2}$	$q^{3\tilde{o}_1}$
Proposed $(2 \nmid q)$	$q^{4\tilde{v}+4\tilde{o}_2-4\tilde{o}_1}$	$q^{3\tilde{v}+3\tilde{o}_2}$	$q^{3\tilde{o}_1}$
Thomae [11] $(2 \mid q)$	$q^{4\tilde{v}+4\tilde{o}_2-4\tilde{o}_1}$	$q^{4\tilde{v}+\tilde{o}_2}$	$q^{3\tilde{o}_1}$
Proposed $(2 \mid q)$	$q^{\tilde{v}+\tilde{o}_2-\tilde{o}_1}$	$q^{\tilde{v}+\tilde{o}_2}$	$q^{\tilde{o}_1}$

Since $\tilde{F} = \hat{T}^{-1} \circ F \circ \hat{S}^{-1}$ with two invertible affine maps $\hat{S} : k^{4\tilde{n}} \to k^{4\tilde{n}}$ derived from M_1, M_2 and $\hat{T} : k^{4\tilde{m}} \to k^{4\tilde{m}}$ derived from the maps $\{F_l^{(1)}\} \mapsto \{F_l^{(2)}\}, \{F_l^{(3)}\} \mapsto \{F_l^{(4)}\}$, inverting $\tilde{F}$ is equivalent to doing F. In order to find $x = (x_1, \ldots, x_{4\tilde{n}})$ with $\tilde{F}(x) = z$ for given $z = (z_1, \ldots, z_{4\tilde{m}})$, we compute as follows.

Step 1. Find $x_1, \ldots, x_{\tilde{n}}$ such that $\tilde{f}_1(x) = z_1, \ldots, \tilde{f}_{\tilde{m}}(x) = z_{\tilde{m}}$.

Step 2. For $x_1, \ldots, x_{\tilde{n}}$ given in Step 1, find $x_{\tilde{n}+1}, \ldots, x_{2\tilde{n}}$ such that $\tilde{f}_{\tilde{m}+1}(x) = z_{\tilde{m}+1}, \ldots, \tilde{f}_{2\tilde{m}}(x) = z_{2\tilde{m}}$.

Step 3. For $x_1, \ldots, x_{2\tilde{n}}$ given in Step 1 and 2, find $x_{2\tilde{n}+1}, \ldots, x_{4\tilde{n}}$ such that $\tilde{f}_{2\tilde{m}+1}(x) = z_{2\tilde{m}+1}, \ldots, \tilde{f}_{4\tilde{m}}(x) = z_{4\tilde{m}}$.

Once $x_1, \ldots, x_{\tilde{n}}$ are fixed, $\tilde{f}_{\tilde{m}+1}(x) = z_{\tilde{m}+1}, \ldots, \tilde{f}_{2\tilde{m}}(x) = z_{2\tilde{m}}$ are linear equations of $\tilde{n}$ variables $x_{\tilde{n}+1}, \ldots, x_{2\tilde{n}}$. Furthermore, once $x_1, \ldots, x_{2\tilde{n}}$ are fixed, $\tilde{f}_{2\tilde{m}+1}(x) = z_{2\tilde{m}+1}, \ldots, \tilde{f}_{4\tilde{m}}(x) = z_{4\tilde{m}}$ are linear equations of $2\tilde{n}$ variables $x_{2\tilde{n}+1}, \ldots, x_{4\tilde{n}}$. Then Step 2 and 3 are computed by the Gaussian eliminations in polynomial time. Recall that finding a solution of $\tilde{f}_1(x) = z_1, \ldots, \tilde{f}_{\tilde{m}}(x) = z_{\tilde{m}}$ is equivalent to generating a dummy signature of the $(\tilde{o}_1, \tilde{o}_2, \tilde{v})$-Rainbow. Thus we conclude that the $(\tilde{o}_1, \tilde{o}_2, \tilde{v})$-Quaternion Rainbow over even characteristic field is as secure as the $(\tilde{o}_1, \tilde{o}_2, \tilde{v})$ original Rainbow, whose complexity against Kipnis-Shamir's attacks is $q^{\tilde{v}+\tilde{o}_2-\tilde{o}_1} \cdot$ (polyn.).

5   Conclusion

In the present paper, we estimate the security of Quaternion Rainbow against Kipnis-Shamir's attack [8,7,3] and the rank attacks [12]. Table 2 summarizes the complexities of the $(\tilde{o}_1, \tilde{o}_2, \tilde{v})$ Quaternion Rainbow.

For both even and odd characteristic cases, the security of Quaternion Rainbow is weaker than expected by the authors of [13]. Especially, the Quaternion Rainbow over even characteristic field is almost as secure as the original Rainbow of 1/4 size. Thus, Quaternion Rainbow is less practical than the original Rainbow. We consider that, since such vulnerabilities emerge from less randomness of the distribution of coefficients in quadratic forms, preserving its randomness will be required to build secure and efficient schemes.

References

1. Bardet, M., Faugère, J.C., Salvy, B., Yang, B.Y.: Asymptotic Expansion of the Degree of Regularity for Semi-Regular Systems of Equations. In: MEGA 2005 (2005)
2. Chen, A.I.-T., Chen, M.-S., Chen, T.-R., Cheng, C.-M., Ding, J., Kuo, E.L.-H., Lee, F.Y.-S., Yang, B.-Y.: SSE Implementation of Multivariate PKCs on Modern x86 CPUs. In: Clavier, C., Gaj, K. (eds.) CHES 2009. LNCS, vol. 5747, pp. 33–48. Springer, Heidelberg (2009)
3. Ding, J., Gower, J.E., Schmidt, D.S.: Multivariate public key cryptosystems. Springer, Heidelberg (2006)
4. Ding, J., Schmidt, D.: Rainbow, a new multivariable polynomial signature scheme. In: Ioannidis, J., Keromytis, A.D., Yung, M. (eds.) ACNS 2005. LNCS, vol. 3531, pp. 164–175. Springer, Heidelberg (2005)
5. Ding, J., Yang, B.-Y., Chen, C.-H.O., Chen, M.-S., Cheng, C.-M.: New Differential-Algebraic Attacks and Reparametrization of Rainbow. In: Bellovin, S.M., Gennaro, R., Keromytis, A.D., Yung, M. (eds.) ACNS 2008. LNCS, vol. 5037, pp. 242–257. Springer, Heidelberg (2008)
6. Faugère, J.C.: A new efficient algorithm for computing Grobner bases (F_4). J. Pure and Applied Algebra 139, 61–88 (1999)
7. Kipnis, A., Patarin, J., Goubin, L.: Unbalanced Oil and Vinegar Signature Schemes. In: Stern, J. (ed.) EUROCRYPT 1999. LNCS, vol. 1592, pp. 206–2006. Springer, Heidelberg (1999)
8. Kipnis, A., Shamir, A.: Cryptanalysis of the Oil & Vinegar Signature Scheme. In: Krawczyk, H. (ed.) CRYPTO 1998. LNCS, vol. 1462, pp. 257–267. Springer, Heidelberg (1998)
9. Petzoldt, A., Bulygin, S., Buchmann, J.: CyclicRainbow – A multivariate signature scheme with a partially cyclic public key. In: Gong, G., Gupta, K.C. (eds.) INDOCRYPT 2010. LNCS, vol. 6498, pp. 33–48. Springer, Heidelberg (2010)
10. Petzoldt, A., Bulygin, S., Buchmann, J.: Selecting Parameters for the Rainbow Signature Scheme. In: Sendrier, N. (ed.) PQCrypto 2010. LNCS, vol. 6061, pp. 218–240. Springer, Heidelberg (2010)
11. Thomae, E.: Quo Vadis Quaternion? Cryptanalysis of Rainbow over Noncommutative Rings. In: Visconti, I., De Prisco, R. (eds.) SCN 2012. LNCS, vol. 7485, pp. 361–373. Springer, Heidelberg (2012)
12. Yang, B.-Y., Chen, J.-M.: Building secure tame-like multivariate public-key cryptosystems: The new TTS. In: Boyd, C., González Nieto, J.M. (eds.) ACISP 2005. LNCS, vol. 3574, pp. 518–531. Springer, Heidelberg (2005)
13. Yasuda, T., Sakurai, K., Takagi, T.: Reducing the Key Size of Rainbow Using Non-commutative Rings. In: Dunkelman, O. (ed.) CT-RSA 2012. LNCS, vol. 7178, pp. 68–83. Springer, Heidelberg (2012)

On Cheater Identifiable Secret Sharing Schemes Secure against Rushing Adversary

Rui Xu[1], Kirill Morozov[2], and Tsuyoshi Takagi[2]

[1] Graduate School of Mathematics, Kyushu University
r-xu@math.kyushu-u.ac.jp
[2] Institute of Mathematics for Industry, Kyushu University

Abstract. At EUROCRYPT 2011, Obana proposed a k-out-of-n secret sharing scheme capable of identifying up to t cheaters with probability $1 - \epsilon$ under the condition $t < k/3$. In that scheme, the share size $|V_i|$ satisfies $|V_i| = |S|/\epsilon$, which is almost optimal. However, Obana's scheme is known to be vulnerable to attacks by rushing adversary who can observe the messages sent by the honest participants prior to deciding her own messages. In this paper, we present a new scheme, which is secure against rushing adversary, with $|V_i| = |S|/\epsilon^{n-t+1}$, assuming $t < k/3$. We note that the share size of our proposal is substantially smaller compared to $|V_i| = |S|(t + 1)^{3n}/\epsilon^{3n}$ in the scheme by Choudhury at PODC 2012 when the secret is a single field element. A modification of the later scheme is secure against rushing adversary under a weaker $t < k/2$ condition. Therefore, our scheme demonstrates an improvement in share size achieved for the price of strengthening the assumption on t.

Keywords: cheater identifiable secret sharing, Shamir secret sharing, rushing adversary.

1 Introduction

Secret sharing, independently introduced by Shamir [1] and Blakley [2], is an important primitive enjoying numerous cryptographic applications such as threshold cryptography [3], secure multiparty computation [4,5], and (perfectly) secure message transmission [6], to mention a few. A typical example is the *threshold* (or k-*out-of-*n) secret sharing scheme that allows a dealer D to distribute a secret s among a set of n participants (or players) $\{P_1, P_2, \ldots, P_n\}$ in such a way that the following two properties hold: (1) *perfect secrecy*: $k - 1$ or less participants can get no information about s from their shares; (2) *correctness*: k or more participants can pool their shares together to reconstruct the secret. In the original setting of secret sharing schemes, it is assumed that all players will provide correct shares when reconstructing the secret. Since this assumption does not model the real life scenario, in which some participants may submit incorrect shares in order to cause the reconstruction of an incorrect secret, a body of work has been done on identifying the cheaters in secret sharing schemes. Next, we will discuss some of the prominent results in this area. If there are more than one

K. Sakiyama and M. Terada (Eds.): IWSEC 2013, LNCS 8231, pp. 258–271, 2013.
© Springer-Verlag Berlin Heidelberg 2013

cheating participant, we will assume a single malicious adversary who controls their behavior. The adversary is called *rushing*, if she is allowed to observe all the messages sent by honest players (in every round) prior to deciding on cheaters' messages.

1.1   Secret Sharing with Cheaters

In this work, we focus on *secret sharing with cheater identification (SSCI)*. In this setting, the dealer is assumed to be honest. At the reconstruction stage, when a qualified subset of participants pool their shares, they will be able to *identify* cheater(s) among them, who submitted a forged share, as long as the number of cheaters is smaller than a certain bound.

The idea of secret sharing with protection against cheating was pioneered by Tompa and Woll [7]. They modified the (k-out-of-n) Shamir secret sharing scheme [1] to enable the *cheater detection* (not identification). The first secret sharing scheme capable of identifying cheaters is due to Rabin and Ben-Or [4]. Later, McEliece and Sarwate [8] showed that Shamir scheme is cheater identifiable by exhibiting its connection to Reed-Solomon codes. Note that this scheme requires the presence of *more than* k participants in order to carry out cheater identification. In contrast, Kurosawa, Obana and Ogata [9] considered the problem of identifying cheaters when only k players take part in the reconstruction. In particular, they gave a lower bound on the share size in this model:

$$|V_i| \geq \frac{|S| - 1}{\epsilon} + 1, \tag{1}$$

where ϵ is the cheaters' success probability.

In this work, we mainly focus on SSCI in the model of Kurosawa *et al.* [9]. They proposed an SSCI scheme identifying $t < k/3$ cheaters in k-out-of-n Shamir secret sharing. Obana [10] improved Kurosawa *et al.*'s scheme by reducing the share size to $|V_i| = |S|/\epsilon$, which is almost optimal, and in addition proposed two (inefficient) SSCI schemes identifying up to $t < k/2$ cheaters.

While the above mentioned schemes can only identify non-rushing cheaters, Choudhury [11] implemented an efficient SSCI scheme which can identify up to $t < k/2$ *rushing* cheaters, achieving the share size $|V_i| = |S|/\epsilon$ when the secret consists of $l = \Omega(n)$ field elements.

Cevallos *et al.* [12] proposed a *robust secret sharing scheme (RSS)* against up to $t < n/2$ rushing cheaters with share size $|V_i| = |S| \left[\log |S| \cdot (t+1)(\frac{\epsilon}{\epsilon})^{\frac{2}{t+1}} \right]^{3n}$ which is close-to-optimal. In their scheme, all the n players are required to take part in the reconstruction phase.

We note that in this work, we focus on *public* cheater identification [10, 11], where reconstruction is performed such that all the shares are treated equally in terms of their trustworthiness by the reconstruction algorithm. It means that this algorithm can be performed even by an external reconstructor. Such type of schemes are only possible for the case of honest majority.

On the contrary, in the schemes with *private* identification [4, 13], the share received by a user from the honest dealer is assumed to be trusted. Such types of schemes do not need honest majority for cheater identification, as explained in [10]. The concept of identifying cheaters without an honest majority is further developed by Ishai *et al.* in [14].

1.2   Related Works

To the best of our knowledge, up to date, the constructions of [11] and [12] are the most efficient secret sharing schemes secure against rushing adversary, in terms of their share size. Both of them are based on the paradigm by Rabin and Ben-Or [4]. In these schemes, a pairwise authentication is applied to identify cheaters in the reconstruction phase. More precisely, every player receives $n-1$ tags computed according to some unconditional Message Authentication Code (MAC) for every share, and the corresponding keys are distributed to the other $n-1$ players, respectively. Therefore, every player can check the validity of shares belonging to other players. Hereby, cheaters' success probability is bounded by the successful substitution attack probability of the used MAC. The schemes [11] and [12] also employ some additional (novel) techniques on top of this generic procedure.

In Choudhury's scheme [11], the shared secret is a vector $s = (s_1, s_2, \ldots, s_l)$ from $\mathbb{F}_p^l$, where $\mathbb{F}_p$ is some finite field and $l \geq 1$. Every player P_i obtains Shamir sharing sh_i of s element-wise. Then the sharing algorithm uses a MAC to authenticate sh_i as $t_{i,j} = MAC(sh_i, k_{j,i})$ where $k_{j,i}$ held by player P_j is the authentication key chosen uniformly and randomly from some finite field. At the reconstruction phase, a majority voting is taken based on the result of verifying each player's tags. Each player whose share is not recognized by the majority is identified as cheater (thus this scheme is public cheater identifiable). Choudhury's scheme is asymptotically optimal when $l = \Omega(n)$.

The sharing phase of Cevallos *et al.* [12] is identical to that of Ben-Or [4] except for the MAC used. Here, it is assumed that $n = 2t + 1$. The sharing algorithm first chooses the secret $s \in \mathbb{F}$ then calculates tags of P_i's Shamir share as $t_{i,j} = MAC(sh_i, k_{j,i})$. At the reconstruction phase player P_i's share sh_i will be accepted as valid only if it is recognized by $t + 1$ players *who hold accepted shares*. After that Reed-Solomon error correction is applied to rule out potential cheaters who are not identified by the majority voting. Due to such an advanced reconstruction phase, shorter keys and tags for the MAC can be used in their scheme, as compared to the straightforward approach. Hereby, a reduction in the share size is achieved.

We point out that in fact, Cevallos *et al.* scheme [12] can identify cheaters. Moreover, it can be modified in a straightforward manner, in order to satisfy the property of SSCI such that only k players can identify up to $t < k/2$ cheaters, since it uses the same message authentication and majority voting strategy as [4] and [11]. However, if there are only k players in the reconstruction phase, then Reed-Solomon error correction will not rule out the potential cheaters who are not identified by the majority voting. Let us explain this point in details: For Reed-Solomon error correction to work, the number of correct symbols must be greater than the degree of the polynomial $f(x)$. In other words, Reed-Solomon error correction can only be used to rule out potential cheaters, if there are at least k honest players in the reconstruction phase. On the other hand, when the number of players taking part in reconstruction is $m < k + t$, Cevallos et al.'s scheme [12] will lose the ability to identify potential cheaters using Reed-Solomon error correction. This will imply that short keys and tags for the MAC will be no longer secure (the security will then come exclusively from the employed MAC). Therefore, if we modify Cevallos et al. [12] to work as a standard SSCI scheme, it will become equivalent to Choudhury's scheme for a single secret.

Let t be the number of cheaters, $|V_i|$ – the size of a share for every player P_i, m – the number of players involved in the reconstruction phase. We summarize the SSCI schemes of [10,11] and the RSS scheme of [12], and compare them to our proposal in Table 1, unifying the parameters of all these schemes, for the convenience sake.

Table 1. Comparison of Our Proposal to Existing SSCI schemes

Scheme	# Cheaters	Share Size	Adversary	# Players at Reconstruction
Obana [10]	$t < k/3$	$|V_i| = |S|/\epsilon$	Non-rushing	$m \geq k$
Choudhury [11]	$t < k/2$	$|V_i| = |S|(t+1)^{3n}/\epsilon^{3n}$	Rushing	$m \geq k$
Cevallos et al. [12]	$t < n/2$	$|V_i| = |S|[\log |S| \cdot (t+1)(\frac{\epsilon}{\epsilon})^{\frac{2}{t+1}}]^{3n}$	Rushing	$m = n$
Our Proposal	$t < k/3$	$|V_i| = |S|/\epsilon^{n-t+1}$	Rushing	$m \geq k$

From Table 1, we observe that Obana's scheme [10] achieved the nearly optimal share size, since the lower bound on the share size is $|V_i| \geq \frac{|S|-1}{\epsilon} + 1$ according to [9]. However, recall that [10] can only deal with non-rushing adversary. Choudhury's scheme [11] is (almost) asymptotically optimal for large secrets, and it has a desirable property of identifying rushing cheaters from the minimal number of shares k. However for a single secret, the share size of this scheme (that is $\frac{|S|(t+1)^{3n}}{\epsilon^{3n}}$) is far from optimal. Cevallos et al. [12] scheme working with a single secret achieves nearly optimal share size. However, their scheme requires more than $k + t$ players to identify t rushing cheaters.

Now, an interesting open question is to introduce a secret sharing scheme (with share size smaller than those of the above schemes) for a single secret with the property that only k players can identify rushing cheaters. Our proposal fills this gap for $t < k/3$.

1.3   Our Result

We present an SSCI scheme with public cheater identification which is a k-out-of-n secret sharing identifying up to $t < k/3$ rushing cheaters. The share size of our SSCI scheme is $|V_i| = |S|/\epsilon^{n-t+1}$, its parameters are summarized in Table 1.

Note that in Table 1, we provide the share size of Choudhury's scheme [11] for the case of a single secret ($l = 1$). As we mentioned before, if the scheme by Cevallos *et al.* [12] is modified to be an SSCI with the property of identifying cheaters from the minimum number of shares, it will turn into Choudhury's [11] scheme with $l = 1$.

We emphasize that all the schemes mentioned in Table 1 are *not* directly comparable, however we list them together since they provide the same functionality. Hereby, it will help the reader to place our contribution in the context of SSCI and related schemes.

Our contribution is to achieve a tradeoff among the existing secret sharing schemes with cheaters, in terms of tolerable cheaters (t), required players at reconstruction (m), and the share size ($|V_i|$). Hereby we fill the following gap: When the number of rushing cheaters is less than $k/3$ and only k players take part in the reconstruction, our SSCI scheme is superior to the existing schemes in terms of share size.

The closest related work is the one by Choudhury [11] so that we will now provide a detailed comparison with this scheme. The share size of our scheme is $\frac{(t+1)^{3n}}{\epsilon^{2n+t-1}}$ times smaller than that of [11] (in the case of a single secret). This advantage comes for the price of strengthening requirements on the number of cheaters, that our scheme can tolerate, to $t < k/3$.

Let us elaborate more on the savings in the share size that we obtain by providing a specific example. Let us consider the bit length to be added for the sake of cheater identification (we will call it *redundancy*) – it will be computed by taking a logarithm of $|V_i|$ and subtracting $\log|S|$. Then the bit length of redundancy in Choudhury's scheme and ours are respectively:

$$Red_{Cho} = 3n \log(t + 1) + 3n \log(\frac{1}{\epsilon}), \qquad (2)$$

$$Red_{Our} = (n - t + 1) \log(\frac{1}{\epsilon}). \qquad (3)$$

From the above equations we can see that asymptotically, our scheme adds at least 3 times less redundancy as compared with Choudhary's scheme if $\frac{1}{\epsilon} >> n$. In Table 2, we compare the redundancy of our scheme to that of [11], fixing the cheater success probability ϵ to be 2^{-80}. For simplicity, we take $t = \lfloor(n-1)/3\rfloor$, although the maximal tolerable number of cheaters t is $\lfloor(k-1)/3\rfloor$ and $\lfloor(k-1)/2\rfloor$ in our scheme and in [11], respectively. We can see from Table 2 that as n grows larger, our scheme needs less and less redundancy as compared with [11]. In particular, even for $n = 4$, our scheme will need 26.7 bytes of redundancy, which is still 4.5 times less than 120.6 bytes needed for Choudhary's scheme. We emphasize again that the reduction of share size comes for the price of sacrifice on the number of tolerable cheaters.

Our scheme (as well as [11] and [12]) has two rounds. In fact, it is round-optimal since Cramer *et al.* [15] showed that two rounds of communication is necessary in the rushing adversary model, if the secret sharing scheme requires

Table 2. Redundancy Needed for Cheater Identification when $t = \lfloor (n-1)/3 \rfloor$, $\epsilon = 2^{-80}$

n	Red_{Cho}	Red_{Our}	Red_{Cho}/Red_{Our}
4	120.6 B	26.7 B	4.5
1024	33.2 KB	6.7 KB	5.0
2^{18}	9.0 MB	1.7 MB	5.4

an agreement among all honest players. Since our scheme is public cheater identifiable, an agreement among all honest players must indeed be achieved.

2 Preliminaries

Let us first fix some notation. Set $[n] = \{1, 2, \ldots, n\}$. The cardinality of the set
X is denoted by $|X|$. Let $\mathbb{F}_p$ be a Galois field of a prime order p satisfying $p > n$.
All computation is done in the specified Galois fields.

2.1 Security Model and Communication Model

Throughout the paper, we consider an active rushing adversary with unbounded
computational power. By being *rushing* we mean that the adversary can observe
the information sent by all the honest players at each communication round,
prior to deciding on her own messages. The adversary can adaptively corrupt
up to t players (which then will be called *cheaters*) during the whole protocol
execution provided that $t < k/3$, where k is the threshold of the secret sharing
scheme. As usual in SSCI schemes, we assume that adversary cannot corrupt
the dealer D.

We assume that the entities are connected pairwise by private and authenticated channels, and also that broadcast channel is available.

2.2 Secret Sharing

The n players are denoted by $\{P_1, P_2, \ldots, P_n\}$. Let s be the secret chosen by D
from some distribution S, and let σ_i be the share distributed to player P_i. The
set of P_i's possible shares is denoted by V_i. By a slight abuse of notation, we
also use S to denote the random variable induced by s and V_i as the random
variable induced by σ_i.

First, we describe k-out-of-n secret sharing scheme by Shamir [1]. A k-out-
of-n secret sharing scheme involves a dealer D and n participants $\{P_1, \ldots, P_n\}$,
and consists of two algorithms: **ShareGen** and **Reconst**. The **ShareGen** al-
gorithm takes a secret $s \in \mathbb{F}_p$ as input and outputs a list $(\sigma_1, \ldots, \sigma_n)$. Each σ_i
is distributed to participant P_i and called her share. The algorithm **Reconst**
takes a list $(\sigma_1, \ldots, \sigma_m)$ as input and outputs the secret s if $m \geq k$. Otherwise,

the **Reconst** outputs $\perp$. Formally, the properties of *correctness* and *perfect secrecy* hold:

1. Correctness: If $m \geq k$, then $\Pr[\textbf{Reconst}(\sigma_1, \ldots, \sigma_m) = s] = 1$;
2. Perfect secrecy: If $m < k$, then $\Pr[S = s | (V_1 = \sigma_1, \ldots, V_m = \sigma_m)] = \Pr[S = s]$ for any $s \in S$.

In Shamir scheme, the above mentioned algorithms proceed as follows:

ShareGen:

1. For a given secret $s \in \mathbb{F}_p$, the dealer D chooses a random polynomial $f(x) \in \mathbb{F}_p[X]$ with degree at most $k - 1$ and $f(0) = s$.
2. For $i \in [n]$, compute $\sigma_i = f(x_i)$ for a fixed $x_i \in \mathbb{F}_p$ (where x_i can be seen as a unique identifier for P_i) and send σ_i privately to participant P_i.

Reconst:

If $m \geq k$ then output the secret s using Lagrange interpolation formula, otherwise output $\perp$.

Remark 1. For simplicity of our presentation, we will henceforth write the identifier of P_i as i, rather than x_i.

Next, we formalize k-out-of-n SSCI schemes. As compared to ordinary secret sharing schemes, we require that the reconstruction algorithm **Reconst** both computes the secret and identifies incorrect shares that point at cheaters among the involved participants. The output of **Reconst** algorithm is a tuple (s', L), where s' is the reconstructed secret and L is the set of cheaters, moreover $s' = s$ except with negligible probability. If a secret can not be reconstructed from the given shares, then s' is set to $\perp$, while $L = \emptyset$ denotes the fact that no cheater is identified.

Definition 1. *A k-out-of-n SSCI scheme Σ is a tuple $(n, k, S, V, \textbf{ShareGen}, \textbf{Reconst})$ consisting of :*

- *A positive integer n called the number of players;*
- *A positive integer k denoting the number of honest shares from which the original secret can be reconstructed;*
- *A finite set S with $|S| \geq 2$, whose elements are called secrets;*
- *A finite set $V = \{V_1, V_2, \ldots, V_n\}$, where V_i is the set of player P_i's shares.*
- *An algorithm **ShareGen**, that takes as input a secret $s \in S$, and outputs a vector of n shares $(\sigma_1, \sigma_2, \ldots, \sigma_n) \in V_1 \times V_2 \times \cdots \times V_n$; and*
- *An algorithm **Reconst**, that takes as input a vector $(\sigma'_{i_1}, \sigma'_{i_2}, \ldots, \sigma'_{i_m}) \in V_{i_1} \times V_{i_2} \times \cdots \times V_{i_m}$, and outputs a tuple (s', L), where s' is the reconstructed secret and L is the set of identified cheaters.*

Remember that t denotes the maximum number of cheaters that a rushing adversary can corrupt. We assume that the players in $A^{(t)} = \{P_{i_1}, P_{i_2}, \ldots, P_{i_t}\}$ are corrupted by the rushing adversary. In the SSCI scheme, a cheater $P_{i_j} (1 \leq j \leq t)$ succeeds if **Reconst** fails to identify P_{i_j} as a cheater when P_{i_j} provides a forged share. Note that if P_{i_j} succeeded in cheating, then the reconstructed secret s' is different from the original secret s. Without loss of generality, we assume that at the reconstruction, the corrupted players P_i's are those with the smallest i's in $[n]$.

Definition 2. *The successful cheating probability of player $P_{i_j} \in A^{(t)}$ against the SSCI scheme $\Sigma = (n, k, S, V, \textbf{ShareGen}, \textbf{Reconst})$ is defined as*

$$\epsilon(\Sigma, A^{(t)}, P_{i_j})$$
$$= \Pr[(s', L) \leftarrow \textbf{Reconst}(\sigma'_{i_1}, \sigma'_{i_2}, \ldots, \sigma'_{i_t}, \sigma_{i_{t+1}}, \ldots, \sigma_{i_k}) \land P_{i_j} \notin L : \sigma'_{i_j} \neq \sigma_{i_j}]. \tag{4}$$

Remark 2. In Definition 2, we set $\epsilon(\Sigma, A^{(t)}, P_{i_j})$ to be the successful cheating probability of an *individual* cheater P_{i_j}. Since at most t players can be corrupted, the overall failure probability for the SSCI scheme (i.e., the probability that at least one cheater in $A^{(t)}$ succeeds) can be upper-bounded using the union bound. We choose the individual successful cheating probability instead of the overall failure probability to be in accordance with the definition of Obana [10].

Definition 3. *A k-out-of-n SSCI scheme $\Sigma = (n, k, S, V, \textbf{ShareGen}, \textbf{Reconst})$ is called (t, ϵ) SSCI scheme if the following properties hold:*

1. *Perfect secrecy: At the end of the algorithm $\textbf{ShareGen}$, any set of players of size at most $k - 1$ have no information about the secret s.*
2. *ϵ-Correctness: $\epsilon(\Sigma, A^{(t)}, P_i) \leq \epsilon$ for any $A^{(t)}$ denoting the set of t or less rushing cheaters, for any cheater $P_i \in A^{(t)}$. If at least k honest players join the reconstruction protocol, the secret will be correctly recovered unless the cheaters remain undetected.*

Remark 3. Note that if at least k honest players take part in the reconstruction protocol, successful identification of cheaters is equivalent to recovering the original secret. The secret is not correctly recovered if and only if one or more cheaters are undetected. However, if less than k honest players are available, our scheme can only identify the cheaters without recovering the original secret. This is an intrinsic limitation of SSCI schemes since we only require k players to identify the cheaters.

Remark 4. Our protocol, as well as the works of [4,10–12], prevents false positive error, i.e., honest participants will never be identified as cheaters.

2.3   Reed-Solomon Error Correction

Let $f(x) \in \mathbb{F}_p[X]$ be a polynomial of degree at most k. Let $x_1, x_2, \ldots, x_n \in \mathbb{F}_p$, for $n > k$, be pairwise distinct interpolation points. Then $C = (f(x_1), f(x_2), \ldots, f(x_n))$ is a codeword of Reed-Solomon error correction code [16]. Reed-Solomon code can correct up to $\frac{n-k}{2}$ erroneous symbols, i.e. when t out of n evaluation points $f(x_i)$ $(1 \leq i \leq n)$ are corrupted, the polynomial can be uniquely determined if and only if $n - k > 2t$. Note that there exist efficient algorithms implementing Reed-Solomon decoding, such as Berlekamp-Welch algorithm [17]. We refer the reader to [18] for details on Reed-Solomon codes.

3   Our Proposal

In this section, we describe our k-out-of-n SSCI scheme secure against $t < k/3$ rushing adversary. We suppose that $m \geq k$ participants take part in the reconstruction phase.

3.1   Overview

Our proposal departs from Obana's scheme [10] and improves it in the following manner. Consider k-out-of-n Shamir secret sharing. Since the maximum number of cheaters is $\lfloor (k-1)/3 \rfloor$ and at least k players will take part in the reconstruction phase, Obana [10] uses a polynomial of degree t to compute authentication tags for each player's share. The degree-t polynomial can be recovered given at least $k \geq 3t+1$ players' tags, t of which might be corrupted, using Reed-Solomon decoding (with probability 1). In this scheme, protection against rushing adversary is not provided, since the latter can see all the tags of the k players and recover the polynomial (since $k \geq t+1$). In other words, the adversary can recover the authentication key, so that she will be able to forge authentication tag for an arbitrary value submitted as her share.

In order to deal with this problem, we split the reconstruction phase into two rounds. In the first round, only the Shamir shares and masked authentication tags are revealed. Then in the second round, the masking key will be submitted by each player. We share the masking key between all the n players using a $(t + 1)$-out-of-n Shamir secret sharing, such that any t corrupted players can neither get any information about the key nor alter it in the reconstruction phase.

Unfortunately, the necessity to share the masking keys takes the share size of our scheme away from the optimal bound. However, we observe that there is no need to mask *all* of the authentication tags: since the knowledge of any t of them gives no advantage to the adversary, it suffices to mask only $n - t$ of them.

3.2   Our Scheme

Let q be a prime power such that $q \geq n \cdot p$ and let $\phi : \mathbb{F}_p \times [n] \to \mathbb{F}_q$ be an injective function.

Our proposed scheme is described below.

Protocol 1 (ShareGen).

Input: Secret $s \in \mathbb{F}_p$.
Output: A list of n shares $\sigma_1, \sigma_2, \ldots, \sigma_n$.

A dealer D performs the following:

1. Generate a random degree-$(k - 1)$ polynomial $f_s(x)$ over $\mathbb{F}_p$, such that $f_s(0) = s$. Compute $v_{s,i} = f_s(i)$, for $i = 1, 2, \ldots, n$.
2. Select a random degree-t polynomial $g(x)$ over $\mathbb{F}_q$. Compute $v_{c,i} = g(\phi(v_{s,i}, i))$.

3. (a) For $i = 1, 2, \ldots, t$: Set $\overline{v_{c,i}} = v_{c,i}$;
   (b) For $i = t+1, t+2, \ldots, n$: Randomly and uniformly generate a key $k_i \in \mathbb{F}_q$, and compute $\overline{v_{c,i}} = v_{c,i} + k_i$.
4. For $i = t+1, t+2, \ldots, n$: Generate a random degree-t polynomial $h_i(x)$ over $\mathbb{F}_q$, such that $h_i(0) = k_i$. Compute $k_{i,j} = h_i(j)$, for $j = 1, 2, \ldots, n$.
5. For $i \in [n]$, set $\sigma_i = \{v_{s,i}, \overline{v_{c,i}}, k_{t+1,i}, \ldots, k_{n,i}\}$ and distribute it privately to player P_i.

Remark 5. Note that in Step 2, we must combine player's share $v_{s,i}$ with her identifier i before authentication, since otherwise a cheater can "steal" a share and its authentication tag from some other player pretending that she has received the same share without being detected. However, when we authenticate the combination of the share and the identifier of a player, which is $\phi(v_{s,i}, i)$, the entities to be authenticated will be distinct for every player even if they received the same share, since $\phi(\cdot, \cdot)$ is an injective function.

Let $CORE = \{i_1, i_2, \ldots, i_m\}$ be the set of identifiers of the m participants who want to recover the secret. Moreover, let $\sigma'_{i_j} = \{v'_{s,i_j}, \overline{v'_{c,i_j}}, k'_{t+1,i_j}, \ldots, k'_{n,i_j}\}$ for each $i_j \in CORE$. Furthermore, at most t out of m shares can be corrupted in a rushing fashion.

Protocol 2 (Reconst).

Input: A list of m shares $(\sigma'_{i_1}, \sigma'_{i_2}, \ldots, \sigma'_{i_m})$, where $m \geq k$.
Output: Either $(\perp, L)$ or (s', L), where L is the list of cheaters.

Communication rounds performed by each player $i_j \in CORE$:

1. Announce $\{v'_{s,i_j}, \overline{v'_{c,i_j}}\}$.
2. Announce $\{k'_{t+1,i_j}, k'_{t+2,i_j}, \ldots, k'_{n,i_j}\}$.

Computation by players in $CORE$:

1. For each $i_j \in CORE \cap \{t+1, t+2, \ldots, n\}$, reconstruct k'_{i_j} from $\{k'_{i_j,i_1}, \ldots, k'_{i_j,i_m}\}$ using Reed-Solomon decoding.
2. For $i_j \in CORE \cap \{1, 2, \ldots, t\}$, set $v'_{c,i_j} = \overline{v'_{c,i_j}}$;
   For $i_j \in CORE \cap \{t+1, t+2, \ldots, n\}$, compute $v'_{c,i_j} = \overline{v'_{c,i_j}} - k'_{i_j}$.
3. Reconstruct $g'(x)$ from $v'_{c,i_1}, v'_{c,i_2}, \ldots, v'_{c,i_m}$ using Reed-Solomon decoding.
4. Check if $v'_{c,i_j} = g'(\phi(v'_{s,i_j}, i_j))$ holds for $1 \leq j \leq m$.
   If $v'_{c,i_j} \neq g'(\phi(v'_{s,i_j}, i_j))$ then i_j is added to the list of cheaters L.
5. If $|L| > m - k$ then output $(\perp, L)$, otherwise:
   Reconstruct $f'_s(x)$ from (k or more) shares v'_{s,i_j} such that $i_j \in CORE \setminus L$ using Lagrange interpolation.
   If $\deg(f'_s) \leq k-1$, output $(f'_s(0), L)$, otherwise output $(\perp, L)$.

Note that the condition $|L| > m - k$ in Step 5 means that the number of honest players is less than k.

Remark 6. For simplicity of our presentation, in the above protocol, we omitted the check similar to that in Step 4, which must be performed in Step 1. In details: When reconstructing in Step 1 the polynomial $h_{i_j}(x)$ (the one used to share the key k_{i_j}) with Reed-Solomon decoding, we must check whether P_i provided a forged share $k'_{i_j,i} \neq k_{i_j,i}$ and put her into the list L, if this is the case. However, we note that in this step, under assumption that $t < k/3$, the cheaters who submitted a forged share will be identified with probability 1.

4   Security Proof

The security of our SSCI scheme is argued in the following theorem.

Theorem 1. *If $t < k/3$ then the scheme described above is a (t, ϵ) SSCI against rushing adversary such that*

$$|S| = p, \ \epsilon = 1/q, \ q \geq n \cdot p, \ |V_i| = p \cdot q^{n-t+1} = |S|/\epsilon^{n-t+1}. \qquad (5)$$

Proof. First, we show that the scheme satisfy perfect secrecy. Suppose that $k-1$ players $\{P_{i_1}, P_{i_1}, \ldots, P_{i_{k-1}}\}$ want to get the secret from their shares. Denote by $\sigma_{i_j} = \{v_{s,i_j}, \overline{v_{c,i_j}}, k_{t+1,i_j}, \ldots, k_{n,i_j}\}$ the share of player P_{i_j}. Due to the secrecy of Shamir scheme, the values $(v_{s,i_1}, v_{s,i_2}, \ldots, v_{s,i_{k-1}})$ do not reveal any information about the secret. Moreover, it is easy to see that the knowledge about $\overline{v_{c,i_j}}$ and $(k_{t+1,i_j}, k_{t+2,i_j}, \ldots, k_{n,i_j})$ does not leak any information about the secret since the polynomial $g(x)$ and the masking keys $(k_{t+1}, k_{t+2}, \ldots, k_n)$ are chosen independently of the secret s.

Next we show that our scheme is ϵ-correct. Our proof follows the lines of [10]. Let us observe the following two facts:

1. For $x_1, \ldots, x_k \in \mathbb{F}_q$, $(g(x_1), g(x_2), \ldots, g(x_k))$ is a codeword of the Reed-Solomon code with minimum distance $k - t$ (since $\deg(g(x)) \leq t$). According to the Reed-Solomon error correction, if $k - t > 2t$ (i.e., $t < k/3$) the degree-t polynomial $g(x)$ can be correctly reconstructed from the k points even if t of them are forged. For the same reason, the masking keys $(k_{t+1}, k_{t+2}, \ldots, k_n)$ can be correctly recovered by k players.

2. The set of functions $\{g(x)|g(x) \in \mathbb{F}_q[X], \deg(g(x)) \leq t\}$ is a class of strongly universal$_{t+1}$ hash functions $\mathbb{F}_q \to \mathbb{F}_q$ [19]; that is, the following equality holds for any distinct $x_1, \ldots, x_t, x_{t+1} \in \mathbb{F}_q$ and the following $y_1, y_2, \ldots, y_t, y_{t+1} \in \mathbb{F}_q$:

$$\Pr[g(x_{t+1}) = y_{t+1}|g(x_1) = y_1, g(x_1) = y_2, \ldots, g(x_t) = y_t] = 1/q. \qquad (6)$$

Let us suppose without loss of generality that the rushing adversary corrupts $P_{i_1}, \ldots, P_{i_t}$ and $CORE \cap \{1, 2, \ldots, t\} = \{i_1, i_2, \ldots, i_l\}$ $(l \leq t)$. Remember that since the adversary is rushing, she can see all the communication of honest players during each round, prior to deciding her own messages. We summarize the view of the adversary in Table 3.

Table 3. Adversary's View in **Reconst**

First Round:	Second Round:
$(v_{s,i_1}, \overline{v_{c,i_1}}, k_{i_{t+1},i_1}, \ldots, k_{i_n,i_1})$	$(v_{s,i_1}, \overline{v_{c,i_1}}, k_{i_{t+1},i_1}, \ldots, k_{i_n,i_1})$
$\ldots$	$\ldots$
$(v_{s,i_t}, \overline{v_{c,i_t}}, k_{i_{t+1},i_t}, \ldots, k_{i_n,i_t})$	$(v_{s,i_t}, \overline{v_{c,i_t}}, k_{i_{t+1},i_t}, \ldots, k_{i_n,i_t})$
$(v_{s,i_{t+1}}, \overline{v_{c,i_{t+1}}})$	$(v_{s,i_{t+1}}, \overline{v_{c,i_{t+1}}}, k_{i_{l+1},i_{t+1}}, \ldots, k_{i_m,i_{t+1}})$
$\ldots$	$\ldots$
$(v_{s,i_m}, \overline{v_{c,i_m}})$	$(v_{s,i_m}, \overline{v_{c,i_m}}, k_{i_{l+1},i_m}, \ldots, k_{i_m,i_m})$

Suppose $P_{i*} \in \{P_{i_1}, P_{i_2}, \ldots, P_{i_t}\}$, who knows the values $\sigma_{i_1}, \sigma_{i_2}, \ldots, \sigma_{i_t}$, submits a forged share $\sigma'_{i*} = (v'_{s,i*}, v'_{c,i*}, k'_{i_l+1,i*}, \ldots, k'_{i_m,i*})$. P_{i*} is not identified as a cheater only if he submits $v'_{c,i*}$ such that $\overline{v'_{c,i*}} = g(\phi(v'_{s,i*}, i*)) + k_{i*}$. At the end of the first round, P_{i*} has to hand in the values $(v'_{s,i*}, \overline{v'_{c,i*}})$. At that time, she can see $(\overline{v_{c,i_1}}, \overline{v_{c,i_2}}, \ldots, \overline{v_{c,i_m}})$, $(k_{i_{t+1},i_1}, k_{i_{t+1},i_2}, \ldots, k_{i_{t+1},i_t})$, $\ldots$, $(k_{i_n,i_1}, k_{i_n,i_2}, \ldots, k_{i_n,i_t})$. From $(k_{j,i_1}, k_{j,i_2}, \ldots, k_{j,i_t})$, $(t+1 \leq j \leq n)$ the cheater P_{i*} can have no information about the masking key k_j since it is shared by the $(t+1)$-out-of-n Shamir scheme. For any $i_j \in CORE \bigcap \{t+1, t+2, \ldots, n\}$, $\overline{v_{c,i_j}} = g(\phi(v_{s,i_j}, i_j)) + k_{i_j}$ looks like a random value to P_{i*}, since k_{i_j} will not be revealed until the second round, and before it serves as a one-time pad. For $i_j \in \{i_1, i_2, \ldots, i_l\}$, P_{i*} will see the values of the function $g(x)$, namely $g(\phi(v_{s,i_j}, i_j)) = v_{c,i_j}$ for $1 \leq j \leq l$, and $l \leq t$. After the second round, all the keys $k_{i_{l+1}}, k_{i_{l+2}}, \ldots, k_{i_m}$ can be correctly reconstructed – since the polynomial $h_j(x)$ hiding k_{i_j} is of degree t and $t < k/3$, we can use Reed-Solomon error correction algorithm to recover the key k_{i_j} despite possibly t corrupted shares of k_{i_j}. By the similar reason, the polynomial $g(x)$ can be correctly reconstructed as well. Since P_{i*} submits the forged share $\sigma'_{i*} = (v'_{s,i*}, \overline{v'_{c,i*}}, k'_{i_l+1,i*}, \ldots, k'_{i_m,i*})$ before he knows the corresponding masking keys, the following holds:

$$\Pr[g(\phi(v'_{s,i*}, i*)) = \overline{v'_{c,i*}} - k_{i*} \mid g(\phi(v_{s,i_j}, i_j)) = v_{c,i_j} : 1 \leq j \leq l \, (l \leq t)] \leq 1/q, \tag{7}$$

where the probability is taken over the random choice of $g(x)$, and $k_{t+1}, k_{t+2}, \ldots, k_n$, and $h_{t+1}(x), h_{t+2}(x), \ldots, h_n(x)$. From the above discussion we can see that any cheater will be identified except with probability at most $1/q$. Therefore, our SSCI scheme satisfies the ϵ-correctness property with $\epsilon = 1/q$.

It is easy to compute the share size as $|V_i| = p \cdot q^{n-t+1} = |S|/\epsilon^{n-t+1}$.

5   Conclusion

We proposed an SSCI scheme capable of identifying $t < k/3$ rushing cheaters. Our scheme is superior to that of Choudhury [11] (for the single secret) and Cevallos et al. [12] (if no more than k players can take part in the reconstruction phase), when the number of cheaters is less than $k/3$.

According to the lower bound (1) from Kurosawa *et al.* [9], our scheme is not optimal in the sense of share size $|V_i|$. It is an interesting open problem to design an SSCI scheme against $t < k/2$ (or at least $t < k/3$) *rushing* cheaters with optimal (or at least constant in n, k and t) size of $|V_i|$, even for sharing of a single field element.

Acknowledgments. R.X. is supported by The China Scholarship Council, No. 201206340057. K.M. is supported by a *kakenhi* Grant-in-Aid for Young Scientists (B) 24700013 from Japan Society for the Promotion of Science.

K.M. and T.T. would like to thank Satoshi Obana for introducing them to this research topic and for some motivating discussions.

The authors would also like to thank Rui Zhang and anonymous reviewers of IWSEC 2013 for some valuable comments that helped to improve this presentation.

References

1. Shamir, A.: How to share a secret. Commun. ACM 22(11), 612–613 (1979)
2. Blakley, G.: Safeguarding cryptographic keys. In: AFIPS:79 National Computer Conference, pp. 313–317. IEEE Computer Society (1979)
3. Desmedt, Y.: Threshold cryptography. European Transactions on Telecommunications 5(4), 449–458 (1994)
4. Rabin, T., Ben-Or, M.: Verifiable secret sharing and multiparty protocols with honest majority (extended abstract). In: STOC, vol. 1989, pp. 73–85 (1989)
5. Cramer, R., Damgård, I., Maurer, U.: General secure multi-party computation from any linear secret-sharing scheme. In: Preneel, B. (ed.) EUROCRYPT 2000. LNCS, vol. 1807, pp. 316–334. Springer, Heidelberg (2000)
6. Dolev, D., Dwork, C., Waarts, O., Yung, M.: Perfectly secure message transmission. J. ACM 40(1), 17–47 (1993)
7. Tompa, M., Woll, H.: How to share a secret with cheaters. J. Cryptology 1(2), 133–138 (1988)
8. McEliece, R., Sarwate, D.: On sharing secrets and reed-solomon codes. Commun. ACM 24(9), 583–584 (1981)
9. Kurosawa, K., Obana, S., Ogata, W.: t-cheater identifiable (k, n) threshold secret sharing schemes. In: Coppersmith, D. (ed.) CRYPTO 1995. LNCS, vol. 963, pp. 410–423. Springer, Heidelberg (1995)
10. Obana, S.: Almost optimum t-cheater identifiable secret sharing schemes. In: Paterson, K.G. (ed.) EUROCRYPT 2011. LNCS, vol. 6632, pp. 284–302. Springer, Heidelberg (2011)
11. Choudhury, A.: Brief announcement: optimal amortized secret sharing with cheater identification. In: PODC 2012, pp. 101–102 (2012)
12. Cevallos, A., Fehr, S., Ostrovsky, R., Rabani, Y.: Unconditionally-secure robust secret sharing with compact shares. In: Pointcheval, D., Johansson, T. (eds.) EUROCRYPT 2012. LNCS, vol. 7237, pp. 195–208. Springer, Heidelberg (2012)
13. Carpentieri, M.: A perfect threshold secret sharing scheme to identify cheaters. Des. Codes Cryptography 5(3), 183–187 (1995)

14. Ishai, Y., Ostrovsky, R., Seyalioglu, H.: Identifying cheaters without an honest majority. In: Cramer, R. (ed.) TCC 2012. LNCS, vol. 7194, pp. 21–38. Springer, Heidelberg (2012)
15. Cramer, R., Damgård, I., Fehr, S.: On the cost of reconstructing a secret, or VSS with optimal reconstruction phase. In: Kilian, J. (ed.) CRYPTO 2001. LNCS, vol. 2139, pp. 503–523. Springer, Heidelberg (2001)
16. Reed, I., Solomon, G.: Polynomial codes over certain finite fields. Journal of the Society for Industrial & Applied Mathematics 8(2), 300–304 (1960)
17. Welch, L., Berlekamp, E.: Error correction for algebraic block codes US Patent 4,633,470 (December 30, 1986)
18. Roth, R.: Introduction to coding theory. Cambridge University Press (2006)
19. Wegman, M., Carter, L.: New hash functions and their use in authentication and set equality. J. Comput. Syst. Sci. 22(3), 265–279 (1981)

One-Round Authenticated Key Exchange without Implementation Trick

Kazuki Yoneyama

NTT Secure Platform Laboratories
3-9-11 Midori-cho Musashino-shi Tokyo 180-8585, Japan
yoneyama.kazuki@lab.ntt.co.jp

Abstract. Fujioka et al. proposed the first generic construction (FSXY construction) of exposure-resilient authenticated key exchange (AKE) from key encapsulation mechanism (KEM) without random oracles. However, the FSXY construction implicitly assumes some intermediate computation result is never exposed though other secret information can be exposed. This is a kind of physical assumption, and an implementation trick (i.e., some on-line computation is executed in a special tamper-proof module) is necessary to achieve the assumption. Unfortunately, such an implementation trick is very costly and should be avoided. In this paper, we introduce a new generic construction without the implementation trick. Our construction satisfies the same security model as the FSXY construction without increasing communication complexity. Moreover, it has another advantage that the protocol can be executed in one-round while the FSXY construction is a sequential two-move protocol. Our key idea is to use KEM with public-key-independent-ciphertext, which allows parties to be able to generate a ciphertext without depending on encryption keys.

Keywords: authenticated key exchange, NAXOS trick, key encapsulation mechanism, exposure-resilience.

1 Introduction

1.1 Background

Authenticated Key Exchange (AKE) is a cryptographic primitive to share a common *session key* among multiple parties through unauthenticated networks such as the Internet. In the ordinary PKI-based setting, each party locally keeps his own *static secret key* (SSK) and publishes a *static public key* (SPK) corresponding to the SSK. Validity of SPKs is guaranteed by a certificate authority. In a key exchange session, each party generates an *ephemeral secret key* (ESK) and sends an *ephemeral public key* (EPK) corresponding to the ESK. A session key is derived from these keys with a *key derivation procedure*. Parties can establish a secure channel with the session key.

An important goal of this area is to achieve exposure-resilience. That means, even if an adversary learns some of secret keys of parties, generated session keys must be protected. For example, SSKs may expose if a party is corrupted, or a device itself (e.g., a smart phone) that records a SSK is physically stolen. On another scenario, ESKs may expose if computations inputting an ESK are executed in a memory area of the smart

K. Sakiyama and M. Terada (Eds.): IWSEC 2013, LNCS 8231, pp. 272–289, 2013.

phone, and a malicious developer steals it via a hidden malware that is embedded in some apps. Therefore, it is desirable to guarantee exposure-resilience in a provably secure way.

There are several studies about modeling exposure-resilience in the AKE setting. Canetti and Krawczyk [1] defined the first security model of AKE capturing exposure of SSKs and session state (i.e., some intermediate computation result), that is called *Canetti-Krawczyk (CK) model*. However, the CK model does not allow an adversary to learn any of SSKs and session state of the target session (called the *test session*). LaMacchia et al. [2] also proposed very strong security models capturing exposure of both SSKs and ESKs, that is called *extended CK (eCK) model*. While the eCK model allows an adversary to directly learn SSKs and ESKs of the test session, exposure of session state is not captured. The CK^+ model [3,4] combines these two models; that is, an adversary can obtain SSKs and ESKs of the test session, and can learn session state of other sessions. Note that the eCK model and the CK^+ model are rigorously incomparable [5,6].

Concrete AKE schemes satisfying these models have been studied. HMQV [3] is one of the most efficient protocols and satisfies the CK^+ model. However, the security proof is given in the random oracle model (ROM) under the knowledge-of-exponent assumption [7] that is a widely criticized assumption [8]. Boyd et al. [9,10] propose a generic construction (BCGNP construction) of AKE from key encapsulation mechanism (KEM), that is secure in the CK model in the standard model (StdM). Because the CK model does not capture exposure of ESKs in the test session, unfortunately, it is unclear whether the BCGNP construction is secure when the ESK of the test session is exposed. Fujioka et al, [4] show that the BCGNP construction is insecure in the CK^+ model, and propose another generic construction (FSXY construction) of AKE from KEM, that is secure in the CK^+ model in the StdM.

1.2 Motivation

The FSXY construction uses a technique to resist exposure of ESKs, which is called the twisted pseudo-random function (PRF) trick [11]. This trick is essentially the same as the NAXOS trick [2] except with/without random oracles (ROs). Roughly, a party uses $H(SSK, ESK)$ to compute an EPK instead of using the ESK directly, where H is some intractable function like ROs. Unless both the SSK and the ESK are exposed, $H(SSK, ESK)$ cannot be computed by an adversary even if the ESK is exposed. Thus, the FSXY construction guarantees the security against exposure of ESKs.

However, such a trick has several problems. First, it needs some *implementation trick*, because it is assumed that exposure of $H(SSK, ESK)$ never occurs while ESKs may be exposed. A typical implementation is that all computations inputting $H(SSK, ESK)$ are executed in a tamper-proof module (TPM) such as a smart card. Without the implementation trick (i.e., $H(SSK, ESK)$ is handled by the same manner as ESKs), the twisted PRF trick is not meaningful, and it may lead to a 'full' ESK exposure attack (i.e., $H(SSK, ESK)$ and ESKs are exposed simultaneously) to the FSXY construction though it is proved in the CK^+ model. The other is an efficiency problem. As discussed above, computations inputting $H(SSK, ESK)$ must be executed in a TPM. In the FSXY construction, $H(SSK, ESK)$ is used as randomness in generating a ciphertext of chosen ciphertext secure (IND-CCA secure) KEM, that is sent to the peer. This computation must be done *on-line* (i.e., any pre-computation is not possible) because the ciphertext is generated *after* the peer of the session is fixed. Therefore, the TPM must process a very heavy on-line computation (i.e., an encryption algorithm of IND-CCA secure KEM) for each session. It is clearly not desirable in practice.

1.3 Our Contribution

First, we clarify that the FSXY construction is insecure in the CK^+ model if the implementation trick is missed (i.e., the outputs of the twisted PRF are handled by the same manner as ESKs) in Section 3. Specifically, we give a simple attack using 'full' ESK exposure. This fact shows that the FSXY construction essentially needs very heavy on-line computations in a TPM or similar implementation tricks.

Next, we introduce a new generic construction of AKE from KEM, that is secure in the CK^+ model without relying on the twisted PRF trick in Section 4. Our key idea is to use *KEM with public-key-independent-ciphertext* (PKIC-KEM) [12]. PKIC-KEM allows that a ciphertext can be generated independently from an encryption key, and a KEM key can be generated with the ciphertext, the encryption key and randomness in generating the ciphertext. While the previous work [12] uses a semantically secure (IND-CPA secure) PKIC-KEM to obtain a one-round AKE scheme, we use IND-CPA secure PKIC-KEM both to resist full ESK exposure and to obtain one-round protocol.[1] A typical example of IND-CPA secure PKIC-KEM is the ElGamal KEM (i.e., an encryption key is g^a, a ciphertext is g^r, and the KEM key is g^{ar}).

Furthermore, though the FSXY construction adapts a strong randomness extractor as a part of the session key derivation procedure, we can replace it with a weaker building block, a *key derivation function* (KDF). The KDF is weaker and more efficient primitive than the strong randomness extractor; the output of the KDF is just guaranteed computationally indistinguishable from random value but the strong randomness extractor guarantees statistical indistinguishability. We can prove the security of our construction only with the computational property; thus, we can improve efficiency of the session key derivation procedure. This technique is proposed in [13,14]

There are some related works [15,16] that achieve exposure-resilient AKE schemes in the StdM without the implementation trick. However, these schemes are specific constructions (i.e., not generic construction), and rely on a strong building block, PRFs with pairwise-independent random sources (πPRF). It is not known how to construct πPRF concretely. Table 1 shows a comparison of exposure-resilient AKE schemes without implementation tricks. HMQV is the most efficient but relies on RO. The schemes in [15,16] is secure in the StdM but relies on πPRF. In addition, the scheme in [16] needs pairing operations. Therefore, our scheme needs less communication cost than the schemes in the StdM, and does not rely on πPRF.

---

[1] One-round means that parties can send their EPKs independently and simultaneously in two-move protocols.

Table 1. Comparison of exposure-resilient AKE without implementation tricks

	Model	Resource	Assumption	Computation (#parings + #[multi,regular]-exp.)	Communication complexity	
[3]	CK$^+$	ROM	gap DH & KEA1	$0 + [2,2]$	$2|p|$	512
[15]	eCK	StdM	DDH & πPRF	$0 + [2,6]$	$9|p|$	2304
[16]	eCK	StdM	DBDH & DLIN & πPRF	$2 + [2,8]$	$12|p|$	3072
Ours	CK$^+$	StdM	DDH	$0 + [4,12]$	$8|p|$	2048

For concreteness the expected ciphertext overhead for a 128-bit implementation is also given. Note that computational costs are estimated without any pre-computation technique. Our protocol is instantiated by the Cramer-Shoup KEM [17] as IND-CCA secure KEM and the ElGamal KEM as IND-CPA secure PKIC-KEM.

2   CK$^+$ Security Model

In this section, we recall the CK$^+$ model [3,4]. We slightly modify the model to specify one-round protocols. It can be trivially extended to any round protocol.

Notations. Throughout this paper we use the following notations. If M is a set, then by $m \in_R$ M we denote that m is sampled uniformly from M. If $\mathcal{R}$ is an algorithm, then by $y \leftarrow \mathcal{R}(x; r)$ we denote that y is output by $\mathcal{R}$ on input x and randomness r (if $\mathcal{R}$ is deterministic, r is empty).

We denote a party by U_P, and party U_P and other parties are modeled as probabilistic polynomial-time (PPT) Turing machines w.r.t. security parameter κ. For party U_P, we denote static secret (public) key by SSK_P (SPK_P) and ephemeral secret (public) key by ESK_P (EPK_P). Party U_P generates its own keys, ESK_P and EPK_P, and the static public key SPK_P is linked with U_P's identity in some systems like PKI.[2]

Session. An invocation of a protocol is called a *session*. Session activation of party U_P is done by an incoming message of the forms $(\Pi, U_P, U_{\bar{P}})$, where we equate Π with a protocol identifier, and $U_{\bar{P}}$ is the party identify of the peer. Party U_P outputs $(\Pi, U_P, U_{\bar{P}}, EPK_P)$, receives an incoming message of the forms $(\Pi, U_{\bar{P}}, U_P, EPK_{\bar{P}})$ from the peer $U_{\bar{P}}$, and then computes the session key SK.

A session of U_P is identified by sid $= (\Pi, U_P, U_{\bar{P}}, EPK_P)$ or sid $= (\Pi, U_P, U_{\bar{P}}, EPK_P, EPK_{\bar{P}})$. We say that U_P is the *owner* of session sid, if the second coordinate of session sid is U_P. We say that U_P is the *peer* of session sid, if the third coordinate of session sid is U_P. We say that a session is *completed* if its owner computes the session key. The *matching session* of $(\Pi, U_P, U_{\bar{P}}, EPK_P, EPK_{\bar{P}})$ is session $(\Pi, U_{\bar{P}}, U_P, EPK_P, EPK_{\bar{P}})$ and vice versa.

Adversary. The adversary $\mathcal{A}$, which is modeled as a PPT machine, controls all communications between parties including session activation by performing the following adversary query.

---

[2] Static public keys must be known to both parties in advance. They can be obtained by exchanging them before starting the protocol or by receiving them from a certification authority. This situation is common for all PKI-based AKE schemes.

- Send(message): The message has one of the following forms: $(\Pi, U_P, U_{\bar{P}})$, or $(\Pi, U_{\bar{P}}, U_P, EPK_{\bar{P}})$. The adversary $\mathcal{A}$ obtains the response from the party.

To capture exposure of secret information, the adversary $\mathcal{A}$ is allowed to issue the following queries.

- SessionKeyReveal(sid): The adversary $\mathcal{A}$ obtains the session key SK for the session sid if the session is completed.
- SessionStateReveal(sid): The adversary $\mathcal{A}$ obtains session state of the owner of session sid if the session is not completed (the session key is not established yet). Session state includes all ESKs and intermediate computation results except for immediately erased information but does not include the SSK.
- Corrupt(U_P): This query allows the adversary $\mathcal{A}$ to obtain all information of the party U_P. If a party is corrupted by a Corrupt(U_P, SPK_P) query issued by the adversary $\mathcal{A}$, then we call the party U_P dishonest. If not, we call the party honest.

Freshness. For the security definition, we need the notion of freshness.

Definition 1 (Freshness). Let sid* $= (\Pi, U_P, U_{\bar{P}}, EPK_P, EPK_{\bar{P}})$ be a completed session between honest users U_P and $U_{\bar{P}}$. If the matching session exists, then let $\overline{\text{sid}^*}$ be the matching session of sid*. We say session sid* is fresh if none of the following conditions hold:

1. The adversary $\mathcal{A}$ issues SessionKeyReveal(sid*), or SessionKeyReveal($\overline{\text{sid}^*}$) if $\overline{\text{sid}^*}$ exists,
2. $\overline{\text{sid}^*}$ exists and the adversary $\mathcal{A}$ makes either of the following queries
   - SessionStateReveal(sid*) or SessionStateReveal($\overline{\text{sid}^*}$),
3. $\overline{\text{sid}^*}$ does not exist and the adversary $\mathcal{A}$ makes the following query
   - SessionStateReveal(sid*).

Security Experiment. For the security definition, we consider the following security experiment. Initially, the adversary $\mathcal{A}$ is given a set of honest users and makes any sequence of the queries described above. During the experiment, the adversary $\mathcal{A}$ makes the following query at once.

- Test(sid*): Here, sid* must be a fresh session. Select random bit $b \in_R \{0, 1\}$, and return the session key held by sid* if $b = 0$, and return a random key if $b = 1$.

The experiment continues until the adversary $\mathcal{A}$ makes a guess b'. The adversary $\mathcal{A}$ wins the game if the test session sid* is still fresh and if the guess of the adversary $\mathcal{A}$ is correct, i.e., $b' = b$. The advantage of the adversary $\mathcal{A}$ in the AKE experiment with the PKI-based AKE protocol Π is defined as

$$\text{Adv}_{\Pi}^{\text{AKE}}(\mathcal{A}) = \Pr[\mathcal{A}\ wins] - \frac{1}{2}.$$

We define the security as follows.

Definition 2 (Security). *We say that a PKI-based AKE protocol Π is secure in the CK^+ model if the following conditions hold:*

1. *If two honest parties complete matching sessions, then, except with negligible probability, they both compute the same session key.*
2. *For any PPT bounded adversary $\mathcal{A}$, $\mathrm{Adv}_{\Pi}^{\mathrm{AKE}}(\mathcal{A})$ is negligible in security parameter κ for the test session sid^*,*
   (a) *if $\overline{\mathsf{sid}^*}$ does not exist, and the static secret key of the owner of sid^* is given to $\mathcal{A}$.*
   (b) *if $\overline{\mathsf{sid}^*}$ does not exist, and the ephemeral secret key of sid^* is given to $\mathcal{A}$.*
   (c) *if $\overline{\mathsf{sid}^*}$ exists, and the static secret key of the owner of sid^* and the ephemeral secret key of sid^* are given to $\mathcal{A}$.*
   (d) *if $\overline{\mathsf{sid}^*}$ exists, and the ephemeral secret key of sid^* and the ephemeral secret key of $\overline{\mathsf{sid}^*}$ are given to $\mathcal{A}$.*
   (e) *if $\overline{\mathsf{sid}^*}$ exists, and the static secret key of the owner of sid^* and the static secret key of the peer of sid^* are given to $\mathcal{A}$.*
   (f) *if $\overline{\mathsf{sid}^*}$ exists, and the ephemeral secret key of sid^* and the static secret key of the peer of sid^* are given to $\mathcal{A}$.*

3 'Full' Ephemeral Key Exposure Attack to FSXY Construction

In this section, we show an attack to the FSXY construction if an adversary can fully expose ESKs of parties. Therefore, it is a realistic attack when the FSXY construction is implemented without a special TPM.

3.1 Protocol of FSXY Construction

First, we recall the protocol of the FSXY construction.

It is a general construction from IND-CCA secure KEM (KeyGen, EnCap, DeCap) and IND-CPA secure KEM (wKeyGen, wEnCap, wDeCap), where the randomness space of encapsulation algorithms is $\mathcal{RS}_E$, the randomness space of key generation algorithms is $\mathcal{RS}_G$ and the KEM key space is $\mathcal{KS}$. Other building blocks are pseudorandom functions (PRFs) and a strong randomness extractor. For a security parameter κ, let $F : \{0,1\}^* \times \mathcal{FS} \to \mathcal{RS}_E$, $F' : \{0,1\}^* \times \mathcal{FS} \to \mathcal{RS}_E$, and $G : \{0,1\}^* \times \mathcal{FS} \to \{0,1\}^\kappa$ be PRFs, where $\mathcal{FS}$ is the key space of PRFs ($|\mathcal{FS}| = \kappa$). Let $Ext : SS \times \mathcal{KS} \to \mathcal{FS}$ be a strong randomness extractor with randomly chosen seed $s \in SS$, where SS is the seed space.

Party U_P randomly selects $\sigma_P \in_R \mathcal{FS}$ and $r \in_R \mathcal{RS}_G$, and runs $(ek_P, dk_P) \leftarrow$ KeyGen$(1^\kappa, r)$. Party U_P's SSK and SPK are $((dk_P, \sigma_P), ek_P)$. Fig. 1 shows the protocol.

3.2 Implementation Trick of FSXY Construction

The FSXY construction uses the twisted PRF trick to compute randomness in generating ciphertext CT_A and CT_B. For instance, randomness is computed as $F_{\sigma_A}(r_A) \oplus F'_{r'_A}(\sigma_A)$

Common public parameter : F, F', G, Ext, s
SSK and SPK for party U_A : $SSK_A := (dk_A, \sigma_A), SPK_A := ek_A$
SSK and SPK for party U_B : $SSK_B := (dk_B, \sigma_B), SPK_B := ek_B$

Party U_A

Party U_B

$r_A, r'_A \in_R \mathcal{FS}; r_{TA} \in_R \mathcal{RS}_G$

$(CT_A, K_A) \leftarrow$

$\underline{\mathsf{EnCap}_{ek_B}(F_{\sigma_A}(r_A) \oplus F'_{r'_A}(\sigma_A))}$

$(ek_T, dk_T) \leftarrow \mathsf{wKeyGen}(1^\kappa, r_{TA})$ $\xrightarrow{\quad U_A, U_B, CT_A, ek_T \quad}$

$r_B, r'_B \in_R \mathcal{FS}; r_{TB} \in_R \mathcal{RS}_E$

$(CT_B, K_B) \leftarrow$

$\underline{\mathsf{EnCap}_{ek_A}(F_{\sigma_B}(r_B) \oplus F'_{r'_B}(\sigma_B))}$

$\xleftarrow{\quad U_A, U_B, CT_B, CT_T \quad}$ $(CT_T, K_T) \leftarrow \mathsf{wEnCap}_{ek_T}(r_{TB})$

$K_B \leftarrow \mathsf{DeCap}_{dk_A}(CT_B)$

$K_T \leftarrow \mathsf{wDeCap}_{dk_T}(CT_T)$

$K_A \leftarrow \mathsf{DeCap}_{dk_B}(CT_A)$

$K'_1 \leftarrow Ext(s, K_A); K'_2 \leftarrow Ext(s, K_B)$ $K'_1 \leftarrow Ext(s, K_A); K'_2 \leftarrow Ext(s, K_B)$

$K'_3 \leftarrow Ext(s, K_T)$ $K'_3 \leftarrow Ext(s, K_T)$

$ST := (U_A, U_B, ek_A, ek_B,$ $ST := (U_A, U_B, ek_A, ek_B,$

$CT_A, ek_T, CT_B, CT_T)$ $CT_A, ek_T, CT_B, CT_T)$

$SK = G_{K'_1}(ST) \oplus G_{K'_2}(ST) \oplus G_{K'_3}(ST)$ $SK = G_{K'_1}(ST) \oplus G_{K'_2}(ST) \oplus G_{K'_3}(ST)$

Fig. 1. FSXY construction

for party U_A. This trick allows that randomness is indistinguishable from a uniformly random value if either of SSK σ_A and ESK r_A is not exposed.

The FSXY construction assumes that the output of the twisted PRF is never exposed. Indeed, though the CK$^+$ model allows an adversary to learn ESKs, the output of the twisted PRF is not contained in the ESK (i.e., The ESK of U_A is only (r_A, r'_A, r_{TA}).) in the security analysis. In order to implement this assumption in the real world, all computations related to the twisted PRF must be executed in a protected area such as a TPM. Specifically, party U_A (resp. U_B) must execute the computation of (CT_A, K_A) $\leftarrow \mathsf{EnCap}_{ek_B}(F_{\sigma_A}(r_A) \oplus F'_{r'_A}(\sigma_A))$ (resp. $(CT_B, K_B) \leftarrow \mathsf{EnCap}_{ek_A}(F_{\sigma_B}(r_B) \oplus F'_{r'_B}(\sigma_B))$) in his TPM on-line. The underlined part of Fig. 1 is computed in TPM. Note that to run a complex operation like encryption algorithms in a TPM is generally very costly and should be avoided. For example, if the Cramer-Shoup KEM [17] is used as IND-CCA secure KEM, the TPM must process 4 exponentiations on-line for each session.

Remark 1. As the case of the twisted PRF, all computations related to SSKs must not be exposed from the session state because SSKs are also assumed to not be learned with ESKs simultaneously. In the FSXY construction, such computations correspond to K_B $\leftarrow \mathsf{DeCap}_{dk_A}(CT_B)$ and $K_A \leftarrow \mathsf{DeCap}_{dk_B}(CT_A)$. However, it is not necessary to execute these computations in TPM. After receiving the message from the peer all computations are executed without stopping, and the session state is immediately erased on finishing the session. Thus, the computations which must be executed in TPM on-line are only the part related to the twisted PRF and the derivation of SK.

3.3 Our Attack

If an implementer misses this assumption, computations related to the twisted PRF may be executed not in a TPM. Then, $F_{\sigma_A}(r_A) \oplus F'_{r'_A}(\sigma_A)$ and $F_{\sigma_B}(r_B) \oplus F'_{r'_B}(\sigma_B)$ can be exposed as same as ESKs (r_A, r'_A, r_{TA}) and (r_B, r'_B, r_{TB}). We show an attack to the FSXY construction without the implementation trick.

An adversary plays the experiment of the CK^+ model in the event corresponding to 2.d in Definition 2 (i.e., Both parties' ESKs are exposed.). In this attack, $F_{\sigma_A}(r_A) \oplus F'_{r'_A}(\sigma_A)$ and $F_{\sigma_B}(r_B) \oplus F'_{r'_B}(\sigma_B)$ is regarded as a part of ESKs. The procedure of the adversary is as follows.

1. specify a session between U_A and U_B as the test session, and learn $ESK_A = (r_A, r'_A, r_{TA}, F_{\sigma_A}(r_A) \oplus F'_{r'_A}(\sigma_A))$ and $ESK_B = (r_B, r'_B, r_{TB}, F_{\sigma_B}(r_B) \oplus F'_{r'_B}(\sigma_B))$.
2. compute K_A, K_B and K_T as $(CT_A, K_A) \leftarrow \mathsf{EnCap}_{ek_B}(F_{\sigma_A}(r_A) \oplus F'_{r'_A}(\sigma_A))$, $(CT_B, K_B) \leftarrow \mathsf{EnCap}_{ek_A}(F_{\sigma_B}(r_B) \oplus F'_{r'_B}(\sigma_B))$ and $(CT_T, K_T) \leftarrow \mathsf{wEnCap}_{ek_T}(r_{TB})$.
3. execute the same key derivation procedure as a party with K_A, K_B and K_T, and derive the session key SK.

Therefore, unless the output of the twisted PRF is strictly protected with an implementation trick, the FSXY construction is insecure against such a 'full' ESK exposure attack.

4 One-Round AKE against Full Ephemeral Key Exposure

In this section, we propose a new generic construction of CK^+-secure AKE from KEM. Our scheme is secure against exposure of all randomness in generating ciphertexts for KEM by avoiding using the twisted PRF trick beside the FSXY construction. Moreover, while the FSXY construction is not one-round protocol, our scheme is one-round protocol by using PKIC-KEM [12].

4.1 Preliminaries

Security Notions of KEM Schemes. Here, we recall the definition of IND-CCA security for KEM, and min-entropy of KEM keys as follows.

Definition 3 (Syntax of KEM). *A KEM scheme consists of the following 3-tuple* (KeyGen, EnCap, DeCap)*:*

$(ek, dk) \leftarrow \mathsf{KeyGen}(1^\kappa, r_g)$: *a key generation algorithm which on inputs 1^κ and $r_g \in \mathcal{RS}_G$, where κ is the security parameter and $\mathcal{RS}_G$ is a randomness space, outputs a pair of keys (ek, dk).*

$(K, CT) \leftarrow \mathsf{EnCap}_{ek}(r_e)$: an encryption algorithm which takes as inputs encapsula-tion key ek and $r_e \in \mathcal{RS}_E$, outputs session key $K \in \mathcal{KS}$ and ciphertext $CT \in \mathcal{CS}$, where $\mathcal{RS}_E$ is a randomness space, $\mathcal{KS}$ is a session key space, and $\mathcal{CS}$ is a cipher-text space.

$K \leftarrow \mathsf{DeCap}_{dk}(CT)$: a decryption algorithm which takes as inputs decapsulation key dk and ciphertext $CT \in \mathcal{CS}$, and outputs session key $K \in \mathcal{KS}$.

Definition 4 (IND-CCA Security for KEM). *A KEM scheme is IND-CCA secure for KEM if the following property holds for security parameter κ; For any PPT adversary $\mathcal{A} = (\mathcal{A}_1, \mathcal{A}_2)$,* $\mathbf{Adv}^{\mathrm{ind-cca}} = |\Pr[r_g \leftarrow \mathcal{RS}_G; (ek, dk) \leftarrow \mathsf{KeyGen}(1^\kappa, r_g); (state) \leftarrow \mathcal{A}_1^{\mathcal{DO}(dk, \cdot)}(ek); b \leftarrow \{0, 1\}; r_e \leftarrow \mathcal{RS}_E; (K_0^*, CT_0^*) \leftarrow \mathsf{EnCap}_{ek}(r_e); K_1^* \leftarrow \mathcal{K}; b' \leftarrow \mathcal{A}_2^{\mathcal{DO}(dk, \cdot)}(ek, (K_b^*, CT_0^*), state); b' = b] - 1/2| \le negl$, *where $\mathcal{DO}$ is the decryption oracle, $\mathcal{K}$ is the space of session key and state is state information that $\mathcal{A}$ wants to preserve from $\mathcal{A}_1$ to $\mathcal{A}_2$. $\mathcal{A}$ cannot submit the ciphertext $CT = CT_0^*$ to $\mathcal{DO}$.*

Definition 5 (Min-Entropy of KEM Key). *A KEM scheme is k-min-entropy KEM if for any ek, distribution D_{KS} of variable K defined by $(K, CT) \leftarrow \mathsf{EnCap}_{ek}(r_e)$, distri-bution D_{other} of public information and random $r_e \in \mathcal{RS}_E$, $H_\infty(D_{KS}|D_{other}) \ge k$ holds, where H_∞ denotes min-entropy.*

Security Notions of KEM with public-key-independent-ciphertext [12]. Here, we recall the syntax for PKIC-KEM, and definitions of IND-CPA security for PKIC-KEM and min-entropy of KEM keys as follows.

Definition 6 (Syntax of PKIC-KEM). *A PKIC-KEM scheme consists of the following 4-tuple (wKeyGen, wEnCapC, wEnCapK, wDeCap):*

$(ek, dk) \leftarrow \mathsf{wKeyGen}(1^\kappa; r_g)$: a key generation algorithm which on inputs 1^κ, where κ is the security parameter and r_g is randomness in space $\mathcal{RS}_G$, outputs a pair of keys (ek, dk).

$CT \leftarrow \mathsf{wEnCapC}(r_e)$: a ciphertext generation algorithm which outputs ciphertext $CT \in \mathcal{CS}$ on inputs public parameters, where r_e is randomness in space $\mathcal{RS}_E$, and $\mathcal{CS}$ is a ciphertext space.

$K \leftarrow \mathsf{wEnCapK}_{ek}(CT; r_e)$: an encryption algorithm which takes as inputs encapsu-lation key ek, ciphertext CT, and randomness r_e, outputs KEM key $K \in \mathcal{KS}$, where r_e is randomness used in wEnCapC, and $\mathcal{KS}$ is a KEM key space.

$K \leftarrow \mathsf{wDeCap}_{dk}(CT)$: a decryption algorithm which takes as inputs decapsulation key dk and ciphertext $CT \in \mathcal{CS}$, and outputs KEM key $K \in \mathcal{KS}$.

Definition 7 (IND-CPA Security for PKIC-KEM). *A PKIC-KEM scheme is IND-CPA secure if the following property holds for security parameter κ; For any PPT adver-sary $\mathcal{A} = (\mathcal{A}_1, \mathcal{A}_2)$,* $\mathbf{Adv}^{\mathrm{ind-cpa}} = |\Pr[r_g \in_R \mathcal{RS}_G; (ek, dk) \leftarrow \mathsf{wKeyGen}(1^\kappa; r_g); state \leftarrow \mathcal{A}_1(ek); b \in_R \{0, 1\}; r_e \in_R \mathcal{RS}_E; CT_0^* \leftarrow \mathsf{wEnCapC}(r_e); K_0^* \leftarrow \mathsf{wEnCapK}_{ek}(CT_0^*; r_e); K_1^* \in_R \mathcal{KS}; b' \leftarrow \mathcal{A}_2(ek, (K_b^*, CT_0^*), state); b' = b] - 1/2| \le negl$, *where state is the state information that $\mathcal{A}$ wants to preserve from $\mathcal{A}_1$ to $\mathcal{A}_2$.*

Definition 8 (Min-Entropy of KEM Key). *We say a PKIC-KEM scheme is k-min-entropy PKIC-KEM if for any ek, distribution D_{KS} of variable K defined by $CT \leftarrow$ wEnCapC(r_e) and $K \leftarrow$ wEnCapK$_{ek}(CT; r_e)$, distribution D_{other} of public information and random $r_e \in RS_E$, $H_\infty(D_{KS}|D_{other}) \geq k$ holds, where H_∞ denotes min-entropy.*

Security Notion of Key Derivation Function. Let $KDF : Salt \times Dom \rightarrow Rng$ be a function with finite domain Dom, finite range Rng, and a space of non-secret random salt $Salt$.

Definition 9 (Key Derivation Function [18]). *We say a function KDF is a key derivation function (KDF) if the following condition holds for a security parameter κ: For any PPT adversary $\mathcal{A}$ and any distribution D_{Rng} over Rng with $H_\infty(D_{Rng}) \geq \kappa$, $|\Pr[y \in_R Rng; 1 \leftarrow \mathcal{A}(y)] - \Pr[x \in_R Dom; s \in_R Salt; y \leftarrow KDF(s, x); 1 \leftarrow \mathcal{A}(y)]| \leq negl$.*

A concrete construction of such a computationally secure KDF is given in [19,20] from a computational extractor and a PRF.

Security Notion of Pseudo-Random Function. Let κ be a security parameter and $F = \{F_\kappa : Dom_\kappa \times \mathcal{F}S_\kappa \rightarrow Rng_\kappa\}_\kappa$ be a function family with a family of domains $\{Dom_\kappa\}_\kappa$, a family of key spaces $\{\mathcal{F}S_\kappa\}_\kappa$ and a family of ranges $\{Rng_\kappa\}_\kappa$.

Definition 10 (Pseudo-Random Function). *We say that function family $F = \{F_\kappa\}_\kappa$ is the PRF family, if for any PPT distinguisher $\mathcal{D}$, $\mathbf{Adv}^{prf} = |\Pr[1 \leftarrow \mathcal{D}^{F_\kappa(\cdot)}] - \Pr[1 \leftarrow \mathcal{D}^{RF_\kappa(\cdot)}]| \leq negl$, where $RF_\kappa : Dom_\kappa \rightarrow Rng_\kappa$ is a truly random function.*

4.2 Our Construction

Design Principle. The goal is to avoid the twisted PRF trick. In the FSXY construction, parties share K_A, K_B and K_T. K_A is protected even if SSK_A and ESK_B are exposed, K_B is protected even if SSK_B and ESK_A are exposed, and K_T is protected even if both SSK_A and SSK_B are exposed. The case that both ESK_A and ESK_B are exposed is solved thanks to the power of the twisted PRF trick; that is, $F_{\sigma_A}(r_A) \oplus F'_{r_A}(\sigma_A)$ and $F_{\sigma_B}(r_B) \oplus F'_{r_B}(\sigma_B)$ look random for an adversary even in this case. To handle this case without the twisted PRF trick, parties must share an additional value that is protected even if both ESK_A and ESK_B are exposed.

Our solution is to change the way to generate SSKs and SPKs. In the FSXY construction, a SSK contains decryption key dk_P of IND-CCA secure KEM and σ_P, and a SPK contains encryption key ek_P. In our construction, party U_P runs $(ek_{SP}, dk_{SP}) \leftarrow$ wKeyGen$(1^\kappa, r')$ and $CT_{SP} \leftarrow$ wEnCapC(r_{SP}) of IND-CPA secure PKIC-KEM, where r' and r_{SP} are randomly chosen. dk_{SP} and r_{SP} are added to the SSK (σ_P is not necessary and is removed), and ek_{SP} and CT_{SP} are added to the SPK. Because wEnCapC can be executed without knowing an encryption key, this key generation phase correctly works.

In each session, parties share K_S in addition to K_A, K_B and K_T. Party U_A generates $K_S \leftarrow$ wDeCap$_{dk_{SA}}(CT_{SB})$, and party U_B generates $K_S \leftarrow$ wEnCapK$_{ek_{SA}}(CT_{SB}; r_{SB})$. The lexicographic order of party identities determines which party to run wDeCap.

$$\boxed{\begin{array}{l}
\textbf{Common public parameter}: G, KDF, s \\
\textbf{SSK and SPK for party } U_A : SSK_A := (dk_A, dk_{SA}, r_{SA}), SPK_A := (ek_A, ek_{SA}, CT_{SA}) \\
\textbf{SSK and SPK for party } U_B : SSK_B := (dk_B, dk_{SB}, r_{SB}), SPK_B := (ek_B, ek_{SB}, CT_{SB})
\end{array}}$$

Party U_A	Party U_B
$r_A \in_R \mathcal{RS}_E; r_{TA} \in_R \mathcal{RS}_G$	$r_B \in_R \mathcal{RS}_E; r_{TB} \in_R \mathcal{RS}_E$
$(CT_A, K_A) \leftarrow \mathsf{EnCap}_{ek_B}(r_A)$	$(CT_B, K_B) \leftarrow \mathsf{EnCap}_{ek_A}(r_B)$
$(ek_T, dk_T) \leftarrow \mathsf{wKeyGen}(1^\kappa, r_{TA})$	$CT_T \leftarrow \mathsf{wEnCapC}(r_{TB})$

$$\xrightarrow{\quad U_A, U_B, CT_A, ek_T \quad}$$
$$\xleftarrow{\quad U_A, U_B, CT_B, CT_T \quad}$$

Party U_A	Party U_B
$K_B \leftarrow \mathsf{DeCap}_{dk_A}(CT_B)$	$K_A \leftarrow \mathsf{DeCap}_{dk_B}(CT_A)$
$K_T \leftarrow \mathsf{wDeCap}_{dk_T}(CT_T)$	$K_T \leftarrow \mathsf{wEnCapK}_{ek_T}(CT_T; r_{TB})$
$K_S \leftarrow \mathsf{wDeCap}_{dk_{SA}}(CT_{SB})$	$K_S \leftarrow \mathsf{wEnCapK}_{ek_{SA}}(CT_{SB}; r_{SB})$
$K'_1 \leftarrow KDF(s, K_A); K'_2 \leftarrow KDF(s, K_B)$	$K'_1 \leftarrow KDF(s, K_A); K'_2 \leftarrow KDF(s, K_B)$
$K'_3 \leftarrow KDF(s, K_T); K'_4 \leftarrow KDF(s, K_S)$	$K'_3 \leftarrow KDF(s, K_T); K'_4 \leftarrow KDF(s, K_S)$
$ST := (U_A, U_B, SPK_A, SPK_B,$	$ST := (U_A, U_B, SPK_A, SPK_B,$
$\qquad CT_A, ek_T, CT_B, CT_T)$	$\qquad CT_A, ek_T, CT_B, CT_T)$
$SK = G_{K'_1}(ST) \oplus G_{K'_2}(ST)$	$SK = G_{K'_1}(ST) \oplus G_{K'_2}(ST)$
$\qquad \oplus G_{K'_3}(ST) \oplus G_{K'_4}(ST)$	$\qquad \oplus G_{K'_3}(ST) \oplus G_{K'_4}(ST)$

Fig. 2. Our construction

From the syntax of PKIC-KEM, K_S is shared non-interactively, and is protected even if both ESK_A and ESK_B are exposed.

Our construction has another advantage that it is one-round protocol. The FSXY construction is not one-round because the responder's EPK depends on the initiator's EPK. We use the technique in [12] using PKIC-KEM to generate K_T.

Also, the session key derivation procedure is more efficient than the FSXY construction because a KDF is used instead of a strong randomness extractor. On input a value having sufficient min-entropy, a strong randomness extractor outputs a value which is *statistically indistinguishable* from a uniformly chosen random value. Indeed, such statistical indistinguishability is not necessary to prove security of our construction. *Computational indistinguishability* is sufficient, and the KDF is suitable.

Protocol. The protocol of our generic construction is shown in Fig. 2.

Public Parameters. Let (KeyGen, EnCap, DeCap) be an IND-CCA secure KEM and (wKeyGen, wEnCapC, wEnCapK, wDeCap) be an IND-CPA secure PKIC-KEM, where the randomness space of encapsulation algorithms is $\mathcal{RS}_E$, the randomness space of key generation algorithms is $\mathcal{RS}_G$ and the KEM key space is $\mathcal{KS}$. Let $G : \{0,1\}^* \times \mathcal{FS} \rightarrow \{0,1\}^\kappa$ be a PRF, where $\mathcal{FS}$ is the key space of PRFs ($|\mathcal{FS}| = \kappa$). Let $KDF : Salt \times \mathcal{KS} \rightarrow \mathcal{FS}$ be a KDF with a non-secret random salt $s \in Salt$, where $Salt$ is the salt space.

Static Secret and Static Public Keys. Party U_P selects $r, r' \in_R \mathcal{RS}_G$ and $r_{SP} \in_R \mathcal{RS}_E$, and generates $(ek_P, dk_P) \leftarrow \mathsf{KeyGen}(1^\kappa, r)$, $(ek_{SP}, dk_{SP}) \leftarrow \mathsf{wKeyGen}(1^\kappa, r')$ and $CT_{SP} \leftarrow$

wEnCapC(r_{SP}). Party U_P's SSK is (dk_P, dk_{SP}, r_{SP}) and SPK is (ek_P, ek_{SP}, CT_{SP}). Note that a party does not use all contents of the SSK to generate K_S in a session.

Session State. The session state of a session owned by U_A contains ephemeral secret keys (r_A, r_{TA}), encapsulated KEM key K_A and ad-hoc decryption key dk_T. Other information that is computed after receiving the message from the peer is immediately erased when the session key is established. Similarly, the session state of a session owned by U_B contains ephemeral secret keys (r_B, r_{TB}) and encapsulated KEM key K_B.

Other intermediate values (e.g., decapsulated KEM keys, and outputs of KDF) are not contained in session state because these values are simultaneously computed with the session key and immediately erased after completing the session.

4.3 Security

We show the following theorem.

Theorem 1. *If* (KeyGen, EnCap, DeCap) *is IND-CCA secure and* κ*-min-entropy KEM,* (wKeyGen, wEnCapC, wEnCapK wDeCap) *is IND-CPA secure and* κ*-min-entropy PKIC-KEM, G is a PRF, and KDF is a KDF, then our generic construction is* CK$^+$*-secure.*

Here, we give an overview of the security proof.

We have to consider the following four exposure patterns in the CK$^+$ security model (matching cases):

$$2\text{-}(c): SSK_A \text{ and } ESK_B, \quad 2\text{-}(d): ESK_A \text{ and } ESK_B,$$
$$2\text{-}(e): SSK_A \text{ and } SSK_B, \quad 2\text{-}(f): ESK_A \text{ and } SSK_B.$$

In case 2-(c), K_A is protected by the security of CT_A because r_A and dk_B are not exposed. In case 2-(d), K_S is protected by the security of CT_S because dk_{SA} and r_{SB} are not exposed. In case 2-(e), K_T is protected by the security of CT_T because dk_T and r_{TB} are not exposed. In case 2-(f), K_B is protected by the security of CT_B because r_B and dk_A are not exposed.

Then, we transform the CK$^+$ security game, and the session key in the test session is randomly distributed in the final game. First, we change the protected KEM key into a random key for each pattern; therefore, the input of *KDF* is randomly distributed and has sufficient min-entropy. Next, we change the output of *KDF* into randomly chosen values. Finally, we change one of the PRFs (corresponding to the protected KEM) into a random function. Therefore, the session key in the test session is randomly distributed; thus, there is no advantage to the adversary. We can show a similar proof in non-matching cases.

Proof. In the experiment of CK$^+$ security, we suppose that sid* is the session identity for the test session, and that there are N users and at most ℓ sessions are activated. Let κ be the security parameter, and let $\mathcal{A}$ be a PPT (in κ) bounded adversary. *Suc* denotes the event that $\mathcal{A}$ wins. We consider the following events that cover all cases of the behavior of $\mathcal{A}$.

- Let E_1 be the event that the test session sid* has no matching session $\overline{\text{sid}}^*$, the owner of sid* is the initiator and the static secret key of the initiator is given to $\mathcal{A}$.
- Let E_2 be the event that the test session sid* has no matching session $\overline{\text{sid}}^*$, the owner of sid* is the initiator and the ephemeral secret key of sid* is given to $\mathcal{A}$.
- Let E_3 be the event that the test session sid* has no matching session $\overline{\text{sid}}^*$, the owner of sid* is the responder and the static secret key of the responder is given to $\mathcal{A}$.
- Let E_4 be the event that the test session sid* has no matching session $\overline{\text{sid}}^*$, the owner of sid* is the responder and the ephemeral secret key of sid* is given to $\mathcal{A}$.
- Let E_5 be the event that the test session sid* has matching session $\overline{\text{sid}}^*$, and both static secret keys of the initiator and the responder are given to $\mathcal{A}$.
- Let E_6 be the event that the test session sid* has matching session $\overline{\text{sid}}^*$, and both ephemeral secret keys of sid* and $\overline{\text{sid}}^*$ are given to $\mathcal{A}$.
- Let E_7 be the event that the test session sid* has matching session $\overline{\text{sid}}^*$, and the static secret key of the owner of sid* and the ephemeral secret key of $\overline{\text{sid}}^*$ are given to $\mathcal{A}$.
- Let E_8 be the event that the test session sid* has matching session $\overline{\text{sid}}^*$, and the ephemeral secret key of sid* and the static secret key of the owner of $\overline{\text{sid}}^*$ are given to $\mathcal{A}$.

To finish the proof, we investigate events $E_i \wedge Suc$ ($i = 1, \ldots, 8$) that cover all cases of event Suc. In this paper, we show the proof of event E_6 because it is most different from that of the FSXY construction. The proof of other events are given in the full paper.

Event $E_6 \wedge Suc$. We change the interface of oracle queries and the computation of the session key. These instances are gradually changed over six hybrid experiments, depending on specific sub-cases. In the last hybrid experiment, the session key in the test session does not contain information of the bit b. Thus, the adversary clearly only output a random guess. We denote these hybrid experiments by $\mathbf{H}_0, \ldots, \mathbf{H}_5$ and the advantage of the adversary $\mathcal{A}$ when participating in experiment $\mathbf{H}_i$ by $\mathbf{Adv}(\mathcal{A}, \mathbf{H}_i)$.

Hybrid Experiment $\mathbf{H}_0$: This experiment denotes the real experiment for CK$^+$ security and in this experiment the environment for $\mathcal{A}$ is as defined in the protocol. Thus, $\mathbf{Adv}(\mathcal{A}, \mathbf{H}_0)$ is the same as the advantage of the real experiment.

Hybrid Experiment $\mathbf{H}_1$: In this experiment, if session identities in two sessions are identical, the experiment halts.

When two ciphertexts from different randomness are identical and two public keys from different randomness are identical, session identities in two sessions are also identical. In any IND-CCA secure KEM and IND-CPA secure PKIC-KEM, such an event occurs with negligible probability. Thus, $|\mathbf{Adv}(\mathcal{A}, \mathbf{H}_1) - \mathbf{Adv}(\mathcal{A}, \mathbf{H}_0)| \leq negl$.

Hybrid Experiment $\mathbf{H}_2$: In this experiment, the experiment selects a party U_A and integer $i \in [1, \ell]$ randomly in advance. If $\mathcal{A}$ poses Test query to a session except i-th session of U_A, the experiment halts.

Since guess of the test session matches with $\mathcal{A}$'s choice with probability $1/N\ell$, $\mathbf{Adv}(\mathcal{A}, \mathbf{H}_2) \geq 1/N\ell \cdot \mathbf{Adv}(\mathcal{A}, \mathbf{H}_1)$. Without loss of generality, we can suppose that the intended peer of the i-th session of U_A is U_B.

Hybrid Experiment $\mathbf{H}_3$: In this experiment, the computation of K_S in the test session is changed. Instead of computing $K_S \leftarrow \mathsf{wDeCap}_{dk_{SA}}(CT_{SB})$ or $K_S \leftarrow \mathsf{wEnCapK}_{ek_{SA}}(CT_{SB}; r_{SB})$, it is changed as choosing $K_S \leftarrow \mathcal{KS}$ randomly.

We construct an IND-CPA adversary S from $\mathcal{A}$ in $\mathbf{H}_2$ or $\mathbf{H}_3$. S performs the following steps.

Init. S receives (ek^*, K_b^*, CT_0^*) as the challenge of IND-CPA game for PKIC-KEM.

Setup. S chooses PRF $G : \{0, 1\}^* \times \mathcal{FS} \to \{0, 1\}^k$, where $\mathcal{FS}$ is the key space of PRFs, and KDF $KDF : Salt \times \mathcal{KS} \to \mathcal{FS}$ with random salt $s \in Salt$, where $Salt$ is the salt space. These are provided as a part of the public parameters. Also, S sets all N users' static secret and public keys except U_A and U_B.

For U_P (except U_A and U_B), S selects $r, r' \in_R \mathcal{RS}_G$ and $r_{SP} \in_R \mathcal{RS}_E$, and generates $(ek_P, dk_P) \leftarrow \mathsf{KeyGen}(1^\kappa, r)$, $(ek_{SP}, dk_{SP}) \leftarrow \mathsf{wKeyGen}(1^\kappa, r')$ and $CT_{SP} \leftarrow \mathsf{wEnCapC}(r_{SP})$. Party U_P's SSK is (dk_P, dk_{SP}, r_{SP}) and SPK is (ek_P, ek_{SP}, CT_{SP}).

For U_A, S selects $r \in_R \mathcal{RS}_G$ and $r_{AP} \in_R \mathcal{RS}_E$, and generates $(ek_A, dk_A) \leftarrow \mathsf{KeyGen}(1^\kappa, r)$ and $CT_{AP} \leftarrow \mathsf{wEnCapC}(r_{AP})$. Party U_A's SSK is $(dk_A, *, r_{AP})$ and SPK is (ek_A, ek^*, CT_{AP}), where $*$ is unknown part for S.

For U_B, S selects $r, r' \in_R \mathcal{RS}_G$, and generates $(ek_B, dk_B) \leftarrow \mathsf{KeyGen}(1^\kappa, r)$ and $(ek_{BP}, dk_{BP}) \leftarrow \mathsf{wKeyGen}(1^\kappa, r')$. Party U_B's SSK is $(dk_B, dk_{BP}, *)$ and SPK is (ek_B, ek_{BP}, CT_0^*), where $*$ is unknown part for S.

Simulation. S maintains the list $\mathcal{L}_{SK}$ that contains queries and answers of SessionKeyReveal. S simulates oracle queries by $\mathcal{A}$ as follows. We suppose that P sorts before $\bar{P}$ lexicographically.

1. Send$(\Pi, U_P, U_{\bar{P}})$: S computes the ephemeral public key $(U_P, U_{\bar{P}}, CT_P, ek_T)$ obeying the protocol, returns it and records $(\Pi, U_P, U_{\bar{P}}, (U_P, U_{\bar{P}}, CT_P, ek_T))$.
2. Send$(\Pi, U_{\bar{P}}, U_P)$: S computes the ephemeral public key $(U_{\bar{P}}, U_P, CT_{\bar{P}}, CT_T)$ obeying the protocol, returns it and records $(\Pi, U_P, U_{\bar{P}}, (U_{\bar{P}}, U_P, CT_{\bar{P}}, CT_T))$.
3. Send$(\Pi, U_{\bar{P}}, U_P, (U_{\bar{P}}, U_P, CT_{\bar{P}}, CT_T))$: If $P = A$, $\bar{P} = B$, the session is i-th session of A, then S sets $K_T := K_b^*$, computes the session key SK^* obeying the protocol, and records $(\Pi, U_A, U_B, (U_A, U_B, CT_A, ek_T), (U_B, U_A, CT_B, CT_T))$ as the completed session and SK^* in the list $\mathcal{L}_{SK}$. Else if $(\Pi, U_P, U_{\bar{P}}, (U_P, U_{\bar{P}}, CT_P, ek_T))$ is not recorded, S records the session $(\Pi, U_P, U_{\bar{P}}, *, (U_{\bar{P}}, U_P, CT_{\bar{P}}, CT_T))$ and waits Send$(\Pi, U_P, U_{\bar{P}})$. Otherwise, S computes the session key SK obeying the protocol, and records $(\Pi, U_P, U_{\bar{P}}, (U_P, U_{\bar{P}}, CT_P, ek_T), (U_{\bar{P}}, U_P, CT_{\bar{P}}, CT_T))$ as the completed session and SK in the list $\mathcal{L}_{SK}$.
4. Send$(\Pi, U_P, U_{\bar{P}}, (U_P, U_{\bar{P}}, CT_P, ek_T))$: If $P = A$, $\bar{P} = B$, the session is the matching session of i-th session of A, then S sets $K_T := K_b^*$, computes the session key SK^* obeying the protocol, and records $(\Pi, U_B, U_A, (U_A, U_B, CT_A, ek_T), (U_B, U_A, CT_B, CT_T))$ as the completed session and SK^* in the list $\mathcal{L}_{SK}$. Else if $(\Pi, U_{\bar{P}}, U_P,$

$(U_{\bar{P}}, U_P, CT_{\bar{P}}, CT_T))$ is not recorded, S records the session $(\Pi, U_{\bar{P}}, U_P, (U_P, U_{\bar{P}}, CT_P, ek_T), *)$ and waits $\mathsf{Send}(\Pi, U_{\bar{P}}, U_P)$. Otherwise, S computes the session key SK obeying the protocol, and records $(\Pi, U_{\bar{P}}, U_P, (U_P, U_{\bar{P}}, CT_P, ek_T), (U_{\bar{P}}, U_P, CT_{\bar{P}}, CT_T))$ as the completed session and SK in the list $\mathcal{L}_{SK}$.

5. $\mathsf{SessionKeyReveal}(\mathsf{sid})$:
   (a) If the session sid is not completed, S returns an error message.
   (b) Otherwise, S returns the recorded value SK.
6. $\mathsf{SessionStateReveal}(\mathsf{sid})$: S responds the ephemeral secret key and intermediate computation results of sid as the definition. Note that the $\mathsf{SessionStateReveal}$ query is not posed to the test session from the freshness definition.
7. $\mathsf{Corrupt}(U_P)$: S responds the static secret key and all unerased session states of U_P as the definition.
8. $\mathsf{Test}(\mathsf{sid})$: S responds to the query as the definition.
9. If $\mathcal{A}$ outputs a guess b', S outputs b'.

Analysis. For $\mathcal{A}$, the simulation by S is same as the experiment $\mathbf{H}_2$ if the challenge is (K_0^*, CT_0^*). Otherwise, the simulation by S is same as the experiment $\mathbf{H}_3$. Also, both K_T in two experiments have κ-min-entropy because (wKeyGen, wEnCapC, wEnCapK, wDeCap) is κ-min-entropy PKIC-KEM. Thus, if the advantage of S is negligible, then $|\mathbf{Adv}(\mathcal{A}, \mathbf{H}_3) - \mathbf{Adv}(\mathcal{A}, \mathbf{H}_2)| \leq negl$.

Hybrid Experiment $\mathbf{H}_4$: In this experiment, the computation of K_4' in the test session is changed. Instead of computing $K_4' \leftarrow KDF(s, K_S)$, it is changed as choosing $K_4' \in \mathcal{F}S$ randomly.

Since K_S is randomly chosen in $\mathbf{H}_3$, it has sufficient min-entropy. Thus, by the definition of the KDF, $|\mathbf{Adv}(\mathcal{A}, \mathbf{H}_4) - \mathbf{Adv}(\mathcal{A}, \mathbf{H}_3)| \leq negl$.

Hybrid Experiment $\mathbf{H}_5$: In this experiment, the computation of SK in the test session is changed. Instead of computing $SK = G_{K_1'}(\mathsf{ST}) \oplus G_{K_2'}(\mathsf{ST}) \oplus G_{K_3'}(\mathsf{ST}) \oplus G_{K_4'}(\mathsf{ST})$, it is changed as $SK = G_{K_1'}(\mathsf{ST}) \oplus G_{K_2'}(\mathsf{ST}) \oplus G_{K_3'}(\mathsf{ST}) \oplus x$ where $x \in \{0, 1\}^\kappa$ is chosen randomly.

We construct a distinguisher $\mathcal{D}'$ between PRF $F^* : \{0, 1\}^* \times \mathcal{F}S \rightarrow \{0, 1\}^\kappa$ and a random function RF from $\mathcal{A}$ in $\mathbf{H}_4$ or $\mathbf{H}_5$. $\mathcal{D}'$ performs the following steps.

Setup. $\mathcal{D}'$ sets $G = F^*$, and chooses KDF $KDF : \mathcal{KS} \rightarrow \mathcal{F}S$. These are provided as a part of the public parameters. Also, $\mathcal{D}'$ sets all N users' static secret and public keys. S selects $r, r' \in_R \mathcal{RS}_G$ and $r_{SP} \in_R \mathcal{RS}_E$, and generates $(ek_P, dk_P) \leftarrow \mathsf{KeyGen}(1^\kappa, r)$, $(ek_{SP}, dk_{SP}) \leftarrow \mathsf{wKeyGen}(1^\kappa, r')$ and $CT_{SP} \leftarrow \mathsf{wEnCapC}(r_{SP})$. Party U_P's SSK is (dk_P, dk_{SP}, r_{SP}) and SPK is (ek_P, ek_{SP}, CT_{SP}).

Simulation. $\mathcal{D}'$ maintains the list $\mathcal{L}_{SK}$ that contains queries and answers of $\mathsf{SessionKeyReveal}$. $\mathcal{D}'$ simulates oracle queries by $\mathcal{A}$ as follows.

1. $\mathsf{Send}(\Pi, U_P, U_{\bar{P}})$: $\mathcal{D}'$ computes the ephemeral public key $(U_P, U_{\bar{P}}, CT_P, ek_T)$ obeying the protocol, returns it and records $(\Pi, U_P, U_{\bar{P}}, (U_P, U_{\bar{P}}, CT_P, ek_T))$.
2. $\mathsf{Send}(\Pi, U_{\bar{P}}, U_P)$: $\mathcal{D}'$ computes the ephemeral public key $(U_{\bar{P}}, U_P, CT_{\bar{P}}, CT_T)$ obeying the protocol, returns it and records $(\Pi, U_P, U_{\bar{P}}, (U_{\bar{P}}, U_P, CT_{\bar{P}}, CT_T))$.

3. Send($\Pi, U_{\bar{P}}, U_P, (U_{\bar{P}}, U_P, CT_{\bar{P}}, CT_T)$): If $P = A$, $\bar{P} = B$, the session is i-th session of A, then $\mathcal{D}'$ poses ST to his oracle (i.e., F^* or a random function RF), obtains $x \in \{0,1\}^\kappa$, computes the session key $SK = G_{K'_1}(\text{ST}) \oplus G_{K'_2}(\text{ST}) \oplus G_{K'_3}(\text{ST}) \oplus x$, and records $(\Pi, U_A, U_B, (U_A, U_B, CT_A, ek_T), (U_B, U_A, CT_B, CT_T))$ as the completed session and SK^* in the list $\mathcal{L}_{SK}$. Else if $(\Pi, U_P, U_{\bar{P}}, (U_P, U_{\bar{P}}, CT_P, ek_T))$ is not recorded, $\mathcal{D}'$ records the session $(\Pi, U_P, U_{\bar{P}}, *, (U_{\bar{P}}, U_P, CT_{\bar{P}}, CT_T))$ and waits Send($\Pi, U_P, U_{\bar{P}}$). Otherwise, $\mathcal{D}'$ computes the session key SK obeying the protocol, and records $(\Pi, U_P, U_{\bar{P}}, (U_P, U_{\bar{P}}, CT_P, ek_T), (U_{\bar{P}}, U_P, CT_{\bar{P}}, CT_T))$ as the completed session and SK in the list $\mathcal{L}_{SK}$.

4. Send($\Pi, U_P, U_{\bar{P}}, (U_P, U_{\bar{P}}, CT_P, ek_T)$): If $P = A$, $\bar{P} = B$, the session is the matching session of i-th session of A, then $\mathcal{D}'$ poses ST to his oracle (i.e., F^* or a random function RF), obtains $x \in \{0,1\}^\kappa$, computes the session key $SK = G_{K'_1}(\text{ST}) \oplus G_{K'_2}(\text{ST}) \oplus G_{K'_3}(\text{ST}) \oplus x$, and records $(\Pi, U_B, U_A, (U_A, U_B, CT_A, ek_T), (U_B, U_A, CT_B, CT_T))$ as the completed session and SK^* in the list $\mathcal{L}_{SK}$. Else if $(\Pi, U_{\bar{P}}, U_P, (U_{\bar{P}}, U_P, CT_{\bar{P}}, CT_T))$ is not recorded, $\mathcal{D}'$ records the session $(\Pi, U_{\bar{P}}, U_P, (U_P, U_{\bar{P}}, CT_P, ek_T), *)$ and waits Send($\Pi, U_{\bar{P}}, U_P$). Otherwise, $\mathcal{D}'$ computes the session key SK obeying the protocol, and records $(\Pi, U_{\bar{P}}, U_P, (U_P, U_{\bar{P}}, CT_P, ek_T), (U_{\bar{P}}, U_P, CT_{\bar{P}}, CT_T))$ as the completed session and SK in the list $\mathcal{L}_{SK}$.

5. SessionKeyReveal(sid):
   (a) If the session sid is not completed, $\mathcal{D}'$ returns an error message.
   (b) Otherwise, $\mathcal{D}'$ returns the recorded value SK.

6. SessionStateReveal(sid): $\mathcal{D}'$ responds the ephemeral secret key and intermediate computation results of sid as the definition. Note that the SessionStateReveal query is not posed to the test session from the freshness definition.

7. Corrupt(U_P): $\mathcal{D}'$ responds the static secret key and all unerased session states of U_P as the definition.

8. Test(sid): $\mathcal{D}'$ responds to the query as the definition.

9. If $\mathcal{A}$ outputs a guess $b' = 0$, $\mathcal{D}'$ outputs that the oracle is the PRF F^*. Otherwise, $\mathcal{D}'$ outputs that the oracle is a random function RF.

Analysis. For $\mathcal{A}$, the simulation by $\mathcal{D}'$ is same as the experiment $\mathbf{H}_4$ if the oracle is the PRF F^*. Otherwise, the simulation by $\mathcal{D}'$ is same as the experiment $\mathbf{H}_5$. Thus, if the advantage of $\mathcal{D}'$ is negligible, then $|\mathbf{Adv}(\mathcal{A}, \mathbf{H}_5) - \mathbf{Adv}(\mathcal{A}, \mathbf{H}_4)| \leq negl$.

In $\mathbf{H}_5$, the session key in the test session is perfectly randomized. Thus, $\mathcal{A}$ cannot obtain any advantage from Test query.

Therefore, $\mathbf{Adv}(\mathcal{A}, \mathbf{H}_5) = 0$ and $\Pr[E_5 \wedge Suc]$ is negligible.

$\square$

Event $E_1 \wedge Suc$. The proof in this case is similar to the event $E_6 \wedge Suc$. There is a difference in the experiment $\mathbf{H}_3$. In the event $E_6 \wedge Suc$, instead of computing $K_S \leftarrow \text{wDeCap}_{dk_{SA}}(CT_{SB})$ or $K_S \leftarrow \text{wEnCapK}_{ek_{SA}}(CT_{SB}; r_{SB})$, it is changed as choosing $K_S \leftarrow \mathcal{KS}$, where we suppose that U_B is the intended partner of U_A in the test session. In the event $E_1 \wedge Suc$, instead of computing $(CT_A, K_A) \leftarrow \text{EnCap}_{ek_B}(r_A)$, it is changed as $K_A \leftarrow \mathcal{KS}$. Since $\mathcal{A}$ cannot obtain r_A and dk_B by the freshness definition in this event, we can construct an adversary S from $\mathcal{A}$ in the similar manner in the proof of the event $E_6 \wedge S\,uc$. Note that if $\mathcal{A}$ poses **Send** query to U_B other than the test session, S simulates K_A by posing the decryption oracle.

Event $E_2 \wedge Suc.$ The proof in this case is almost same as the event $E_6 \wedge S\,uc$.

Event $E_3 \wedge Suc.$ The proof in this case is similar to the event $E_6 \wedge S\,uc$. There is a difference in the experiment $\mathbf{H_3}$. In the event $E_6 \wedge S\,uc$, instead of computing $K_S \leftarrow \mathsf{wDeCap}_{dk_{S_A}}(CT_{SB})$ or $K_S \leftarrow \mathsf{wEnCapK}_{ek_{S_A}}(CT_{SB}; r_{SB})$, it is changed as choosing $K_S \leftarrow \mathcal{K}S$, where we suppose that U_B is the intended partner of U_A in the test session. In the event $E_3 \wedge S\,uc$, instead of computing $(CT_B, K_B) \leftarrow \mathsf{EnCap}_{ek_A}(r_B)$, it is changed as $K_B \leftarrow \mathcal{K}S$. Since $\mathcal{A}$ cannot obtain r_B and dk_A by the freshness definition in this event, we can construct an adversary S from $\mathcal{A}$ in the similar manner in the proof of the event $E_6 \wedge S\,uc$. Note that if $\mathcal{A}$ poses **Send** query to U_A other than the test session, S simulates K_B by posing the decryption oracle.

Event $E_4 \wedge Suc.$ The proof in this case is almost same as the event $E_6 \wedge S\,uc$.

Event $E_5 \wedge Suc.$ The proof in this case is similar to the event $E_6 \wedge S\,uc$. There is a difference in the experiment $\mathbf{H_3}$. In the event $E_6 \wedge S\,uc$, instead of computing $K_S \leftarrow \mathsf{wDeCap}_{dk_{S_A}}(CT_{SB})$ or $K_S \leftarrow \mathsf{wEnCapK}_{ek_{S_A}}(CT_{SB}; r_{SB})$, it is changed as choosing $K_S \leftarrow \mathcal{K}S$, where we suppose that U_B is the intended partner of U_A in the test session. In the event $E_5 \wedge S\,uc$, instead of computing $K_T \leftarrow \mathsf{wDeCap}_{dk_T}(CT_T)$ or $K_T \leftarrow \mathsf{wEnCapK}_{ek_T}(CT_T; r_{TB})$, it is changed as $K_T \leftarrow \mathcal{K}S$. Since $\mathcal{A}$ cannot obtain r_{TA} and r_{TB} by the freshness definition in this event, we can construct an adversary S from $\mathcal{A}$ in the similar manner in the proof of the event $E_6 \wedge S\,uc$.

Event $E_7 \wedge Suc.$ The proof in this case is almost same as the event $E_1 \wedge S\,uc$.

Event $E_8 \wedge Suc.$ The proof in this case is almost same as the event $E_2 \wedge S\,uc$.

References

1. Canetti, R., Krawczyk, H.: Analysis of key-exchange protocols and their use for building secure channels. In: Pfitzmann, B. (ed.) EUROCRYPT 2001. LNCS, vol. 2045, pp. 453–474. Springer, Heidelberg (2001)
2. LaMacchia, B.A., Lauter, K., Mityagin, A.: Stronger security of authenticated key exchange. In: Susilo, W., Liu, J.K., Mu, Y. (eds.) ProvSec 2007. LNCS, vol. 4784, pp. 1–16. Springer, Heidelberg (2007)
3. Krawczyk, H.: HMQV: A High-Performance Secure Diffie-Hellman Protocol. In: Shoup, V. (ed.) CRYPTO 2005. LNCS, vol. 3621, pp. 546–566. Springer, Heidelberg (2005)
4. Fujioka, A., Suzuki, K., Xagawa, K., Yoneyama, K.: Strongly Secure Authenticated Key Exchange from Factoring, Codes, and Lattices. In: Fischlin, M., Buchmann, J., Manulis, M. (eds.) PKC 2012. LNCS, vol. 7293, pp. 467–484. Springer, Heidelberg (2012)

5. Cremers, C.J.F.: Session-state Reveal Is Stronger Than Ephemeral Key Reveal: Attacking the NAXOS Authenticated Key Exchange Protocol. In: Abdalla, M., Pointcheval, D., Fouque, P.-A., Vergnaud, D. (eds.) ACNS 2009. LNCS, vol. 5536, pp. 20–33. Springer, Heidelberg (2009)
6. Cremers, C.J.F.: Examining Indistinguishability-Based Security Models for Key Exchange Protocols: The case of CK, CK-HMQV, and eCK. In: ASIACCS 2011, pp. 80–91 (2011)
7. Damgård, I.: Towards Practical Public Key Systems Secure Against Chosen Ciphertext Attacks. In: Feigenbaum, J. (ed.) CRYPTO 1991. LNCS, vol. 576, pp. 445–456. Springer, Heidelberg (1992)
8. Naor, M.: On Cryptographic Assumptions and Challenges. In: Boneh, D. (ed.) CRYPTO 2003. LNCS, vol. 2729, pp. 96–109. Springer, Heidelberg (2003)
9. Boyd, C., Cliff, Y., Gonzalez Nieto, J.M., Paterson, K.G.: Efficient One-Round Key Exchange in the Standard Model. In: Mu, Y., Susilo, W., Seberry, J. (eds.) ACISP 2008. LNCS, vol. 5107, pp. 69–83. Springer, Heidelberg (2008)
10. Boyd, C., Cliff, Y., González Nieto, J.M., Paterson, K.G.: One-round key exchange in the standard model. IJACT 1(3), 181–199 (2009)
11. Okamoto, T.: Authenticated Key Exchange and Key Encapsulation in the Standard Model. In: Kurosawa, K. (ed.) ASIACRYPT 2007. LNCS, vol. 4833, pp. 474–484. Springer, Heidelberg (2007)
12. Yoneyama, K.: One-Round Authenticated Key Exchange with Strong Forward Secrecy in the Standard Model against Constrained Adversary. In: Hanaoka, G., Yamauchi, T. (eds.) IWSEC 2012. LNCS, vol. 7631, pp. 69–86. Springer, Heidelberg (2012)
13. Yoneyama, K.: One-Round Authenticated Key Exchange with Strong Forward Secrecy in the Standard Model against Constrained Adversary. IEICE Transactions 96-A(6), 1124–1138 (2013)
14. Yoneyama, K.: Generic Construction of Two-Party Round-Optimal Attribute-Based Authenticated Key Exchange without Random Oracles. IEICE Transactions 96-A(6), 1112–1123 (2013)
15. Moriyama, D., Okamoto, T.: An eCK-Secure Authenticated Key Exchange Protocol without Random Oracles. In: Pieprzyk, J., Zhang, F. (eds.) ProvSec 2009. LNCS, vol. 5848, pp. 154–167. Springer, Heidelberg (2009)
16. Yang, Z., Schwenk, J.: Strongly Authenticated Key Exchange Protocol from Bilinear Groups without Random Oracles. In: Takagi, T., Wang, G., Qin, Z., Jiang, S., Yu, Y. (eds.) ProvSec 2012. LNCS, vol. 7496, pp. 264–275. Springer, Heidelberg (2012)
17. Cramer, R., Shoup, V.: A Practical Public Key Cryptosystem Provably Secure Against Adaptive Chosen Ciphertext Attack. In: Krawczyk, H. (ed.) CRYPTO 1998. LNCS, vol. 1462, pp. 13–25. Springer, Heidelberg (1998)
18. Gennaro, R., Shoup, V.: A Note on An Encryption Scheme of Kurosawa and Desmedt. In: Cryptology ePrint Archive: 2004/194 (2004)
19. Krawczyk, H.: Cryptographic Extraction and Key Derivation: The HKDF Scheme. In: Rabin, T. (ed.) CRYPTO 2010. LNCS, vol. 6223, pp. 631–648. Springer, Heidelberg (2010)
20. Dachman-Soled, D., Gennaro, R., Krawczyk, H., Malkin, T.: Computational Extractors and Pseudorandomness. In: Cramer, R. (ed.) TCC 2012. LNCS, vol. 7194, pp. 383–403. Springer, Heidelberg (2012)

Attacks to the Proxy Re-Encryption Schemes from IWSEC2011

Toshiyuki Isshiki[1], Manh Ha Nguyen[2,*], and Keisuke Tanaka[2,*]

[1] NEC Corporation, Japan
t-issiki@bx.jp.nec.com
[2] Tokyo Institute of Technology, Japan
{nguyen9,keisuke}@is.titech.ac.jp

Abstract. Proxy re-encryption (PRE) allows a proxy to convert a ciphertext encrypted for Alice (delegator) into a ciphertext for Bob (delegatee) by using a re-encryption key generated by Alice. In PRE, non-transferability is a property that colluding proxies and delegatees cannot re-delegate decryption rights to a malicious user. In IWSEC 2011, Hayashi, Matsushita, Yoshida, Fujii, and Okada introduced the unforgeability of re-encryption keys against collusion attack (UFReKey-CA), which is a relaxed notion of the non-transferability. They also proposed a stronger security notion, the *strong* unforgeability of re-encryption keys against collusion attack (sUFReKey-CA). Since sUFReKey-CA implies UFReKey-CA and sUFReKey-CA is simpler (i.e. easier to treat) definition than UFReKey-CA, sUFReKey-CA is useful to prove UFReKey-CA. They then proposed two concrete constructions of PRE and claimed that they meet both replayable-CCA security and sUFReKey-CA under two new variants of the Diffi-Hellman inversion assumption. In this paper, we present two concrete attacks to their PRE schemes. The first attack is to the sUFReKey-CA property on their two schemes. The second attack is to the assumptions employed in the security proofs for sUFReKey-CA of their two schemes.

Keywords: Proxy re-encryption, non-transferability, unforgeability of re-encryption keys.

1   Introduction

Proxy re-encryption (PRE), introduced by Blaze, Bleumer, and Strauss [2] in EURO-CRYPT'98, allows a semi-trust proxy to translate a ciphertext intended for Alice into another ciphertext intended for Bob. The proxy, however, cannot learn anything about the underlying messages. According to the direction of transformation, PRE can be categorized into *bidirectional* PRE, in which the proxy can transform from Alice to Bob and vice versa, and *unidirectional* PRE, in which the proxy cannot transform ciphertexts in the opposite direction. PRE can also be categorized to *multi-hop* PRE, in which the ciphertexts can be transformed from Alice to Bob and then to Charlie and so on, and *single-hop* PRE, in which the ciphertexts can only be transformed once.

---

* Supported by Ministry of Education, Culture, Sports, Science and Technology.

K. Sakiyama and M. Terada (Eds.): IWSEC 2013, LNCS 8231, pp. 290–302, 2013.
© Springer-Verlag Berlin Heidelberg 2013

Recently, as cloud computing emerges, PRE gains much more attention as one of the key security components to provide secure cloud services. The security against corrupted proxies is especially important in such applications since the proxies may be out of control of honest users and the proxies are more likely to be attacked than those in on-premise systems. In [1], Ateniese, Fu, Green, and Hohenberger mentioned the security notion, *non-transferability*, with respect to the security against malicious proxies, which is described as *"The (malicious) proxy and a set of colluding delegatees cannot re-delegate decryption rights."* They also note that *"achieving a proxy scheme that is non-transferable, in the sense that the only way for Bob to transfer offline decryption capabilities to Carol is to expose his own secret key, seems to be the main open problem left for proxy re-encryption."*

Until now, some attempts for non-transferable PRE have been taken. For example, in the scheme proposed by Libert and Vergnaud [5], a delegator can identify the malicious proxies by analyzing a re-encryption key to convert ciphertexts of the delegator into some malicious user's generated (forged) by the colluding proxies and delegatees. Although it is one possible approach to the non-transferable PRE, it still cannot prevent colluding proxies and delegatees from re-delegating the decryption rights. Further, the scheme is less efficient in the sense that the ciphertext size depends on the number of delegations and it is only proved to be secure against chosen plaintext attack (CPA). In the scheme by Wang et al. [8] which is an ID-based PRE scheme, a trusted third party (privatekey generator, PKG) takes part in generating re-encryption keys. This approach, however, is not a complete solution to non-transferable PRE because, as pointed out by He, Chim, Hui, and Yiu [4], it is just a transformation of "delegatee-proxy-collusion transferable problem" to "PKG alone transferable problem." In the ID-based scheme by He, Chim, Hui, and Yiu [4], the delegator and the delegatee communicate and send some information to each other to generate the re-encryption key (The delegator also communicates with PKG). Therefore, colluding proxies and delegatees cannot generate a new re-encryption key without delegator's help.

In IWSEC 2011, Hayashi, Matsushita, Yoshida, Fujii, and Okada [3] made an important step toward the non-transferability by introducing a security notion called *the unforgeability of re-encryption keys against collusion attack (UFReKey-CA)*, which is a relaxed notion (necessary condition) of the non-transferability, and its formal definition. Roughly speaking, UFReKey-CA means that even colluding proxies and delegatees cannot generate a re-encryption key for some user. They also proposed a stronger security notion, the *strong* unforgeability of re-encryption keys against collusion attack (sUFReKey-CA). Since sUFReKey-CA implies UFReKey-CA and sUFReKey-CA is simpler (i.e. easier to treat) definition than UFReKey-CA, sUFReKey-CA is useful to prove UFReKey-CA. We note that sUFReKey-CA does not imply the non-transferability and vice-versa. They then proposed a concrete PRE scheme and its variant supporting temporary delegation, which can limit the lifetime of re-encryption keys within a certain time interval. They claimed that their schemes meet both the replayable CCA (RCCA) security and sUFReKey-CA in the standard model. To prove the above schemes meet sUFReKey-CA, they proposed two new variants of Diffie-Hellman inversion problem and assumed their hardness.

In this paper, we present two concrete attacks to their PRE schemes.

The first attack is to the sUFReKey-CA property on their two schemes (in Section 4). In particular, we identify the weakness in their schemes by using the linearity of the exponents of the re-encryption key.

The second attack is to the assumptions employed in the security proofs for sUFReKey-CA of their two schemes (in Section 5). In particular, we show that the two computational problems called the 2-DHIwRA and the m-2-DHIwRA problems can be solved efficiently.

2 Preliminaries

We use $x \xleftarrow{\text{R}} \mathcal{S}$ to denote that an element x is chosen uniformly at random from $\mathcal{S}$.

2.1 Bilinear Maps

Groups $(\mathbb{G}, \mathbb{G}_T)$ of prime order p are called *bilinear map groups* if there is a mapping $\mathbf{e} : \mathbb{G} \times \mathbb{G} \to \mathbb{G}_T$ with the following properties:

1. bilinearity: $\mathbf{e}(g^a, h^b) = \mathbf{e}(g, h)^{ab}$ for any $(g, h) \in \mathbb{G} \times \mathbb{G}$ and $a, b \in \mathbb{Z}_p$;
2. efficient computability for any input pair;
3. non-degeneracy: $\mathbf{e}(g, h) \neq 1_{\mathbb{G}_T}$ whenever $g, h \neq 1_{\mathbb{G}}$.

2.2 Unidirectional Proxy Re-Encryption

Definition 1 (Unidirectional, Single-Hop PRE [6]). *A unidirectional, single-hop PRE scheme is a tuple of algorithms* $\Pi = (\mathbf{Setup}, \mathbf{KGen}, \mathbf{ReKey}, \mathbf{Enc}, \mathbf{ReEnc}, \mathbf{Dec})$ *for message space* $\mathcal{M}$:

- **Setup**$(1^\lambda) \to PP$. *On input security parameter* 1^λ, *the setup algorithm outputs the public parameters* PP.
- **KGen**$(\lambda, PP) \to (pk, sk)$. *On input parameters, the key generation algorithm outputs a public key* pk *and a secret key* sk.
- **ReKey**$(PP, sk_i, pk_j) \to rk_{i \to j}$. *Given a secret key* sk_i *and a public key* pk_j, *where* $i \neq j$, *the re-encryption key generation algorithm outputs a unidirectional re-encryption key* $rk_{i \to j}$. *The restriction that* $i \neq j$ *is provided as re-encrypting a message to the original recipient is impractical.*
- **Enc**$_1(PP, pk_i, m) \to CT_i$. *On input a public key* pk_i *and a message* $m \in \mathcal{M}$, *the encryption algorithm outputs a first level ciphertext* CT_i *that cannot be re-encrypted for another party.*
- **Enc**$_2(PP, pk_i, m) \to CT_i$. *On input a public key* pk_i *and a message* $m \in \mathcal{M}$, *the encryption algorithm outputs a second level ciphertext* CT_i *that can be re-encrypted into a first level one (intended for a possibly different receiver) using a suitable re-encryption key.*
- **ReEnc**$(PP, rk_{i \to j}, CT_i) \to CT_j$. *Given a re-encryption key* $rk_{i \to j}$ *and an original ciphertext* CT_i *for* i, *the re-encryption algorithm outputs a first level ciphertext* CT_j *for* j *or a distinguished message 'invalid'.*

- $\mathbf{Dec}_1(PP, sk_i, CT_i) \rightarrow m$. *Given a secret key sk_i and a first level ciphertext CT_i, the decryption algorithm outputs a message $m \in \mathcal{M}$ or a distinguished message 'invalid'.*
- $\mathbf{Dec}_2(PP, sk_i, CT_i) \rightarrow m$. *Given a secret key sk_i and a second level ciphertext CT_i, the decryption algorithm outputs a message $m \in \mathcal{M}$ or a distinguished message 'invalid'.*

To lighten notations, from now, we will omit the public parameters PP as the input of the algorithms.

For all $m \in \mathcal{M}$ and all pair (pk_i, sk_i), (pk_j, sk_j) these algorithms should satisfy the following conditions of correctness:

$$\mathbf{Dec}_1(sk_i, \mathbf{Enc}_1(pk_i, m)) = m;$$
$$\mathbf{Dec}_2(sk_i, \mathbf{Enc}_2(pk_i, m)) = m;$$
$$\mathbf{Dec}_1(sk_j, \mathbf{ReEnc}(\mathbf{ReKey}(sk_i, pk_j), \mathbf{Enc}_2(pk_i, m))) = m.$$

2.3 Unidirectional Proxy Re-Encryption with Temporary Delegation

In this section, we review the syntactic definition of unidirectional proxy re-encryption with temporary delegation [6]. In the PRE with temporary delegation, it only allows the proxy to re-encrypt messages from Alice to Bob during a limited time period.

The model of unidirectional PRE supporting temporary delegation is almost the same as that in Definition 1 except that re-encryption key generation, encryption, and re-encryption algorithms take a period $\ell \in \{1, \ldots, L\}$ as input. Intuitively, the re-encryption key generated by **ReKey** with a period ℓ, can be used to re-encrypt the ciphertext generaed by **Enc**$_2$ with the same period ℓ. Note that the public and the secret keys are common to all time periods.

2.4 Unforgeability of Re-Encryption Keys against Collusion Attack

In this section, we recall the security definitions proposed by Hayashi et al. [3] called the (strong) unforgeability of re-encryption keys against collusion attack. More detailed explanation can be found in [3].

In the following definitions, keys subscripted by $*$, h, j, and c_i are those for a target honest delegator, a honest user, a malicious user, and a corrupted delegatee, respectively, and $i \in \{1, \ldots L\}$ where L is polynomially bounded.

Definition 2 (Unforgeability of Re-Encryption Keys against Collusion Attack, UFReKey-CA [3]). *A unidirectional single-hop proxy re-encryption scheme meets the unforgeability of re-encryption keys against collusion attack if there exists a polynomial time algorithm $\mathcal{P}$ such that*

$$\Pr[(pk_*, sk_*) \leftarrow \mathbf{KGen}(\lambda); (pk_h, sk_h) \leftarrow \mathbf{KGen}(\lambda); \{(pk_{c_i}, sk_{c_i}) \leftarrow \mathbf{KGen}(\lambda)\};$$

$$(pk_j, sk_j) \leftarrow \mathbf{KGen}(\lambda); \{rk_{* \to c_i} \leftarrow \mathbf{ReKey}(sk_*, pk_{c_i})\};$$

$$\{rk_{h \to c_i} \leftarrow \mathbf{ReKey}(sk_h, pk_{c_i})\}; m \xleftarrow{\text{R}} \mathcal{M}; C^* \leftarrow \mathbf{Enc}_2(pk_*, m);$$

$$\{m_i \xleftarrow{\text{R}} \mathcal{M}\}; \{C_i \leftarrow \mathbf{Enc}_2(pk_{c_i}, m_i)\}; \{m_i' \xleftarrow{\text{R}} \mathcal{M}\}; \{C_i' \leftarrow \mathbf{Enc}_1(pk_{c_i}, m_i')\};$$

$$\{m_i'' \xleftarrow{\text{R}} \mathcal{M}\}; \{C_i'' \leftarrow \mathbf{ReEnc}(rk_{h \to c_i}, \mathbf{Enc}_2(pk_h, m_i''))\};$$

$$X \leftarrow \mathcal{C}(pk_*, \{(pk_{c_i}, sk_{c_i})\}, \{rk_{* \to c_i}\}); rk^\dagger_{* \to j} \leftarrow \mathcal{J}(X, (pk_j, sk_j));$$

$$m_{\mathcal{P}} \leftarrow \mathcal{P}(X, (pk_j, sk_j), \{C_i\}, \{C_i'\}, \{C_i''\})$$

$$: m \neq \mathbf{Dec}_1(sk_j, \mathbf{ReEnc}(rk^\dagger_{* \to j}, C^*)) \vee m_{\mathcal{P}} \in \{m_i\} \cup \{m_i'\} \cup \{m_i''\}]$$

is overwhelming for any polynomial time algorithm $\mathcal{C}, \mathcal{J}$, and polynomial L.

Intuitively, this definition states that it is impossible for $\mathcal{C}$ (the colluding proxies and delegatees) to re-delegate the decryption rights of the target honest delegator $*$ to $\mathcal{J}$ (a malicious user) by giving the information to forge the re-encryption key for $\mathcal{J}$ without delegating any right related to secret keys of any member in $\mathcal{C}$ to $\mathcal{P}$ (the malicious user). As compared with the non-transferability, the way to re-delegate the decryption rights is limited to the forgery of the re-encryption key in the definition of UFReKey-CA.

In the above definition, the adversary tries to return a forged re-encryption key $rk^\dagger_{* \to j}$ such that

$$m = \mathbf{Dec}_1(sk_j, \mathbf{ReEnc}(rk^\dagger_{* \to j}, \mathbf{Enc}_2(pk_*, m))) \tag{1}$$

where $m \xleftarrow{\text{R}} \mathcal{M}$. The adversary always wins if she returns the *well-formed* re-encryption key, which is one of the outputs of $\mathbf{ReKey}(sk_*, pk_j)$. On the other hand, the adversary does not have to output the well-formed re-encryption key to win the game. That is, the adversary also wins if she forges a re-encryption key which satisfies the equation (1) and it is an ill-formed re-encryption key, which is never returned from $\mathbf{ReKey}(sk_*, pk_j)$.

In the definition of UFReKey-CA, there exists the adversary $\mathcal{P}$ which extracts the plaintext from the information X. Generally speaking, it is difficult (complicated) to prove the (non-)existence of the plaintext extractor $\mathcal{P}$. For convenience, [3] proposed a simple and useful security notion to prove UFReKey-CA as the following.

Definition 3 (Strong Unforgeability of Re-Encryption Keys against Collusion Attack, sUFReKey-CA [3]) *A unidirectional single-hop proxy re-encryption scheme meets* the strong unforgeability of re-encryption keys against collusion attack *if*

$$\Pr[(pk_*, sk_*) \leftarrow \mathbf{KGen}(\lambda); \{(pk_{c_i}, sk_{c_i}) \leftarrow \mathbf{KGen}(\lambda)\}; (pk_j, sk_j) \leftarrow \mathbf{KGen}(\lambda);$$

$$\{rk_{* \to c_i} \leftarrow \mathbf{ReKey}(sk_*, pk_{c_i})\}; m \xleftarrow{\text{R}} \mathcal{M}; C^* \leftarrow \mathbf{Enc}_2(pk_*, m);$$

$$rk^\dagger_{* \to j} \leftarrow \mathcal{A}(pk_*, \{(pk_{c_i}, sk_{c_i})\}, (pk_j.sk_j), \{rk_{* \to c_i}\})$$

$$: m = \mathbf{Dec}_1(sk_j, \mathbf{ReEnc}(rk^\dagger_{* \to j}, C^*))]$$

is negligible for any polynomial time algorithm $\mathcal{A}$, and polynomial L.

Intuitively, this definition states that it is impossible for $\mathcal{C}$ (colluding proxies and delegatees) to re-delegate the decryption rights of the target honest delegator $*$ to a malicious user $\mathcal{A}$ by giving the forged re-encryption keys for the malicious user, where the secret key(s) of the colluding proxies and delegatees may be revealed to the malicious user, and the secret key of the malicious user may be revealed to to the colluding proxies and delegatees. It is easy to see that the scheme which satisfies sUFReKey-CA also meets UFReKey-CA. Since there exists no plaintext extractor in the definition of sUFReKey-CA, the proof of sUFReKey-CA is simpler than that of UFReKey-CA.

Note that we can consider several variations of UFReKey-CA by changing goals of $\mathcal{C}$. For example, it may be defined that $\mathcal{C}$ tries to generate a secret key sk_{c_i} itself, or generate a forged secret key $sk^\dagger$ such that $\mathbf{Dec}_1(sk^\dagger, C) = \mathbf{Dec}_1(sk_{c_i}, C)$. We also note that even when such kinds of $\mathcal{C}$ are defined in UFReKey-CA, sUFReKey-CA still implies UFReKey-CA and sUFReKey-CA is still useful to prove UFReKey-CA.

3   Review of the PRE Schemes by Hayashi et al.

In this section, we review the PRE schemes proposed by Hayashi et al. in [3]. There are two schemes. The first one is the main scheme which is based on the scheme in [6]. The other is its variant which supports temporary delegation.

Before describing them, we review the strong one-time signature which is employed to construct the schemes. One-time signature $\mathbf{Sig} = (\mathcal{G}, \mathcal{S}, \mathcal{V})$ consists if a triple of algorithms. The algorithm $\mathcal{G}$ takes a security parameter λ and returns a pair of signing/verification keys (ssk, svk). Then, for any message M, $\mathcal{V}(\sigma, svk, M)$ returns 1 whenever $\sigma = \mathcal{S}(ssk, M)$ and 0 otherwise. We say that $\mathbf{Sig}$ is a strong one-time signature if no polynomial time adversary can create a new signature for a previously signed message (See [6] for the formal security definition).

3.1   The Main Scheme

The main scheme proposed by Hayashi et al. [3] is as follows.

Setup(λ): given a security parameter λ, choose bilinear map groups $(\mathbb{G}, \mathbb{G}_T)$ of prime order $p > 2^\lambda$, generators $g, g_1(= g^\alpha), g_2(= g^\beta), u, v \xleftarrow{R} \mathbb{G}$, and a one-time signature scheme $\mathbf{Sig} = (\mathcal{G}, \mathcal{S}, \mathcal{V})$. The public parameters are $PP := \{p, \mathbb{G}, \mathbb{G}_T, g, g_1, g_2, u, v, \mathbf{Sig}\}$. The message space $\mathcal{M}$ is equal to $\mathbb{G}_T$.

KGen(λ, PP): user i chooses $x_i, y_i, z_i \xleftarrow{R} \mathbb{Z}_p^*$. The secret key is $sk_i = (x_i, y_i, z_i)$. The public key is $pk_i = (X_i, Y_{1i}, Y_{2i}, Z_i, Z_{1i})$, where $X_i \leftarrow g^{x_i}, Y_{1i} \leftarrow g_1^{y_i}, Y_{2i} \leftarrow g_2^{y_i}, Z_i \leftarrow g^{z_i}, Z_{1i} \leftarrow g_1^{z_i}$.

Enc$_1(PP, pk_j, m)$: to encrypt a message $m \in \mathbb{G}_T$ under the public key pk_j at the first level, the sender proceeds as follows:

1. Select a one-time signature key pair $(svk, ssk) \leftarrow \mathcal{G}(\lambda)$ and set $C_1 = svk$.

2. Pick $r, s, t, k, \gamma \xleftarrow{R} \mathbb{Z}_p^*$ and compute,
$$C'_{2X} = Y_{2j}^s, C''_{2X} = Y_{2j}^{rs}, C'_{2Y} = X_j^t, C''_{2Y} = X_j^{rt}, C'_{2Z} = Y_{2j}^k,$$
$$C''_{2Z} = Y_{2j}^{rk}, C'_{2Z1} = X_j^k, C''_{2Z1} = X_j^{rk}, C_3 = e(g_1 g_2, g)^r \cdot m,$$
$$C_4 = (u^{svk} \cdot v)^r, C_{5X} = (g_1 \cdot g^\gamma)^{\frac{1}{s}}, C_{5Y} = g^{\frac{\gamma+1}{t}}, C_{5Z} = (g_1 \cdot g^{\gamma+1})^{\frac{1}{k}}.$$

3. Generate a one-time signature $\sigma \leftarrow \mathcal{S}(ssk, (C_3, C_4))$ on (C_3, C_4).

The (first level) ciphertext is $C_j = (C_1, C'_{2X}, C''_{2X}, C'_{2Y}, C''_{2Y}, C'_{2Z}, C''_{2Z}, C'_{2Z1}, C''_{2Z1}, C_3, C_4, C_{5X}, C_{5Y}, C_{5Z}, \sigma)$.

Enc$_2(PP, pk_i, m)$: to encrypt a message $m \in \mathbb{G}_T$ under the public key pk_i at the second level, the sender proceeds as follows:

1. Select a one-time signature key pair $(svk, ssk) \leftarrow \mathcal{G}(\lambda)$ and set $C_1 = svk$.

2. Pick $r \xleftarrow{R} \mathbb{Z}_p^*$ and compute,
$$C_{2X} = X_i^r, C_{2Y} = Y_{1i}^r, C_{2Z} = Z_i^r, C_{2Z1} = Z_{1i}^r,$$
$$C_3 = e(g_1 g_2, g)^r \cdot m, C_4 = (u^{svk} \cdot v)^r.$$

3. Generate a one-time signature $\sigma \leftarrow \mathcal{S}(ssk, (C_3, C_4))$ on (C_3, C_4).

The (first level) ciphertext is $C_i = (C_1, C_{2X}, C_{2Y}, C_{2Z}, C_{2Z1}, C_3, C_4, \sigma)$.

ReKey(PP, sk_i, pk_j): given user i's secret key sk_i and user j's public key pk_j, generate the re-encryption key $rk_{i \to j} = (rk_{ij1}, rk_{ij2}, rk_{ij3})$, where $\gamma \leftarrow \mathbb{Z}_p^*$ and
$$rk_{ij1} = (X_j \cdot g^\gamma)^{1/x_i} = g^{\frac{x_j + \gamma}{x_i}}, rk_{ij2} = (Y_{2j} \cdot g^\gamma)^{1/y_i} = g^{\frac{\beta y_j + \gamma}{y_i}},$$
$$rk_{ij3} = (X_j \cdot Y_{2j} \cdot g^\gamma)^{1/z_i} = g^{\frac{x_j + \beta y_j + \gamma}{z_i}}.$$

ReEnc$(PP, rk_{i \to j}, C_i)$: on input of the re-encryption key $rk_{i \to j}$ and a second level ciphertext C_i, check the validity of the ciphertext by testing:
$$e(C_{2X}, u^{C1} \cdot v) = e(X_i, C_4), e(C_{2Y}, u^{C1} \cdot v) = e(Y_{1i}, C_4),$$
$$e(C_{2Z}, u^{C1} \cdot v) = e(Z_i, C_4), e(C_{2Z1}, u^{C1} \cdot v) = e(Z_{1i}, C_4), \tag{2}$$
$$\mathcal{V}(C_1, \sigma, (C_3, C_4)) = 1.$$

If the relations (2) hold (well-formed), C_i is re-encrypted by choosing $s, t, k \xleftarrow{R} \mathbb{Z}_p^*$ and computing
$$C'_{2X} = X_i^s, C''_{2X} = X_i^{rs}, C'_{2Y} = Y_{1i}^t, C''_{2Y} = C_{2Y}^t = Y_{1i}^{rt}, C'_{2Z} = Z_i^k,$$
$$C''_{2Z} = Z_i^{rk}, C'_{2Z1} = Z_{1i}^k, C''_{2Z1} = C_{2Z1}^k = Z_{1i}^{rk}, C_{5X} = rk_{ij1}^{\frac{1}{s}}, C_{5Y} = rk_{ij2}^{\frac{1}{t}}, C_{5Z} =$$
$rk_{ij3}^{\frac{1}{k}}$, and a re-encrypted ciphertext $C_j = (C_1, C'_{2X}, C''_{2X}, C'_{2Y}, C''_{2Y}, C'_{2Z}, C''_{2Z}, C'_{2Z1}, C''_{2Z1}, C_3, C_4, C_{5X}, C_{5Y}, C_{5Z}, \sigma)$ is returned. Otherwise, 'invalid' is returned.

Dec$_1(sk_j, C_j)$: the validity of the first level ciphertext C_j is checked by testing:
$$e(C''_{2X}, u^{C1} \cdot v) = e(C'_{2X}, C_4), e(C''_{2Y}, u^{C1} \cdot v) = e(C'_{2Y}, C_4),$$
$$e(C''_{2Z}, u^{C1} \cdot v) = e(C'_{2Z}, C_4), e(C''_{2Z1}, u^{C1} \cdot v) = e(C'_{2Z1}, C_4),$$
$$e(C_{5Z}, C'_{2Z}) = e(C_{5X}, C'_{2X}) \cdot e(Y_{2j}, g), \tag{3}$$
$$e(C_{5Z}, C'_{2Z1}) = e(C_{5Y}, C'_{2Y}) \cdot e(X_j, g_1),$$
$$\mathcal{V}(C_1, \sigma, (C_3, C_4)) = 1.$$

If the relations (3) hold (well-formed), the plaintext
$$m = C_3 \Big/ \left\{ \left(\frac{e(C_{5Z}, C''_{2Z})}{e(C_{5X}, C''_{2X})} \right)^{\frac{1}{y_j}} \cdot \left(\frac{e(C_{5Z}, C''_{2Z1})}{e(C_{5Y}, C''_{2Y})} \right)^{\frac{1}{x_j}} \right\},$$

is returned. Otherwise (ill-formed), the algorithm outputs 'invalid'.

Dec$_2(sk_i, C_i)$: if the second level ciphertext C_i satisfies the relations (2) (well-formed), the plaintext $m = C_3 / e(g_1 g_2, C_{2X})^{\frac{1}{x_i}}$ is returned. Otherwise, 'invalid' is returned.

3.2 The Scheme with Temporary Delegation

The PRE scheme supporting temporary delegation in [3] is as follows.

Setup(λ): the same as that in Section 3.1.

KGen(λ, PP): user i chooses $x_i, y_i, z_i, w_i \xleftarrow{R} \mathbb{Z}_p^*$. The secret key is $sk_i = (x_i, y_i, z_i, w_i)$. The public key is $pk_i = (X_i, Y_{1i}, Y_{2i}, Z_i, Z_{1i}, W_i)$, where $X_i \leftarrow g^{x_i}, Y_{1i} \leftarrow g_1^{y_i}, Y_{2i} \leftarrow g_2^{y_i}, Z_i \leftarrow g^{z_i}, Z_{1i} \leftarrow g_1^{z_i}, W_i \leftarrow g^{w_i}$. A function $F_i : \{1, \ldots, L\} \rightarrow \mathbb{G}$ is implicitly defined as $F_i(\ell) = g^{\ell} \cdot W_i = g^{\ell + w_i}$.

Enc$_1(PP, pk_j, m, \ell)$: to encrypt a message $m \in \mathbb{G}_T$ under the public key pk_j at the first level during period ℓ, the sender proceeds as follows:

1. Select a one-time signature key pair $(svk, ssk) \leftarrow \mathcal{G}(\lambda)$ and set $C_1 = svk$.

2. Pick $r, s, t, k, \gamma, \delta_x, \delta_y \xleftarrow{R} \mathbb{Z}_p^*$ and compute,
$$C'_{2X} = Y_{2j}^s, C''_{2X} = Y_{2j}^{rs}, C'_{2Y} = X_j^t, C''_{2Y} = X_j^{rt}, C'_{2Z} = Y_{2j}^k,$$
$$C''_{2Z} = Y_{2j}^{rk}, C'_{2Z1} = X_j^k, C''_{2Z1} = X_j^{rk}, C_3 = e(g_1 g_2, g)^r \cdot m,$$
$$C_4 = (u^{svk} \cdot v)^r, C_{5X} = (g_1 \cdot g^\gamma \cdot F_j(\ell)^{\delta_y})^{\frac{1}{s}} = g^{\frac{\alpha + \gamma + (\ell + w_j)\delta_y}{s}},$$
$$C_{5Y} = (g^{\gamma+1} \cdot F_j(\ell)^{\delta_x})^{\frac{1}{t}} = g^{\frac{1 + \gamma + (\ell + w_j)\delta_y}{t}},$$
$$C_{5Z} = (g_1 \cdot g^{\gamma+1})^{\frac{1}{k}} = g^{\frac{\alpha + 1 + \gamma}{k}}, C_{5FX} = (Y_{2j})^{\frac{\delta_y}{k}}, C_{5FY} = (X_j)^{\frac{\delta_x}{k}}$$

3. Generate a one-time signature $\sigma \leftarrow \mathcal{S}(ssk, (\ell, C_3, C_4))$ on (ℓ, C_3, C_4). The (first level) ciphertext $C_j = (C_1, C'_{2X}, C''_{2X}, C'_{2Y}, C''_{2T}, C'_{2Z}, C''_{2Z}, C'_{2Z1}, C''_{2Z1}, C_3, C_4, C_{5X}, C_{5Y}, C_{5Z}, C_{5FX}, C_{5FY}, \sigma)$.

Enc$_2(PP, pk_i, m, \ell)$: to encrypt a message $m \in \mathbb{G}_T$ under the public key pk_i at the second level, the sender proceeds as follows:

1. Select a one-time signature key pair $(svk, ssk) \leftarrow \mathcal{G}(\lambda)$ and set $C_1 = svk$.

2. Pick $r \xleftarrow{R} \mathbb{Z}_p^*$ and compute,
$$C_{2X} = X_i^r, C_{2Y} = Y_{1i}^r, C_{2Z} = Z_i^r, C_{2Z1} = Z_{1i}^r, C_{2F} = F_i(\ell)^r,$$
$$C_3 = e(g_1 g_2, g)^r \cdot m, C_4 = (u^{svk} \cdot v)^r.$$

3. Generate a one-time signature $\sigma \leftarrow \mathcal{S}(ssk, (\ell, C_3, C_4))$ on (ℓ, C_3, C_4).
The (first level) ciphertext is $C_i = (\ell, C_1, C_{2X}, C_{2Y}, C_{2Z}, C_{2Z1}, C_{2F}, C_3, C_4, \sigma)$.

ReKey(PP, sk_i, pk_j, ℓ): given a period number ℓ, user i's secret key sk_i and user j's public key pk_j, generate the re-encryption key $rk_{i \rightarrow j|\ell} = (rk_{ij\ell1}, rk_{ij\ell2}, rk_{ij\ell3}, rk_{ij\ell4}, rk_{ij\ell5})$, where $\gamma, \delta_x, \delta_y \leftarrow \mathbb{Z}_p^*$ and
$$rk_{ij\ell1} = (X_j \cdot g^\gamma)^{1/x_i} \cdot F_i(\ell)^{\delta_x} = g^{\frac{x_j + \gamma}{x_i} + (\ell + w_i)\delta_x},$$
$$rk_{ij\ell2} = (Y_{2j} \cdot g^\gamma)^{1/y_i} \cdot F_i(\ell)^{\delta_y} = g^{\frac{\beta y_j + \gamma}{y_i} + (\ell + w_i)\delta_y},$$
$$rk_{ij\ell3} = (X_j \cdot Y_{2j} \cdot g^\gamma)^{1/z_i} = g^{\frac{x_j + \beta y_j + \gamma}{z_i}}, rk_{ij\ell4} = X_i^{\delta_x}, rk_{ij\ell5} = Y_{1i}^{\delta_y}.$$

ReEnc$(PP, rk_{ijl}, C_i, \ell)$: on input of the re-encryption key $rk_{i \rightarrow j}$ for period ℓ and a second level ciphertext C_i, check the validity of the ciphertext by testing:

$$e(C_{2X}, u^{C_1} \cdot v) = e(X_i, C_4), e(C_{2Y}, u^{C_1} \cdot v) = e(Y_{1i}, C_4),$$
$$e(C_{2Z}, u^{C_1} \cdot v) = e(Z_i, C_4), e(C_{2Z1}, u^{C_1} \cdot v) = e(Z_{1i}, C_4), \qquad (4)$$
$$e(C_{2F}, u^{C_1} \cdot v) = e(F_i(\ell), C_4), \mathcal{V}(C_1, \sigma, (C_3, C_4)) = 1.$$

If the relations (4) hold (well-formed), C_i is re-encrypted by computing

$$C'_{2X} = X_i^s, C''_{2X} = X_i^{rs}, C'_{2Y} = Y_{1i}^t, C''_{2Y} = C_{2Y}^t = Y_{1i}^{rt}, C'_{2Z} = Z_i^k,$$
$$C''_{2Z} = Z_i^{rk}, C'_{2Z1} = Z_{1i}^k, C''_{2Z1} = C_{2Z1}^k = Z_{1i}^{rk},$$
$$C'_{2F} = F_i(\ell)^h, C''_{2F} = C_{2F}^h = F_i(\ell)^{rh},$$
$$C_{5X} = rk_{ij\ell1}^{\frac{1}{s}}, C_{5Y} = rk_{ij\ell2}^{\frac{1}{t}}, C_{5Z} = rk_{ij\ell3}^{\frac{1}{k}}, C_{FX} = rk_{ij\ell4}^{\frac{1}{h}}, C_{FY} = rk_{ij\ell5}^{\frac{1}{h}},$$

where $s, t, k, h \xleftarrow{R} \mathbb{Z}_p^*$, and re-encrypted ciphertext $C_j = (C_1, C'_{2X}, C''_{2X}, C'_{2Y},$
$C''_{2T}, C'_{2Z}, C''_{2Z}, C'_{2Z1}, C''_{2Z1}, C_3, C_4, C_{5X}, C_{5Y}, C_{5Z}, C_{5FX}, C_{5FY}, \sigma)$ is returned.
Otherwise (ill-formed), the algorithm outputs 'invalid'.

$\mathbf{Dec}_1(sk_j, C_j)$: the validity of the first level ciphertext C_j is checked by testing:

$$e(C''_{2X}, u^{C1} \cdot v) = e(C'_{2X}, C_4), e(C''_{2Y}, u^{C1} \cdot v) = e(C'_{2Y}, C_4),$$
$$e(C''_{2Z}, u^{C1} \cdot v) = e(C'_{2Z}, C_4), e(C''_{2Z1}, u^{C1} \cdot v) = e(C'_{2Z1}, C_4),$$
$$e(C''_{2F}, u^{C1} \cdot v) = e(C'_{2F}, C_4), \mathcal{V}(C_1, \sigma, (C_3, C_4)) = 1, \tag{5}$$
$$e(C_{5Z}, C'_{2Z}) = e(C_{5FX}, C'_{2X}) \cdot e(Y_{2j}, g),$$
$$e(C_{5Z}, C'_{2Z1}) = e(C_{5FY}, C'_{2Y}) \cdot e(X_j, g_1).$$

If the relations (5) hold (well-formed), the plaintext

$$m = C_3 \bigg/ \left\{ \left(\frac{e(C_{5Z}, C''_{2Z}) \cdot e(C_{5FX}, C''_{2F})}{e(C_{5X}, C''_{2X})} \right)^{\frac{1}{y_j}} \cdot \left(\frac{e(C_{5Z}, C''_{2Z1} \cdot e(C_{5FY}, C''_{2F})}{e(C_{5Y}, C''_{2Y})} \right)^{\frac{1}{x_j}} \right\},$$

is returned. Otherwise (ill-formed), the algorithm outputs 'invalid'.

$\mathbf{Dec}_2(sk_i, C_i)$: if the second level ciphertext C_i satisfies the relations (4) (well-formed),
the plaintext $m = C_3/e(g_1g_2, C_{2X})^{\frac{1}{x_i}}$ is returned. Otherwise, 'invalid' is returned.

4   Security Analysis of the PRE Schemes by Hayashi et al.

In this section, we present concrete attacks to sUFReKey-CA of both schemes. Before
presenting their details, we first identify the potential weakness in their schemes; that
is, the linearity of the exponents of the re-encryption key, i.e. the linearity of x_j, y_j, γ
in $\frac{x_j + \gamma}{x_i}, \frac{\beta y_j + \gamma}{y_i}, \frac{x_j + \beta y_j + \gamma}{z_i}$.

For example, given two re-encryption keys $rk_{* \to c_i} = (rk_{i1}, rk_{i2}, rk_{i3})$, and $rk_{* \to c_j}$
$= (rk_{j1}, rk_{j2}, rk_{j3})$ one can compute a re-encryption key $rk_{* \to c_t} = (rk_{t1}, rk_{t2}, rk_{t3})$,
where the secret key sk_t of the user t is a linear combination of sk_i and sk_j (i.e. $x_t = ax_i + bx_j, y_t = ay_i + by_j$, for some $a, b \in \mathbb{Q}$) as follows:

- $rk_{t1} = rk_{i1}^a \cdot rk_{j1}^b = g^{\frac{ax_i + a\gamma_1}{x_*}} \cdot g^{\frac{bx_j + b\gamma_2}{x_*}} = g^{\frac{(ax_i + bx_j) + (a\gamma_1 + b\gamma_2)}{x_*}} = g^{\frac{x_t + \gamma}{x_*}}$, where
  $\gamma := a\gamma_1 + b\gamma_2$,
- $rk_{t2} = rk_{i2}^a \cdot rk_{j2}^b = g^{\frac{\beta(ay_i + by_j) + (a\gamma_1 + b\gamma_2)}{y_*}} = g^{\frac{\beta y_t + \gamma}{y_*}}$,
- $rk_{t3} = rk_{i3}^a \cdot rk_{j3}^b = g^{\frac{(ax_i + bx_j) + \beta(ay_i + by_j) + (a\gamma_1 + b\gamma_2)}{z_*}} = g^{\frac{x_t + \beta y_t + \gamma}{z_*}}$.

Then, there exists an adversary who can break sUFReKey-CA of the main scheme
(it is the same in the case of the PRE scheme with temporary delegation) as follows:
given $sk_{c_1} = (x_1, y_1, z_1), sk_{c_2} = (x_2, y_2, z_2), sk_j = (x_j, y_j, z_j), rk_{* \to c_1}, rk_{* \to c_2}$, the

adversary can first compute $a, b \in \mathbb{Q}$ such that $x_j = ax_1 + bx_2, y_j = ay_1 + by_2$, then uses it to compute a forged re-encryption key $rk_{* \to j}$ as the above example.

In the next sections, we give the details of attacks to sUFReKey-CA of the main scheme and the scheme supporting temporary delegation, respectively.

4.1 Attack to sUFReKey-CA of the Main Scheme

Given $pk_*, \{(pk_{c_i}, sk_{c_i})\}, (pk_j, sk_j), \{rk_{* \to c_i}\}$ and the public parameters $PP = \{p, \mathbb{G}, \mathbb{G}_T, g, g_1, g_2, u, v, \mathbf{Sig}\}$, the adversary $\mathcal{A}$ breaks sUFReKey-CA of the scheme as follows:

1. $\mathcal{A}$ chooses from $\{sk_{c_i}\}$ two secret keys $sk_{c_{i_1}} = (x_1, y_1, z_1)$ and $sk_{c_{i_2}} = (x_2, y_2, z_2)$ such that $\frac{x_1}{y_1} \neq \frac{x_2}{y_2}$. It is easy to see that this is possible since x_i, y_i are independently chosen from $\mathbb{Z}_p^*$. Let $rk_{* \to c_{i_1}} = (rk_{i_1 1}, rk_{i_1 2}, rk_{i_1 3})$ and $rk_{* \to c_{i_2}} = (rk_{i_2 1}, rk_{i_2 2}, rk_{i_2 3})$.

2. $\mathcal{A}$ solves the following system of linear equations (with variables u and v)

$$\begin{cases} u \cdot x_1 + v \cdot x_2 = x_j \\ u \cdot y_1 + v \cdot y_2 = y_j \end{cases}$$

Since $\frac{x_1}{y_1} \neq \frac{x_2}{y_2}$, the above system has a pair of root $(a, b) \in \mathbb{Q}^2$. It means that $\mathcal{A}$ easily computes $(a, b) \in \mathbb{Q}^2$ such that

$$\begin{cases} a \cdot x_1 + b \cdot x_2 = x_j \\ a \cdot y_1 + b \cdot y_2 = y_j \end{cases}$$

3. $\mathcal{A}$ outputs a re-encryption key $rk_{* \to j}^{\dagger} = (R_1, R_2, R_3)$, where $R_1, R_2,$ and R_3 are computed as follows:

   - $R_1 = rk_{i_1 1}^a \cdot rk_{i_2 1}^b = \left(g^{\frac{x_1 + \gamma_1}{x_*}}\right)^a \cdot \left(g^{\frac{x_2 + \gamma_2}{x_*}}\right)^b = g^{\frac{(ax_1 + bx_2) + (a\gamma_1 + b\gamma_2)}{x_*}} = g^{\frac{x_j + \gamma}{x_*}},$

   where $\gamma := a\gamma_1 + b\gamma_2$.

   - $R_2 = rk_{i_1 2}^a \cdot rk_{i_2 2}^b = \left(g^{\frac{\beta y_1 + \gamma_1}{y_*}}\right)^a \cdot \left(g^{\frac{\beta y_2 + \gamma_2}{y_*}}\right)^b = g^{\frac{\beta(ay_1 + by_2) + (a\gamma_1 + b\gamma_2)}{y_*}} =$

   $g^{\frac{\beta y_j + \gamma}{y_*}},$

   - $R_3 = rk_{i_1 3}^a \cdot rk_{i_2 3}^b = \left(g^{\frac{x_1 + \beta y_1 + \gamma_1}{z_*}}\right)^a \cdot \left(g^{\frac{x_2 + \beta y_2 + \gamma_2}{z_*}}\right)^b$

   $= g^{\frac{(ax_1 + bx_2) + \beta(ay_1 + by_2) + (a\gamma_1 + b\gamma_2)}{z_*}} = g^{\frac{x_j + \beta y_j + \gamma}{z_*}}.$

It is easy to see that the above re-encryption key $rk_{* \to j}^{\dagger}$ is a well-formed re-encryption key if $\gamma = a\gamma_1 + b\gamma_2 \in \mathbb{Z}_p^*$; otherwise, it is an ill-formed re-encryption key satisfying the equation (1) since outputs of the algorithms $\mathbf{Dec}_1, \mathbf{ReEnc},$ and $\mathbf{Enc}_2$ are not depend on whether $\gamma \in \mathbb{Z}_p^*$. Therefore, the above re-encryption key $rk_{* \to j}^{\dagger}$ is really a forged re-encryption key.

4.2 Attack to sUFReKey-CA of the Scheme with Temporary Delegation

Before describing the attack, we recall the definition of sUFReKey-CA for the scheme with temporary delegation proposed by Hayashi et al. [3]. It is defined as follows:

The adversary is given the same public/secret keys as those in Definition 3, the target time period $\ell^* \xleftarrow{R} \{1, \ldots, L\}$ where L is polynomially bounded, re-encryption keys $rk_{*\to c|\ell}$ for any corrupted delegatee $c(\neq j)$ at any period $1 \leq \ell \leq L$, and re-encryption keys $rk_{*\to c|\ell}$ for the malicious user j at period $\ell \neq \ell^*$. Then the adversary tries to compute $rk^{\dagger}_{*\to c|\ell^*}$ such that

$$m = \mathbf{Dec}_1(sk_j, \mathbf{ReEnc}(rk^{\dagger}_{*\to j}, \mathbf{Enc}_2(pk_*, m, \ell^*), \ell^*)). \tag{6}$$

THE ATTACK DETAILS. Given $pk_*, \{(pk_{c_i}, sk_{c_i})\}, (pk_j, sk_j), \ell^*, \{rk_{*\to c_i|\ell}\}, \{rk_{*\to j|\ell\neq\ell^*}\}$, and the public parameters $PP = \{p, \mathbb{G}, \mathbb{G}_T, g, g_1, g_2, u, v, \mathbf{Sig}\}$, the adversary $\mathcal{A}$ breaks sUFReKey-CA of the scheme as follows:

1. $\mathcal{A}$ chooses from $\{sk_{c_i}\}$ two secret keys $sk_{c_{i_1}} = (x_1, y_1, z_1, w_1)$ and $sk_{c_{i_2}} = (x_2, y_2, z_2, w_2)$ such that $\frac{x_1}{y_1} \neq \frac{x_2}{y_2}$. It is easy to see that this is possible since x_i, y_i are independently chosen from $\mathbb{Z}_p^*$. Let $rk_{*\to c_{i_1}|\ell^*} = (rk_{i_1 1}, rk_{i_1 2}, rk_{i_1 3}, rk_{i_1 4}, rk_{i_1 5})$ and $rk_{*\to c_{i_2}|\ell^*} = (rk_{i_2 1}, rk_{i_2 2}, rk_{i_2 3}, rk_{i_2 4}, rk_{i_2 5})$.

2. $\mathcal{A}$ solves the following system of linear equations (with variables u and v)

$$\begin{cases} u \cdot x_1 + v \cdot x_2 = x_j \\ u \cdot y_1 + v \cdot y_2 = y_j \end{cases}$$

Since $\frac{x_1}{y_1} \neq \frac{x_2}{y_2}$, the above system has a pair of root $(a, b) \in \mathbb{Q}^2$. It means that $\mathcal{A}$ easily computes $(a, b) \in \mathbb{Q}^2$ such that

$$\begin{cases} a \cdot x_1 + b \cdot x_2 = x_j \\ a \cdot y_1 + b \cdot y_2 = y_j \end{cases}$$

3. $\mathcal{A}$ outputs a re-encryption key $rk^{\dagger}_{*\to j|\ell^*} = (R_1, R_2, R_3, R_4, R_5)$, where $R_1, R_2, R_3, R_4,$ and R_5 are computed as follows:

   - $R_1 = rk_{i_1 1}^a \cdot rk_{i_2 1}^b = \left(g^{\frac{x_1+\gamma_1}{x_*}+(\ell^*+w_*)\delta_{x1}}\right)^a \cdot \left(g^{\frac{x_2+\gamma_2}{x_*}+(\ell^*+w_*)\delta_{x2}}\right)^b = g^{\frac{(ax_1+bx_2)+(a\gamma_1+b\gamma_2)}{x_*}+(\ell^*+w_*)(a\delta_{x1}+b\delta_{x2})} = g^{\frac{x_j+\gamma}{x_*}+(\ell^*+w_*)\delta_x}$, where $\gamma := a\gamma_1 + b\gamma_2$ and $\delta_x := a\delta_{x1} + b\delta_{x2}$,

   - $R_2 = rk_{i_1 2}^a \cdot rk_{i_2 2}^b = \left(g^{\frac{\beta y_1+\gamma_1}{y_*}+(\ell^*+w_*)\delta_{y1}}\right)^a \cdot \left(g^{\frac{\beta y_2+\gamma_2}{y_*}+(\ell^*+w_*)\delta_{y2}}\right)^b = g^{\frac{\beta(ay_1+by_2)+(a\gamma_1+b\gamma_2)}{y_*}+(\ell^*+w_*)(a\delta_{y1}+b\delta_{y2})} = g^{\frac{\beta y_j+\gamma}{y_*}+(\ell^*+w_*)\delta_y}$, where $\delta_y := a\delta_{y1} + b\delta_{y2}$.

   - $R_3 = rk_{i_1 3}^a \cdot rk_{i_2 3}^b = \left(g^{\frac{x_1+\beta y_1+\gamma_1}{z_*}}\right)^a \cdot \left(g^{\frac{x_2+\beta y_2+\gamma_2}{z_*}}\right)^b = g^{\frac{(ax_1+bx_2)+\beta(ay_1+by_2)+(a\gamma_1+b\gamma_2)}{z_*}} = g^{\frac{x_j+\beta y_j+\gamma}{z_*}}$,

   - $R_4 = rk_{i_1 4}^a \cdot rk_{i_2 4}^b = \left(X_*^{\delta_{x1}}\right)^a \cdot \left(X_*^{\delta_{x2}}\right)^b = X_*^{\delta_x}$,

   - $R_5 = rk_{i_1 5}^a \cdot rk_{i_2 5}^b = \left(Y_{1*}^{\delta_{y1}}\right)^a \cdot \left(Y_{1*}^{\delta_{y2}}\right)^b = Y_{1*}^{\delta_y}$.

It is easy to see that the above re-encryption key $rk^{\dagger}_{*\to j}$ is a well-formed re-encryption key if $\gamma, \delta_x, \delta_y \in \mathbb{Z}_p^*$; otherwise, it is an ill-formed re-encryption key satisfying the equation (6) since outputs of the algorithms $\mathbf{Dec}_1, \mathbf{ReEnc},$ and $\mathbf{Enc}_2$ are not depend on whether $\gamma, \delta_x, \delta_y \in \mathbb{Z}_p^*$. Therefore, the above re-encryption key $rk^{\dagger}_{*\to j}$ is really a forged re-encryption key.

5   Attack to the Assumptions by Hayashi et al.

In this section, we review the problems proposed by Hayashi et al. [3] and show that they are not really hard to compute.

5.1   Review of the Problems

To prove the PRE schemes meet sUFReKey-CA, Hayashi et al. proposed two new variants of the Diffie-Hellman inversion problem and assumed their hardness. These new problems are as follows:

Definition 4   (2-DHIwRA problem [3]). *The 2-Diffie-Hellman inversion with randomized answers problem is computing* $g^{1/(a+c)}$ *given the following:*

- *input 1:* g, g^a, g^{a^2}, c, *where* $a, c \xleftarrow{\text{R}} \mathbb{Z}_p^*$;
- *input 2:* $(x_i, y_i, D_i, E_i, F_i) = (x_i, y_i, g^{\frac{x_i+\gamma_i}{a(a+c)}}, g^{\frac{ay_i+\gamma_i}{a+c}}, g^{\frac{x_i+\gamma_i}{a}})$, *where* $x_i, y_i, \gamma_i \xleftarrow{\text{R}}$ $\mathbb{Z}_p^*$ *for* $i \in \{1, \ldots, L\}$ *and* L *is polynomially bounded.*

The 2-DHIwRA problem without the input 2 is a variant of the 2-DHI problem [7], where $c = 0$.

Definition 5   (m-2-DHIwRA problem [3]). *The modified 2-Diffie-Hellman inversion with randomized answers problem is computing* $g^{1/(a+c)}$ *given the inputs 1 and 2 of 2-DHIwRA problem and*

- *input 3:* $(y', \mu, D', E', F', G', H') = (y', \mu, g^{\frac{cy'+\gamma'}{a(a+c)}+\delta_x'}, g^{\frac{a\mu y'+\gamma'}{a+c}+\delta_y'}, g^{\frac{cy'+\gamma'}{a}},$ $g^{a(a+c)\delta_x'}, g^{a(a+c)\delta_y'})$, *where* $y', \mu, \gamma', \delta_x', \delta_y' \xleftarrow{\text{R}} \mathbb{Z}_p^*$.

Remark. The m-2-DHIwRA problem is the same as the 2-DHIwRA problem except for the additional input 3, so a solving method for the 2-DHIwRA problem implies that for the m-2-DHIwRA problem.

5.2   Solving the Problems

By combining D_i, E_i, F_i from the input 2 and c from the input 1, we can remove the element γ_i and obtain $g^{\frac{1}{a+c}}$ (see the following theorem).

Theorem 1 *The 2-DHIwRA and the m-2-DHIwRA problems are not hard.*

Proof. Using x_i, y_i, D_i, E_i, F_i from input 2, and c from input 1 of the 2-DHIwRA problem, we compute N_i as follows.

$$N_i = \frac{F_i \cdot g^{y_i}}{D_i^c \cdot E_i} = \frac{g^{\frac{x_i+\gamma_i}{a}} \cdot g^{y_i}}{g^{\frac{x_i+\gamma_i}{a(a+c)}c} \cdot g^{\frac{ay_i+\gamma_i}{a+c}}} = \frac{g^{\frac{x_i+ay_i+\gamma_i}{a}}}{g^{\frac{(x_i+\gamma_i)c+(ay_i+\gamma_i)a}{a(a+c)}}}$$

$$= \frac{g^{\frac{(a+c)x_i+a(a+c)y_i+(a+c)\gamma_i}{a(a+c)}}}{g^{\frac{cx_i+a^2 y_i+(a+c)\gamma_i}{a(a+c)}}}$$

$$= g^{\frac{ax_i+acy_i}{a(a+c)}} = g^{\frac{x_i+cy_i}{a+c}}.$$

$$\Rightarrow (N_i)^{\frac{1}{x_i+cy_i}} = \left(\frac{F_i \cdot g^{y_i}}{D_i^c \cdot E_i}\right)^{\frac{1}{x_i+cy_i}} = g^{\frac{1}{a+c}}.$$

Note that in the above computation we do not use input 3 of the m-2-DHIwRA problem. Therefore, since the m-2-DHIwRA problem is the same as the 2-DHIwRA problem, except for the additional input 3, we can solve the m-2-DHIwRA problem in the same way.

$\square$

6   Concluding Remarks

We have presented two concrete attacks to the PRE schemes proposed by Hayashi et al. [3]. The first attack is to the sUFReKey-CA property on their two schemes. The second attack is to the assumptions employed in the security proofs for sUFReKey-CA of their two schemes. We stress that the work of Hayashi et al. [3] is still considered as an important step in this research area. Namely, the formal definition of UFReKey-CA and its stronger variant sUFReKey-CA are really significant steps toward the non-transferability. Moreover, due to their schemes, we can figure out of main difficulties for constructing PRE which meets UFReKey-CA. It is an open problem of constructing UFReKey-CA secure PRE schemes.

References

1. Ateniese, G., Fu, K., Green, M., Hohenberger, S.: Improved proxy re-encryption schemes with applications to secure distributed storage. In: NDSS (2005)
2. Blaze, M., Bleumer, G., Strauss, M.: Divertible protocols and atomic proxy cryptography. In: Nyberg, K. (ed.) EUROCRYPT 1998. LNCS, vol. 1403, pp. 127–144. Springer, Heidelberg (1998)
3. Hayashi, R., Matsushita, T., Yoshida, T., Fujii, Y., Okada, K.: Unforgeability of re-encryption keys against collusion attack in proxy re-encryption. In: Iwata, T., Nishigaki, M. (eds.) IWSEC 2011. LNCS, vol. 7038, pp. 210–229. Springer, Heidelberg (2011)
4. He, Y., Chim, T., Hui, L., Yiu, S.: Non-transferable proxy re-encryption. In: Cryptology ePrint Archive (2010), http://eprint.iacr.org/2010/192
5. Libert, B., Vergnaud, D.: Tracing malicious proxies in proxy re-encryption. In: Galbraith, S.D., Paterson, K.G. (eds.) Pairing 2008. LNCS, vol. 5209, pp. 332–353. Springer, Heidelberg (2008)
6. Libert, B., Vergnaud, D.: Unidirectional chosen-ciphertext secure proxy re-encryption. In: Cramer, R. (ed.) PKC 2008. LNCS, vol. 4939, pp. 360–379. Springer, Heidelberg (2008)
7. Mitsunari, S., Sakai, R., Kasahara, M.: A new traitor tracing. IEICE Transactions on Fundamentals of Electronics, Communications and Computer Sciences E85-A(2), 481–484 (2002)
8. Wang, L., Wang, L., Mambo, M., Okamoto, E.: New identity-based proxy re-encryption schemes to prevent collusion attacks. In: Joye, M., Miyaji, A., Otsuka, A. (eds.) Pairing 2010. LNCS, vol. 6487, pp. 327–346. Springer, Heidelberg (2010)

Game-Theoretic Security for Bit Commitment*

Haruna Higo[1], Keisuke Tanaka[1], and Kenji Yasunaga[2]

[1] Tokyo Institute of Technology, Japan
{higo9,keisuke}@is.titech.ac.jp
[2] Kanazawa University, Japan
yasunaga@se.kanazawa-u.ac.jp

Abstract. Higo, Tanaka, Yamada, and Yasunaga (ACISP 2012) studied oblivious transfer (OT) from a game-theoretic viewpoint in the malicious model. Their work can be considered as an extension of the study on two-party computation in the fail-stop model by Asharov, Canetti, and Hazay (EUROCRYPT 2011).

This paper focuses on bit commitment, and continues to study it from a perspective of game theory. In a similar manner to the work on OT, we consider bit commitment in the malicious model. In order to naturally capture the security properties of bit commitment, we characterize them with a single game where both parties are rational. In particular, we define a security notion from a game theoretic viewpoint, and prove the equivalence between it and the standard security notion.

Keywords: Cryptography, game theory, bit commitment.

1 Introduction

1.1 Motivations

Cryptographic protocols are designed for some parties to accomplish some purposes. When defining their security, we consider situations among honest parties and adversaries. Honest parties always follow the protocol description, while adversaries may deviate from it to attack others, e.g., dig out secrets of others. We usually say a protocol is secure if no adversary can damage the honest parties. The adversaries are assumed to be interested in attacking, however, not interested in protecting their own secret. Also, we assume there is at least one honest party. That is, we do not consider situations where all parties conduct some sort of attack.

Game theory mathematically analyzes decision making of multiple parties. In particular, non-cooperative game theory deals with the situations where the parties act independently. The parties are called rational, since they only care about their own preferences and act to achieve their best satisfactions. If a party

---

* This research was supported in part by a grant of I-System Co. Ltd., and JSPS Grant-in-Aid for Scientific Research Numbers 23500010, 24240001, 25106509, and 23700010.

K. Sakiyama and M. Terada (Eds.): IWSEC 2013, LNCS 8231, pp. 303–318, 2013.

has two or more preferences, he considers the trade-offs among them and aims to obtain the most reasonable result.

As described, both non-cooperative game theory and cryptography study the situations where parties act. However, they capture situations from different perspectives. In reality, even adversaries may be reluctant to reveal their secrets. Also, for example, if a party is sure that there is no danger, he may try to obtain more information than expected. That is, all parties may not be completely honest. In a game-theoretic framework, we can formalize such realistic perspectives.

There is a line of work using game-theoretic concepts to study cryptographic protocols. For a survey on the joint work of cryptography and game theory, we refer to [15,13]. Halpern and Teague [9] introduced such approach of study on secret sharing. Their work has been followed in many subsequent work called rational secret sharing (see [3] and the references therein for the subsequent work). They study it in the presence of rational parties, seeking for secure protocols in a game-theoretic framework. Besides secret sharing, there are several studies using game-theoretic frameworks for cryptographic protocols, e.g., two-party computation [1,7], leader election [6], byzantine agreement [8], oblivious transfer (OT) [11], and public-key encryption [17]. As an extension of the work by Asharov, Canetti, and Hazay [1] and Higo, Tanaka, Yamada, and Yasunaga [11], we are interested in whether the standard security notions of cryptographic protocols are reasonable in such a realistic model. In order to investigate it, we employ a game-theoretic framework.

In this work, we focus on bit commitment. Two parties, called the *sender* and the *receiver*, interact to implement it. They conduct two phases in series. In the first phase, called the *commit phase*, the sender who has a bit b interacts with the receiver. After that, the receiver obtains a commitment string c, and the sender obtains c and a decommitment string d. In the latter phase, called the *open phase*, the sender persuade the receiver that the committed bit is b through an interaction using c and d . Finally, the receiver outputs a bit representing whether she accepts that b is the committed bit.

In cryptography, we usually require three properties, *hiding property*, *binding property*, and *correctness*, as the security properties for bit commitment. Hiding property guarantees that no receiver can learn the committed bit before starting the open phase. Binding property guarantees that no sender can generate a pair of decommitment strings to open the commitment to both 0 and 1. These two properties are required to protect the sender and the receiver respectively. Correctness guarantees that if two parties honestly follow the protocol description, they can open the bit that was committed in the commit phase. Note that, in cryptography, each of the three properties is defined individually. Thus, for example, we do not consider parties who want to break hiding property and to protect binding property at the same time.

1.2   Previous Studies on Game-Theoretic Security

Asharov et al. [1] studied two-party protocols in the fail-stop model from a game-theoretic viewpoint. Fail-stop adversaries are allowed to abort the protocol rather than continuing at each round, but they cannot conduct other deviation, such as sending illegal messages to the others. They focus on the properties of privacy, correctness, and fairness. They characterized them individually in a game-theoretic manner using a concept called computational Nash equilibrium. For privacy and correctness, they showed the equivalence between the corresponding cryptographic and the game-theoretic notions. For fairness, they showed that their game-theoretic notion is strictly weaker than existing cryptographic ones, and proposed a new cryptographic notion that is equivalent to the game-theoretic one. Groce and Katz [7] continued their consideration on fairness, and showed a way to circumvent impossibility results in cryptography in a game-theoretic framework.

Higo et al. [11] studied two-message oblivious transfer (OT) from a game-theoretic viewpoint, characterizing its security using computational Nash equilibrium. They restrict the target protocol from general two-party computation to OT. However, the characterization of Higo et al. [11] has several advantages. First, they investigated the security in the malicious model, where the adversaries can arbitrarily deviate from the protocol description. Second, both parties are rational in their game while a game defined in [1] is essentially played between a rational party and an honest party. Finally, they characterized all security properties by a single game, whereas each security property is defined in an individual game in [1]. Specifically, Higo et al. [11] listed three preferences for each party. Since parties may have different strength of preferences, they formalize them as a weighted sum of the probabilities where each preference is satisfied. This way of formalization was introduced in order to make the model closer to the reality. With this model, they showed the equivalence between their game-theoretic security and the standard cryptographic security.

1.3   This Work

In this paper, we study bit commitment in a game-theoretic framework. In particular, we define a security notion from a game theoretic viewpoint, and examine the relation between it and the standard security notion. As summarized in Table 1., our work has various advantages compared to the previous studies.

We consider bit commitment in the malicious model. In order to naturally capture its security properties, we define a single game where both parties are rational. In other words, we take over the advantages of [11] over [1].

Since both bit commitment and OT are types of two-party computation, one might think that we can simply extend the result of [11] to the case of bit commitment. However, this is not the case. Bit commitment and OT have an essential difference in the functions they compute. The function of OT is a single function, that is, it has a single pair of inputs and a single pair of outputs. On the other hands, what bit commitment computes is a type of reactive functionalities [10,12], which have multiple phases in their computations. Bit commitment

Table 1. Results of [1], [11], and this work

	Asharov et al. [1]	Higo et al. [11]	This work
Target protocol	Two-party computation	Two-message OT	Bit commitment
# of phases	1	1	2
# of messages	Not restricted	2	Not restricted
Adversary model	Fail-stop model	Malicious model	Malicious model
# of rational parties	1 out of 2 parties	Both of 2 parties	Both of 2 parties
Properties	1	3	3
Utility functions	Fixed	Weighted	General

has two phases, with a pair of inputs and outputs for each phase, where the second input may depend on the result of the first phase. Moreover, Higo et al. [11] focused on two-message OT, whose interaction has only one round. For bit commitment, we do not only consider multiple phases, but also get rid of the limitation on the number of rounds. When we consider from a game-theoretic perspective, this difference makes the characterization and the analysis more complicated than those in the case of OT.

Generalized utility functions and a simpler solution concept. In the field of game theory, *utility function* mathematically represents the preferences of each party. We formalize the preferences of each party in bit commitment into a form of utility function.

We do not employ a fixed form such as a fixed value in [1] or a weighted sum in [11]. Our utility functions are said to be more general than the ones in the previous work.

Moreover, we reform the way of perceiving the preferences. Since protocols may be used repeatedly, the users are not just interested in a good outcome of a game but prefer to use a good protocol. We characterize the preferences of the parties not over the outcomes of single executions of a protocol, but over the algorithms used by the parties. Although it is not an essential difference, it contributes to employ Nash equilibrium rather than computational Nash equilibrium. As a result, we obtain a simple description of the theorem and its proof.

Non-triviality of our theorem. We prove that our security is equivalent to the standard cryptographic one. The implications between the two security notions are not trivial. Actually, they are, in general, not comparable. In the cryptographic security, we define the three properties individually, whereas rational parties pay attention to the trade-offs among them. That is, if there is a way of attacking some property of a protocol, it is not secure in cryptography. However since rational parties may not perform the attack to the protocol in case this attack together derives a negative result, it may satisfy the game-theoretic security. In this sense, the cryptographic security seems stronger. However, when we focus on the number of non-honest parties, the other seems stronger. Considering security in cryptography, we generally assume that there is at least one

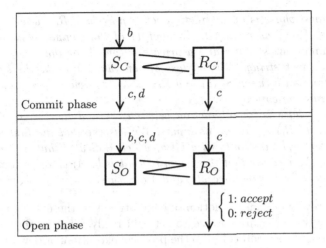

Fig. 1. Bit commitment protocol

honest party, but all parties are rational in game theory. That is, everyone is allowed to take arbitrary action.

2   Preliminaries

In this section, we review some cryptographic definitions and game-theoretic concepts.

First, we review some basic definitions. We say a function $\mu : \mathbb{N} \to \mathbb{R}$ is *negligible* if for any polynomial p, there exists $N \in \mathbb{N}$ such that for any $n > N$ it holds that $\mu(n) < 1/p(n)$. We describe a negligible function as negl$(\cdot)$. An algorithm is *PPT* if it runs in probabilistic polynomial time. In this paper, all the parties are assumed to use PPT algorithms in the security parameter n. Formally, each party has an input 1^n, but we omit this part. For two algorithms A and B, denote the view of A during the interaction with B by $\mathsf{view}_A(B)$, and the output of A after the interaction with B by $\mathsf{out}_A(B)$.

2.1   Bit Commitment in Cryptography

In this section, we review security of bit commitment in a cryptographic framework as defined in [5,2]. Bit commitment (Fig. 1.) has two phases, the *commit phase* and the *open phase*, which are executed in series. Note that this definition allows interactions in both phases.

Definition 1 (Bit commitment protocol). *A bit commitment protocol* Com *is a tuple of four PPT interactive algorithms, denoted by* Com $= ((S_C, S_O), (R_C, R_O))$.

- *The commit phase is an interaction between S_C and R_C, where S_C receives a bit $b \in \{0, 1\}$ as an input. The output of the commit phase consists of the commitment string c and a private output d for the sender, called the decommitment string. Without loss of generality, let c be the transcript of the interaction between $S_C(b)$ and R_C, and d the view of S_C, including the private random coin of S_C.*
- *The open phase is an interaction between S_O and R_O, where S_O receives (b, c, d), and R_O receives c as inputs. We assume that the first message by the sender explicitly contains a bit b, which indicates that the sender is to persuade the receiver that the committed bit is b. After the interaction, R_O outputs 1 if the receiver accepts, and 0 otherwise.*

Next, we review a security notion of commitment in the malicious model. In this model, adversaries are allowed to act arbitrarily. That is, they may follow the description of the protocol, stop the protocol execution, or deviate from it. A protocol is called secure if it satisfies three properties, hiding property, binding property, and correctness. Since we derive a new security notion in terms of game theory in the next section, this one is called the *cryptographic security*.

Definition 2 (Cryptographic security). *Let* $\mathsf{Com} = ((S_C, S_O), (R_C, R_O))$ *be a bit commitment protocol. We say* Com *is cryptographically secure if it satisfies the following three properties.*

Hiding Property: *For any $b \in \{0, 1\}$, PPT cheating receiver R_C^*, and PPT distinguisher D, it holds that*

$$\Pr[D(\mathrm{view}_{R_C^*}(S_C(b))) = b] \le 1/2 + \mathrm{negl}(n).$$

Binding Property: *For any $b \in \{0, 1\}$, PPT cheating sender (S_C^*, S_O^*), and PPT decommitment finder F, it holds that*

$$\Pr[\mathrm{out}_{R_O(c^*)}(S_O^*(0, c^*, d_0)) = \mathrm{out}_{R_O(c^*)}(S_O^*(1, c^*, d_1)) = 1] \le \mathrm{negl}(n),$$

where c^* is the transcript between $S_C^*(b)$ and R_C, (d_0, d_1) is the output of $F(\mathrm{view}_{S_C^*(b)}(R_C))$.

Correctness: *For any $b \in \{0, 1\}$, it holds that*

$$\Pr[\mathrm{out}_{R_O(c)}(S_O(b, c, d)) = 1] \ge 1 - \mathrm{negl}(n),$$

where c is the transcript between $S_C(b)$ and R_C, and $d = \mathrm{view}_{S_C(b)}(R_C)$.

2.2   Game Theory

Game theory [4,16] studies actions of some parties aiming at their own goals. We characterize the situations as a *game* in terms of game theory. The parties of the game have their own preferences. In games, parties choose the best actions from their alternatives to obtain the most preferable outcome. The series of actions of each party is collectively called *strategies*. When we analyze cryptographic protocols from a game-theoretic viewpoint, the tuple of algorithms of each party accounts for his strategy.

Utility functions stands for the preferences of the parties. A utility function maps from a tuple of strategies of parties to a real number. When all parties choose their strategies, the outcome of the game is (probabilistically) determined. The values of utility functions usually represent the degree of its preference over the outcome. Higher rate represents stronger preference. Each party guesses the actions of the others, and estimate his own utility to choose his best strategy. Every party chooses the algorithm that delivers him the highest utility.

We are interested in how the parties act in the game. Solution concepts characterize which tuples of strategies are likely to be chosen by the parties. While there are many solution concepts introduced, we employ *Nash equilibrium*, which is one of the most commonly used. When all parties choose the Nash equilibrium strategies, no party can gain his utility by changing his strategy unilaterally. Namely, if parties are assumed to choose the Nash equilibrium strategies, no party have any motivation to change his strategy.

3   Bit Commitment in Game Theory

In this section, we introduce game-theoretic definitions with respect to bit commitment. First, we define a game to execute a protocol. Then, we consider the natural preferences of the sender and the receiver. The solution concept we employ is Nash equilibrium [15,13]. Finally, we characterize the required properties for bit commitment using these notions in the game-theoretic framework.

Game. Given a bit commitment protocol $\mathsf{Com} = ((S_C, S_O), (R_C, R_O))$, we define a game between a sender and a receiver. A sender has three PPT algorithms (S'_C, S'_O, F), and a receiver has two PPT algorithms (R'_C, D) in our game. Here is an informal description of the game. (See also Fig. 2.)

First, the sender and the receiver execute a commit phase by using S'_C and R'_C together with a random bit b as the input for the sender. Then, a distinguisher D of the receiver tries to guess the committed bit b using her view in the commit phase. After that, a decommitment finder F of the sender tries to generate two decommitment strings d_0 and d_1, where d_b is used for opening b as the committed bit. Using S'_O and R_O, two open phases are executed, whether d_0 and d_1 are correctly used to open the commitment generated in the commit phase. Note that the receiver has to use R_O as the open phase algorithm. Since otherwise, the receiver can even accept/reject all the commitment, and such strategies should be excluded from her choice.

Now we formally define a bit commitment game.

Definition 3 (Game). *For a bit commitment protocol* $\mathsf{Com} = ((S_C, S_O), (R_C, R_O))$, *and PPT algorithms* S'_C, S'_O, F, R'_C, *and* D, *the game* $\Gamma^{\mathsf{Com}}((S'_C, S'_O, F), (R'_C, D))$ *is executed as follows.*

1. *Choose a bit b uniformly at random and set* $\mathsf{guess} = \mathsf{amb} = \mathsf{suc} = \mathsf{abort} = 0$.

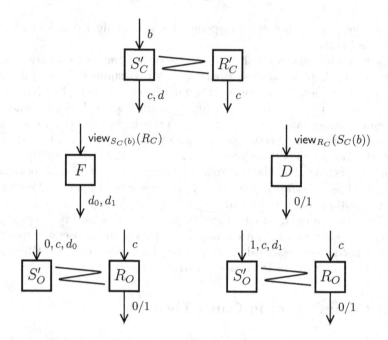

Fig. 2. Bit commitment game

2. *Observe an interaction between $S'_C(b)$ and R'_C, and c denotes the transcript during the interaction. Set* abort $= 1$ *if some party aborts the protocol.*
3. *Set* guess $= 1$ *if* $b = D(\text{view}_{R'_C}(S'_C(b)))$.
4. *Run* $F(\text{view}_{S'_C(b)}(R'_C))$ *and get* (d_0, d_1) *as output.*
5. *Observe an interaction between $S'_O(0, c, d_0)$ and $R_O(c)$, and between $S'_O(1, c, d_1)$ and $R_O(c)$. Set* abort $= 1$ *if some party aborts.*
6. *Set* amb $= 1$ *if* $\text{out}_{R_O(c)}(S'_O(0, c, d_0)) = \text{out}_{R_O(c)}(S'_O(1, c, d_1)) = 1$, *and* suc $= 1$ *if either* $\text{out}_{R_O(c)}(S'_O(b, c, d_b)) = 1$ *or* abort $= 1$.

The tuple (guess, amb, suc) is the outcome of this game, and is explained as follows.

After the commit phase, the receiver tries to learn the committed bit b beforehand. If she succeeded in guessing, then guess $= 1$. Otherwise, guess $= 0$. The sender tries to find two decommitment strings d_0 and d_1 in order that $d_{b'}$ can be opened to b'. Acceptance of both bits implies that he can choose the bit to be opened. If the sender succeed in finding such values, then amb $= 1$. Otherwise, amb $= 0$. If the receiver can accept the commitment for the committed bit b, or one of the parties aborts the protocol, then suc $= 1$. Otherwise, suc $= 0$.

Utility functions. We consider that each party of bit commitment has multiple goals listed as the following preferences.

We consider that the sender has the following two preferences:

– He does not prefer the receiver to know the committed bit b before executing the open phase.

- On executing the open phase, he prefers to be able to choose a bit to be opened.

Next, the receiver is considered to have the following three preferences:

- She prefers to learn the committed bit b before executing the open phase.
- She does not prefer the sender to change the bit to be opened in the open phase.
- She prefers to open the committed bit b in the open phase unless the protocol was aborted.

We formalize these preferences as utility functions. Similar to the work of Higo et al. [11], each party has a single utility function that represents all the preferences in a lump. However, our utility functions are not in a fixed form such as weighted sum used in [11]. Moreover, to describe the preferences over the algorithms used in the game, the arguments of utility functions are the algorithms. They are evaluated using the prescribed three random variables guess, amb and suc that represents the outcome of the game.

For simplicity, we use the following notations. We denote by $a \prec b$ or $b \succ a$ for $a(n), b(n) \in \mathbb{R}$, if it holds that $a(n) < b(n) - \epsilon(n)$ for some non-negligible function ϵ. Also, $a \approx b$ denotes that $|a(n) - b(n)| \leq \mathrm{negl}(n)$.

Definition 4 (Utility functions). *For a bit commitment protocol* Com, *and PPT algorithms* $S_C, S_O, R_C, S'_C, S'_O, R'_C, D,$ *and* F, *let* (guess, amb, suc) *and* (guess', amb', suc') *be the random variables representing the outcome of* $\Gamma^{\mathsf{Com}}((S_C, S_O, F), (R_C, D))$ *and* $\Gamma^{\mathsf{Com}}((S'_C, S'_O, F), (R'_C, D))$, *respectively. The utility function* U_S^{Com} *for the sender satisfies* $U_S^{\mathsf{Com}}((S_C, S_O, F), (R_C, D)) > U_S^{\mathsf{Com}}((S'_C, S'_O, F), (R'_C, D))$ *if one of the following conditions holds.*

S-1. $|\mathrm{Pr}[\mathsf{guess} = 1] - 1/2| \prec |\mathrm{Pr}[\mathsf{guess}' = 1] - 1/2|$ *and* $\mathrm{Pr}[\mathsf{amb} = 1] \approx \mathrm{Pr}[\mathsf{amb}' = 1]$.
S-2. $\mathrm{Pr}[\mathsf{guess} = 1] \approx \mathrm{Pr}[\mathsf{guess}' = 1]$ *and* $\mathrm{Pr}[\mathsf{amb} = 1] \succ \mathrm{Pr}[\mathsf{amb}' = 1]$.

The utility function U_R^{Com} *for the receiver satisfies* $U_R^{\mathsf{Com}}((S_C, S_O, F), (R_C, D)) > U_R^{\mathsf{Com}}((S'_C, S'_O, F), (R'_C, D))$ *if one of the following conditions holds.*

R-1. $|\mathrm{Pr}[\mathsf{guess} = 1] - 1/2| \succ |\mathrm{Pr}[\mathsf{guess}' = 1] - 1/2|$, $\mathrm{Pr}[\mathsf{amb} = 1] \approx \mathrm{Pr}[\mathsf{amb}' = 1]$, *and* $\mathrm{Pr}[\mathsf{suc} = 1] \approx \mathrm{Pr}[\mathsf{suc}' = 1]$.
R-2. $\mathrm{Pr}[\mathsf{guess} = 1] \approx \mathrm{Pr}[\mathsf{guess}' = 1]$, $\mathrm{Pr}[\mathsf{amb} = 1] \prec \mathrm{Pr}[\mathsf{amb}' = 1]$, *and* $\mathrm{Pr}[\mathsf{suc} = 1] \approx \mathrm{Pr}[\mathsf{suc}' = 1]$.
R-3. $\mathrm{Pr}[\mathsf{guess} = 1] \approx \mathrm{Pr}[\mathsf{guess}' = 1]$, $\mathrm{Pr}[\mathsf{amb} = 1] \approx \mathrm{Pr}[\mathsf{amb}' = 1]$, *and* $\mathrm{Pr}[\mathsf{suc} = 1] \succ \mathrm{Pr}[\mathsf{suc}' = 1]$.

Note that we use the value $|\mathrm{Pr}[\mathsf{guess} = 1] - 1/2|$ rather than $\mathrm{Pr}[\mathsf{guess} = 1]$. After a single execution of the game, the sender prefers guess to be 0, and the receiver 1. However, focusing on what the parties hope the algorithm to be, we consider that the sender prefers guess to be close to $1/2$, and the receiver prefers it to be far from $1/2$.

Nash equilibrium. As mentioned in Section 2.2., we use Nash equilibrium as the solution concept in this paper. When a pair of strategies in a Nash equilibrium is chosen by the parties, neither party can gain more no matter how he changes his strategy unilaterally. Although all strategies we consider are polynomially bounded, we do not need to use the extended notion named computational Nash equilibrium as is used in the previous work [1,11]. This conversion is attributed to the reformation of the utility. Since our utility functions describe the preferences over the strategies not over the outcomes of the games, the discussion of computability is done with evaluating utility functions.

Definition 5 (Nash equilibrium). *Let* Com *be a bit commitment protocol. A tuple of PPT strategies* $((S_C, S_O), R_C)$ *is in a* Nash equilibrium, *if for any PPT algorithms* S_C^*, S_O^*, R_C^*, D, *and* F, *neither of the followings hold.*

- $U_S^{\mathsf{Com}}((S_C, S_O, F), (R_C, D)) < U_S^{\mathsf{Com}}((S_C^*, S_O^*, F), (R_C, D))$
- $U_R^{\mathsf{Com}}((S_C, S_O, F), (R_C, D)) < U_R^{\mathsf{Com}}((S_C, S_O, F), (R_C^*, D))$

Note that the strategies of the parties are (S_C, S_O) and R_C. D and F are excluded from strategies. That is because, informally, the parties always choose the best D and F to improve their utilities.

Game-theoretic security. We characterize the required properties for bit commitment using the prescribed notions. If a protocol is in a Nash equilibrium, it means that the parties will prefer to take the strategies according to the protocol. In other words, the parties do not have a motivation to deviate from the protocol. We call such protocols *game-theoretically secure*.

Definition 6 (Game-theoretic security). *Let* Com $= ((S_C, S_O), (R_C, R_O))$ *be a bit commitment protocol. We say* Com *is game-theoretically secure if the tuple of the strategies* $((S_C, S_O), R_C)$ *is in a Nash equilibrium.*

4   Equivalence between the Two Security Notions

In this section, we prove the equivalence between the cryptographic security (Definition 2) and the game-theoretic security (Definition 6). In other words, we show that a protocol is cryptographically secure if and only if the protocol itself is in a Nash equilibrium.

Theorem 1. *Let* Com *be a bit commitment protocol.* Com *is cryptographically secure if and only if* Com *is game-theoretically secure.*

As mentioned in Section 1.3., this relationship is not trivial. We provide both directions of implication one by one.

First, we show that the cryptographic security implies the game-theoretic security.

Lemma 1. *If a bit commitment protocol* Com *is cryptographically secure, then* Com *is game-theoretically secure.*

We prove the contrapositive of this statement. If a protocol is not game-theoretically secure, that is, it is not in a Nash equilibrium, at least one party can gain with using some alternative strategies rather than the protocol description. From the definitions of the utility functions, it is natural that the alternative strategies break some of the cryptographic property, which implies that the protocol is not cryptographically secure. Actually, the definition of Nash equilibrium makes the proof a little complicated. The formal proof is as follows.

Proof. To prove this lemma, we assume that $\mathsf{Com} = ((S_C, S_O), (R_C, R_O))$ is not game-theoretically secure, and show that Com is not cryptographically secure. Namely, Com does not satisfy at least one of the three properties, hiding property, binding property, and correctness.

Suppose Com is not game-theoretically secure. Then, there exist a tuple $((S_C^*, S_O^*), R_C^*)$ of PPT strategies, a PPT distinguisher D and a PPT decommitment finder F such that at least one of the following two inequalities holds:

$$U_S^{\mathsf{Com}}((S_C, S_O, F), (R_C, D)) < U_S^{\mathsf{Com}}((S_C^*, S_O^*, F), (R_C, D)), \tag{1}$$
$$U_R^{\mathsf{Com}}((S_C, S_O, F), (R_C, D)) < U_R^{\mathsf{Com}}((S_C, S_O, F), (R_C^*, D)). \tag{2}$$

First, assume that Equality (1) holds. It implies that the sender can get a higher utility by changing his strategy from (S_C, S_O) to (S_C^*, S_O^*). There are two possibilities for the cause of this increase:

Case S-1: $|\Pr[\mathsf{guess} = 1] - 1/2|$ decreases with the change of the strategy.
Case S-2: $\Pr[\mathsf{amb} = 1]$ increases with the change of the strategy.

Case S-1 implies that $|\Pr[\mathsf{guess} = 1] - 1/2| \succ 0$ holds when both parties choose its honest strategy. This means that Com does not satisfy hiding property for R_C.

Case S-2 implies that $\Pr[\mathsf{amb} = 1] \succ 0$ holds for the strategy tuple $((S_C^*, S_O^*), R_C)$. Hence, Com does not satisfy binding property for (S_C^*, S_O^*).

Next, assume that Equality (2) holds. It implies that the receiver can get a higher utility by changing her strategy from R_C to R_C^*. There are three possibilities for the cause of this increase:

Case R-1: $|\Pr[\mathsf{guess} = 1] - 1/2|$ increases with the change of the strategy.
Case R-2: $\Pr[\mathsf{amb} = 1]$ decreases with the change of the strategy.
Case R-3: $\Pr[\mathsf{suc} = 1]$ increases with the change of the strategy.

Case R-1 implies that $|\Pr[\mathsf{guess} = 1] - 1/2| \succ 0$ holds for the strategy tuple $((S_C, S_O), R_C^*)$. This means that Com does not satisfy hiding property for R_C^*.

Case R-2 implies that $\Pr[\mathsf{amb} = 1] \succ 0$ holds when both parties choose their honest strategies. Hence, Com does not satisfy binding property for (S_C, S_O).

Case R-3 implies that $\Pr[\mathsf{suc} = 1] \prec 1$ holds when both parties choose their honest strategy. This means that Com does not satisfy correctness.

In every case, we have shown that Com is not cryptographically secure. Therefore, the statement follows. □

Next, we show that the game-theoretic security implies the cryptographic security.

Lemma 2. *If a bit commitment protocol* Com *is game-theoretically secure, then* Com *is cryptographically secure.*

The proof of this direction is more technical than that of Lemma 1. We prove it by showing that the contrapositive is true. Assume that a protocol is not cryptographically secure, at least one of the security properties, hiding property, binding property and correctness, does not hold. Provided that an algorithm breaks one of the properties, we cannot simply say that the protocol is not in a Nash equilibrium. That is because, the parties consider the tradeoffs among the preferences. If the algorithms together leads to some negative result, the party cannot gain his utility by using this algorithm. This cannot be the reason of the protocol being not game-theoretically secure. This lemma seems not trivial at this point.

Despite this point, the lemma holds because the definition of Nash equilibrium requires the inequality to hold for any D and F. If an algorithm breaks some property, then some D and F makes a situation where only the probability related to the broken property ($\Pr[\mathsf{guess} = 1]$, $\Pr[\mathsf{amb} = 1]$ or $\Pr[\mathsf{suc} = 1]$) changes by using the algorithm rather than following the protocol. That is, when at least one of the security properties does not hold, some tuple of algorithm makes the protocol not in Nash equilibrium.

Here, we provide a formal proof.

Proof. Suppose that Com $= ((S_C, S_O), (R_C, R_O))$ is not cryptographically secure. We consider the following five cases, and show that Com is not game-theoretically secure in each case.

Case 1: Com does not satisfy correctness.

Case 2: Com satisfies correctness and does not satisfy binding property for (S_C, S_O).

Case 3: Com satisfies correctness and binding property for (S_C, S_O), and does not satisfy binding property for some $(S_C^*, S_O^*) \neq (S_C, S_O)$.

Case 4: Com satisfies correctness and binding property, and does not satisfy hiding property for R_C.

Case 5: Com satisfies correctness, binding property, and hiding property for R_C, and does not satisfy hiding property for some $R_C^* \neq R_C$.

In Case 1, even if both parties follow the protocol description, the probability that they cannot open the committed bit is non-negligible. That is, for some $b \in \{0, 1\}$, it holds that

$$\Pr[\mathsf{out}_{R_O(c)}(S_O(b, c, d)) = 1] \prec 1,$$

where c is the transcript between $S_C(b)$ and R_C, and $d = \text{view}_{S_C(b)}(R_C)$. Let D^{rand} be an algorithm that outputs 0 or 1 uniformly at random, F^{honest} an algorithm that outputs (d_0, d_1) where $d_b = \text{out}_{S_C(b)}(R'_C)$ and $d_{1-b} = \perp$, where R'_C is an algorithm of the receiver in the commit phase, and R_C^{abort} a strategy of sending the abort message right after starting the protocol. Note that the three algorithms, D^{rand}, F^{honest}, and R_C^{abort}, are PPT algorithms. We denote the outcome of the games $\Gamma^{\text{Com}}((S_C, S_O, F^{\text{honest}}), (R_C, D^{\text{rand}}))$ and $\Gamma^{\text{Com}}((S_C, S_O, F^{\text{honest}}), (R_C^{\text{abort}}, D^{\text{rand}}))$ by $(\text{guess}, \text{amb}, \text{suc})$ and $(\text{guess}', \text{amb}', \text{suc}')$, respectively. Now we obtain the following equalities:

- $|\Pr[\text{guess} = 1] - 1/2| \approx |\Pr[\text{guess}' = 1] - 1/2| \approx 0$,
- $\Pr[\text{amb} = 1] = \Pr[\text{amb}' = 1] = 0$,
- $\Pr[\text{suc} = 1] \prec \Pr[\text{suc}' = 1] = 1$.

Hence, it holds that $U_R^{\text{Com}}((S_C, S_O, F^{\text{honest}}), (R_C, D^{\text{rand}})) < U_R^{\text{Com}}((S_C, S_O, F^{\text{honest}}), (R_C^{\text{abort}}, D^{\text{rand}}))$, which implies that the tuple $((S_C, S_O), R_C)$ is not in a Nash equilibrium.

In Case 2, the sender can break binding property with the honest strategy (S_C, S_O). That is, for some PPT decommitment finder F and $b \in \{0, 1\}$, it holds that

$$\Pr[\text{out}_{R_O(c)}(S_O(0, c, d_0)) = \text{out}_{R_O(c)}(S_O(1, c, d_1)) = 1] \succ 0,$$

where (d_0, d_1) is the output of $F(\text{view}_{S_C(b)}(R_C))$. We denote the outcome of the games $\Gamma^{\text{Com}}((S_C, S_O, F), (R_C, D^{\text{rand}}))$ and $\Gamma^{\text{Com}}((S_C, S_O, F), (R_C^{\text{abort}}, D^{\text{rand}}))$ by $(\text{guess}, \text{amb}, \text{suc})$ and $(\text{guess}', \text{amb}', \text{suc}')$, respectively. Now we obtain the following equalities:

- $|\Pr[\text{guess} = 1] - 1/2| \approx |\Pr[\text{guess}' = 1] - 1/2| \approx 0$,
- $\Pr[\text{amb} = 1] \succ \Pr[\text{amb}' = 1] = 0$,
- $\Pr[\text{suc} = 1] = \Pr[\text{suc}' = 1] = 1$.

Hence, it holds that $U_R^{\text{Com}}((S_C, S_O, F), (R_C, D^{\text{rand}})) < U_R^{\text{Com}}((S_C, S_O, F), (R_C^{\text{abort}}, D^{\text{rand}}))$, which implies that the tuple $((S_C, S_O), R_C)$ is not in a Nash equilibrium.

In Case 3, the sender cannot break binding property with honest strategy (S_C, S_O) but with some strategy $(S_C^*, S_O^*) \neq (S_C, S_O)$. That is, for some PPT decommitment finder F and $b \in \{0, 1\}$, it holds that

$$\Pr[\text{out}_{R_O(c)}(S_O^*(0, c^*, d_0)) = \text{out}_{R_O(c)}(S_O^*(1, c^*, d_1)) = 1] \succ 0,$$

where c^* is the transcript between $S_C^*(b)$ and R_C, and (d_0, d_1) is the output of $F(\text{view}_{S_C^*(b)}(R_C))$. For the same F and b, it holds that

$$\Pr[\text{out}_{R_O(c)}(S_O(0, c, d_0)) = \text{out}_{R_O(c)}(S_O(1, c, d_1)) = 1] \approx 0.$$

We denote the outcome of the games $\Gamma^{\text{Com}}((S_C, S_O, F), (R_C, D^{\text{rand}}))$ and $\Gamma^{\text{Com}}((S_C^*, S_O^*, F), (R_C, D^{\text{rand}}))$ by (guess, amb, suc) and (guess', amb', suc'), respectively. Now we obtain the following equalities:

- $|\Pr[\text{guess} = 1] - 1/2| \approx |\Pr[\text{guess}' = 1] - 1/2| \approx 0$,
- $0 = \Pr[\text{amb} = 1] \prec \Pr[\text{amb}' = 1]$.

Hence, it holds that $U_S^{\text{Com}}((S_C, S_O, F), (R_C, D^{\text{rand}})) <$ $U_S^{\text{Com}}((S_C^*, S_O^*, F), (R_C, D^{\text{rand}}))$, which implies that the tuple $((S_C, S_O), R_C)$ is not in a Nash equilibrium.

In Case 4, the receiver can break hiding property with the honest strategy R_C. That is, for some PPT distinguisher D, it holds that

$$\Pr[D(\text{view}_{R_C}(S_C(b))) = b] \succ 1/2.$$

Let S_C^{abort} be a strategy of sending the abort message right after starting the protocol. We denote the outcome of the games $\Gamma^{\text{Com}}((S_C, S_O, F^{\text{honest}}), (R_C, D))$ and $\Gamma^{\text{Com}}((S_C^{\text{abort}}, S_O, F^{\text{honest}}), (R_C, D))$ by (guess, amb, suc) and (guess', amb', suc'), respectively. Now we obtain the following equalities:

- $|\Pr[\text{guess} = 1] - 1/2| \succ |\Pr[\text{guess}' = 1] - 1/2| \approx 0$,
- $\Pr[\text{amb} = 1] = \Pr[\text{amb}' = 1] = 0$.

Hence, it holds that $U_S^{\text{Com}}((S_C, S_O, F^{\text{honest}}), (R_C, D)) <$ $U_S^{\text{Com}}((S_C^{\text{abort}}, S_O, F^{\text{honest}}), (R_C, D))$, which implies that the tuple $((S_C, S_O), R_C)$ is not in a Nash equilibrium.

In Case 5, the receiver can not break hiding property with honest strategy R_C but with some strategy $R_C^* \neq R_C$. That is, for some PPT distinguisher D, it holds that

$$\Pr[D(\text{view}_{R_C^*}(S_C(b))) = b] \succ 1/2, \text{ and } \Pr[D(\text{view}_{R_C}(S_C(b))) = b] \approx 1/2.$$

Let $\tilde{R}_C^*$ be a strategy of following R_C^* in the commit phase and not participating in the open phase. Then, it holds that $\Pr[D(\text{view}_{\tilde{R}_C^*}(S_C(b))) = b] \succ 1/2$. We denote the outcome of the games $\Gamma^{\text{Com}}((S_C, S_O, F^{\text{honest}}), (R_C, D))$ and $\Gamma^{\text{Com}}((S_C, S_O, F^{\text{honest}}), (\tilde{R}_C^*, D))$ by (guess, amb, suc) and (guess', amb', suc'), respectively. Now we obtain the following equalities:

- $0 \approx |\Pr[\text{guess} = 1] - 1/2| \prec |\Pr[\text{guess}' = 1] - 1/2|$,
- $\Pr[\text{amb} = 1] = \Pr[\text{amb}' = 1] = 0$,
- $\Pr[\text{suc} = 1] = \Pr[\text{suc}' = 1] = 1$.

Hence, it holds that $U_R^{\text{Com}}((S_C, S_O, F^{\text{honest}}), (R_C, D)) <$ $U_R^{\text{Com}}((S_C, S_O, F^{\text{honest}}), (\tilde{R}_C^*, D))$, which implies that the tuple $((S_C, S_O), R_C)$ is not in a Nash equilibrium.

In every case, we show that the tuple $((S_C, S_O), R_C)$ is not in a Nash equilibrium. Therefore, the statement follows.

5 Concluding Remarks

This paper has focused on bit commitment and characterized its security in a game-theoretic manner. Our work is based on the work of OT by Higo et al. [11]. Since bit commitment and OT computes different numbers of functions in their protocols, the characterization of bit commitment is more complicated. In this paper, we have defined a game in which parties execute a bit commitment protocol, and picked up the natural preferences of the sender and the receiver. Using Nash equilibrium as a solution concept, we have defined the notion of game-theoretic security. We have shown the equivalence between the game-theoretic security and the cryptographic security.

Although we have introduced game-theoretic concepts as a formalization of realistic perspectives, no practical application has been known. Further work is expected in this area to describe some practical implication or limitations.

References

1. Asharov, G., Canetti, R., Hazay, C.: Towards a game theoretic view of secure computation. In: Paterson, K.G. (ed.) EUROCRYPT 2011. LNCS, vol. 6632, pp. 426–445. Springer, Heidelberg (2011)
2. Chung, K.-M., Liu, F.-H., Lu, C.-J., Yang, B.-Y.: Efficient string-commitment from weak bit-commitment. In: Abe, M. (ed.) ASIACRYPT 2010. LNCS, vol. 6477, pp. 268–282. Springer, Heidelberg (2010)
3. Fuchsbauer, G., Katz, J., Naccache, D.: Efficient rational secret sharing in standard communication networks. In: Micciancio (ed.) [14], pp. 419–436
4. Fudenberg, D., Tirole, J.: Game theory (3. pr.). MIT Press (1991)
5. Goldreich, O.: The foundations of cryptography, vol. 2. Basic applications. Cambridge University Press (2004)
6. Gradwohl, R.: Rationality in the full-information model. In: Micciancio (ed.) [14], pp. 401–418
7. Groce, A., Katz, J.: Fair computation with rational players. In: Pointcheval, D., Johansson, T. (eds.) EUROCRYPT 2012. LNCS, vol. 7237, pp. 81–98. Springer, Heidelberg (2012)
8. Groce, A., Katz, J., Thiruvengadam, A., Zikas, V.: Byzantine agreement with a rational adversary. In: Czumaj, A., Mehlhorn, K., Pitts, A., Wattenhofer, R. (eds.) ICALP 2012, Part II. LNCS, vol. 7392, pp. 561–572. Springer, Heidelberg (2012)
9. Halpern, J.Y., Teague, V.: Rational secret sharing and multiparty computation: extended abstract. In: Babai, L. (ed.) STOC, pp. 623–632. ACM (2004)
10. Hazay, C., Lindell, Y.: Efficient secure two-party protocols: techniques and constructions, 1st edn. Springer-Verlag New York, Inc., New York (2010)
11. Higo, H., Tanaka, K., Yamada, A., Yasunaga, K.: A game-theoretic perspective on oblivious transfer. In: Susilo, W., Mu, Y., Seberry, J. (eds.) ACISP 2012. LNCS, vol. 7372, pp. 29–42. Springer, Heidelberg (2012)
12. Jeffs, R.A., Rosulek, M.: Characterizing the cryptographic properties of reactive 2-party functionalities. In: Sahai, A. (ed.) TCC 2013. LNCS, vol. 7785, pp. 263–280. Springer, Heidelberg (2013)

13. Katz, J.: Bridging game theory and cryptography: Recent results and future directions. In: Canetti, R. (ed.) TCC 2008. LNCS, vol. 4948, pp. 251–272. Springer, Heidelberg (2008)
14. Micciancio, D. (ed.): TCC 2010. LNCS, vol. 5978. Springer, Heidelberg (2010)
15. Nisan, N., Roughgarden, T., Tardos, E., Vazirani, V.V.: Algorithmic game theory. Cambridge University Press, New York (2007)
16. Osborne, M.J., Rubinstein, A.: A course in game theory. MIT Press (1994)
17. Yasunaga, K.: Public-key encryption with lazy parties. In: Visconti, I., De Prisco, R. (eds.) SCN 2012. LNCS, vol. 7485, pp. 411–425. Springer, Heidelberg (2012)

Author Index